# The Object-Oriented Approach to Problem Solving and Machine Learning with Python

This book is a comprehensive guide suitable for beginners and experienced developers alike. It teaches readers how to master object-oriented programming (OOP) with Python and use it in real-world applications.

Start by solidifying your OOP foundation with clear explanations of core concepts such as use cases and class diagrams. This book goes beyond theory as you get practical examples with well-documented source code available in the book and on GitHub.

This book doesn't stop at the basics. Explore how OOP empowers fields such as data persistence, graphical user interfaces (GUIs), machine learning, and data science, including social media analysis. Learn about machine learning algorithms for classification, regression, and unsupervised learning, putting you at the forefront of AI innovation.

Each chapter is designed for hands-on learning. You'll solidify your understanding with case studies, exercises, and projects that apply your newfound knowledge to real-world scenarios. The progressive structure ensures mastery, with each chapter building on the previous one, reinforced by exercises and projects.

Numerous code examples and access to the source code enhance your learning experience. This book is your one-stop shop for mastering OOP with Python and venturing into the exciting world of machine learning and data science.

**Sujith Samuel Mathew** holds a PhD in computer science from the University of Adelaide, Australia. He is an associate professor at Zayed University, UAE. He specializes in ubiquitous and distributed computing, focusing on the Internet of Things and related Smart City applications.

**Mohammad Amin Kuhail** holds an MSc in software engineering from the University of York and a PhD in software development from IT University of Copenhagen. He is an associate professor at Zayed University, specializing in human–computer interaction and software engineering, and he researches chatbot technology, user behavior, and education.

**Maha Hadid** holds an MSc in Information Sciences and Systems from the University of Marseille in France. She is an instructor at Zayed University, UAE, with experience in undergraduate courses and instructional design and delivery for blended and classroom-based courses.

**Shahbano Farooq** holds an MSc in computer science from the University of Calgary. She is an instructor at Zayed University, UAE, specializing in human-computer interaction and machine learning.

# The Object-Oriented Approach to Problem Solving and Machine Learning with Python

Sujith Samuel Mathew
Mohammad Amin Kuhail
Maha Hadid
Shahbano Farooq

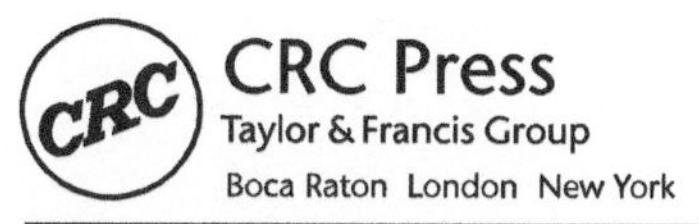

CRC Press is an imprint of the
Taylor & Francis Group, an informa business
A CHAPMAN & HALL BOOK

Designed cover image: Shutterstock_2081373361

First edition published 2025
by CRC Press
2385 NW Executive Center Drive, Suite 320, Boca Raton FL 33431

and by CRC Press
4 Park Square, Milton Park, Abingdon, Oxon, OX14 4RN

*CRC Press is an imprint of Taylor & Francis Group, LLC*



*Library of Congress Cataloging-in-Publication Data*
Names: Mathew, Sujith Samuel, author. | Kuhail, Mohammad Amin, 1982-author. | Hadid, Maha, author. | Farooq, Shahbano, author.
Title: The object-oriented approach to problem solving and machine learning with Python / Sujith Samuel Mathew, Mohammad Amin Kuhail, Maha Hadid and Shahbano Farooq.
Description: First edition. | Boca Raton, FL : CRC Press, 2025. | Includes bibliographical references and index. |
Identifiers: LCCN 2024033649 (print) | LCCN 2024033650 (ebook) | ISBN 9781032668338 (hbk) | ISBN 9781032668314 (pbk) | ISBN 9781032668321 (ebk)
Subjects: LCSH: Object-oriented programming (Computer science) | Python (Computer program language) | Machine learning.
Classification: LCC QA76.64 .M3845 2025 (print) | LCC QA76.64 (ebook) | DDC 005.1/17--dc23/eng/20241120
LC record available at https://lccn.loc.gov/2024033649
LC ebook record available at https://lccn.loc.gov/2024033650

ISBN: 978-1-032-66833-8 (hbk)
ISBN: 978-1-032-66831-4 (pbk)
ISBN: 978-1-032-66832-1 (ebk)

DOI: 10.1201/9781032668321

Typeset in Times
by SPi Technologies India Pvt Ltd (Straive)

Support material available at https://github.com/Object-Oriented-Programming-2024/Object-Oriented-Programming

# Contents

# 1 Introduction to Object-Oriented Programming

## 1.1 THE PARADIGM SHIFT – PROCEDURAL TO OO PROGRAMMING

Software is a collection of interoperable computer programs written in high-level programming languages, such as Python. A program is a logical sequence of instructions written to achieve a specific task. An instruction or a command is translated or compiled into machine language that computers can understand, and machine language is a combination of bits (0s and 1s).

Programming paradigms categorize high-level programming languages based on their capabilities and structure. One of the earliest programming paradigms was the unstructured programming paradigm.

### 1.1.1 Unstructured Programming

In the past, computer programs were written using the unstructured programming paradigm, where the code was a continuous block of instructions, and GOTO statements were used for redirecting the flow of control. A GOTO statement was used to jump from one line of code to another to execute the program logic (see Figure 1.1).

The data used in these programs was global data, i.e., it was accessible from anywhere in the program for its entire life cycle. As software became more prevalent in everyday systems, the size of programs grew and managing them became difficult. This approach was found to be harmful as computer programs failed and affected people's lives.

**Advantages of Unstructured Programs:**

- Freedom to express and test logical execution sequences of very small programs.
- Typically used as a scratchpad to check functionality.

**Disadvantages of Unstructured Programs:**

- Data is global, and this creates the possibility of unintentional changes to the state of the code.
- Reusability of code is not supported.
- Very difficult to manage, modify, scale, and debug programs.
- Large programs tend to turn into spaghetti code.

When an unstructured program is small, it is easy to read and understand. However, as the program's size increases, it becomes spaghetti code, a computer program that is hard to understand – like a bowl of tangled spaghetti. While the program may work, it isn't easy to follow, read, or manage.

DOI: 10.1201/9781032668321-1

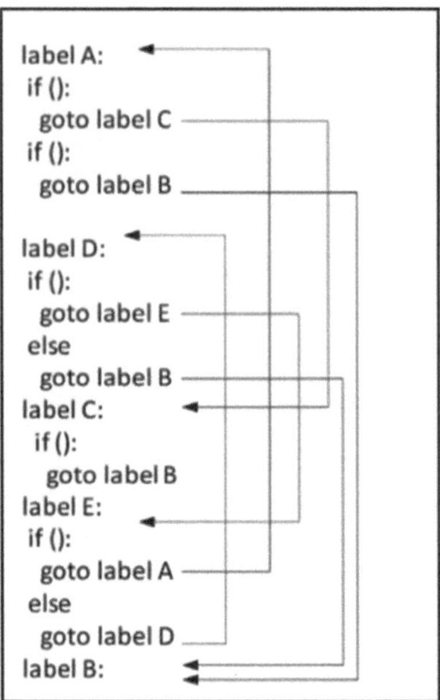


**FIGURE 1.1** A depiction of control flow using the unstructured programming paradigm.

### 1.1.2 Structured Programming

As the dependency on technology increased and program size increased, it became challenging to remember the flow of the program. Managing, debugging, and modifying programs became impossible when programs were unstructured. Structured programming was a solution to these problems.

Structured programming is based on the top-down approach in which a program is further divided into subprograms. Corrado Böhm and Giuseppe Jacopini formalized the structured programming concept in 1966. In structured programming, programmers use subprograms and loops instead of the inefficient GOTO statements, making the code more readable, efficient, and reusable. Structured programs are modular, decomposing a program into a cohesive set of subprograms. Structured programming focuses on breaking down a program into a collection of variables, data structures, and subprograms. These subprograms are often referred to as subroutines, procedures, or methods. Many programming languages like Python and C support the structured programming paradigm.

The data used in a structured program can be local data or global data.

- **Local Data:** The variables that are declared within a method body. Their scope is limited to within the method. Local variables cannot be accessed outside the method.
- **Global Data:** These are variables that are declared outside all methods. Their scope is within the whole program and accessible within the methods.

**Advantages of Structured Programming**

- Easy to manage, modify, and debug.
- Data can be either global or local.
- Reusability of code is supported.

**Disadvantages of Structured Programming**

- With code size constantly increasing, maintaining the code becomes difficult.

- Any significant changes required to the functionality may result in many changes in the code.
- Very large programs may still turn into spaghetti code if good coding standards are not followed and the code is poorly documented.
- It is not easy to represent real-life entities and relationships.

### 1.1.3 Object-Oriented Programming (OOP)

Real-world systems are becoming more and more reliant on software. For example, software systems used in banking, transportation, navigation, health care, education, and security heavily rely on software. Software has become very complex for all real-world systems, and the relative size has increased enormously. The need to manage software has become very critical in all areas of life. Therefore, reducing software complexity is important in reducing cost, reducing the effort to build and maintain software, and, often, life-saving because software controls real-world systems.

One major reason for software complexity is the difficulty in understanding the software model representing a software system. Having a good software model helps translate ideas (software requirements) into working software and helps to easily change or update software based on new requirements. Another reason for software complexity is the inability of programming languages to map real-world entities into programs easily and to manage the growing size and complexity of writing programs. These issues gave rise to the Object-Oriented Programming (OOP) paradigm that helps overcome these issues.

OOP is a powerful programming paradigm that centers around the concept of an object. These objects are essentially independent units of code structured to mimic real-world entities' behaviors and attributes. This approach creates a clear and intuitive way to understand relationships between objects. OOP utilizes a bottom-up approach, which involves piecing together smaller code units to create larger systems. This method allows developers to describe the smaller parts of the system in detail before integrating them into the larger program. By creating a collection of interacting objects, OOP provides an effective way to manage, modify, scale, and debug programs. One of the benefits of OOP is its data-hiding features, which enhance data security (Table 1.1).

## 1.2 THE CLASS AND THE OBJECT

An object in any software context is a uniquely identifiable representation of a component required in the software. The object may represent a real-world entity like a student or a car or a virtual entity

**TABLE 1.1**
**Structured vs. Object-Oriented Programming Paradigm**

| Structured Programming | Object-Oriented Programming |
|---|---|
| In structured programming, the program is divided into smaller subprograms called methods. | Object-Oriented Programming divides the program into small independent parts called objects. |
| Structured programming follows a top-down approach, dividing a program into methods. | Object-Oriented Programming follows a bottom-up approach. Here, the objects are first created and then assembled. |
| Structured programming does not have an access specifier for variables, so there is no easy way to hide data, making it less secure. | Object-Oriented Programming (OOP) uses access specifiers like private, public, and protected to provide a certain level of data security. OOP also provides data hiding, making it more secure. |
| Code reusability is comparatively limited in structured programming. | Code reusability is a major benefit of object-oriented programming. |

like a button or menu item on the page of a mobile application. In other words, any entity that must be represented in a program is considered an object. For example, Sarah, who is an employee in an organization; Samir, who is a student at the university; or my sports car, are all examples of objects that can be represented in software.

To represent these objects in software, we need to create a class. A class is the template of an object, i.e., it is used to create or model an object. For example,

- A class named "Car" could create multiple instances of software representations of real-world cars. A red-colored sports car or a black-colored convertible could be objects of the class Car.
- A class named "Student" can be used to create multiple instances of students. Mariam or Samir, university students, could be objects of the class Student.
- A class "Employee" can be used to create multiple instances of an employee. Sarah, a manager, or Carson, a programmer, could be objects of the class Employee.

### 1.2.1 Identifying and Creating Objects

OOP follows the bottom-up approach to programming, and therefore, the first step in building an OO program is to identify the objects that constitute the program. So, how do we represent a real-world entity like a car in an OOP?

- First, identify the object, its attributes, and its behaviors (functionalities).
- Next, create classes (templates) to represent all the objects.
- Identify relationships between the objects.
- Finally, use the class to create the objects to participate in the flow of the program.

An intuitive method for identifying objects would be to define use cases and scenarios about the program or software being analyzed. For example, a use case for vehicle software is "Turn on the vehicle ignition", and a scenario would be "James turns on the ignition of his car with his key", Scenarios and use cases are usually written in a natural language or with formal standard modeling tools like unified modeling language (UML) diagrams to indicate the various software requirements. These concepts are discussed in detail in the following unit of this book. In the example above, James's *car* and *car key* are relevant to the scenario. Considering that the car is one of the objects to represent in the software, we would need a class as a template to represent the object. Let's call the class Vehicle.

Consider the Python script below, which shows the creation of the object "james_car", an instance of the class *Vehicle*.

**CODE 1.1 CODE TO CREATE A VEHICLE OBJECT BASED ON THE VEHICLE CLASS**

```
class Vehicle:
    """Class to represent any vehicle"""
    pass


# Create an object of class Vehicle
james_car = Vehicle()
```

In Code 1.1 snippet:

- In Line 1, the keyword "class" defines the class that would model or define any vehicle in the program.
- Notice the indentation in Line 2, which indicates that the line is included within the class. The line is called the "doc_string" and provides information specific to the class.
- Line 3 has the keyword "pass", a placeholder for code that will be written later.
- Lines 4 and 5 are empty.
- Line 6 is a comment statement in Python.
- Line 7 is where the object is created. Compiling this line creates an object named "james_car" as an instance of the class Vehicle.

### 1.2.2 The Class Constructor and the Class Initializer

In Python, creating an object of a class leads to the execution of two built-in methods. The "__new__()" method, which is the constructor that creates an instance of the class, and the "__init__()" method, which is the initializer, are used to assign values to the class's attributes when an object is created. When Python executes Line 12 in Code 1.2, "james_car = Vehicle()", it first calls the "Vehicle.__new__()" method and then calls the "Vehicle.__init__()" method.

Programmers often define an "__init__()" method for their classes, where the class will override the default behavior of the "__init__()" method. However, programmers usually do not write the "__new__()" method, as overriding "__new__()" is uncommon.

The "__init__()" method initializes the data attributes and is always called once an object is created.

**CODE 1.2 INITIALIZING VARIABLES OF AN OBJECT USING THE __INIT__() METHOD**

```
1. class Vehicle:
2.     """Class to represent any vehicle"""
3.
4.     # Defining the attributes
5.     def __init__(self):
6.         self.manufacturer = "" # default value for string
7.         self.num_Doors = 0 # default value for an integer
8.         self.fuel_capacity = 0.0 # default value for a float
9.
10.
11. # Create objects of class Vehicle
12. james_car = Vehicle()
13. her_car = Vehicle()
```

In Code 1.2, the class "Vehicle" defines the "__init__()" method to initialize the attributes that define a vehicle. When the two objects are created (Lines 12 and 13), each will have its own copy of the three attributes, i.e., "manufacturer", "num_doors", and "fuel_capacity". These data attributes are initialized to their default values.

The parameter "self" in the "__init__()" method refers to each of the objects constructed by the "__new__()" method, and by convention, it is called "self". Though self is included in the "__init__()" method as a parameter, note that it is not referred to when the object is created in Lines 12 and 13.

The initializer can be parameterized, i.e., values can be passed to the method to be assigned to the data attributes. In this case, when the object is created, the first argument remains to be *self* as a reference to the instance constructed, and the programmer does not provide a value for it. The rest of the arguments are provided by the programmer.

**CODE 1.3 INITIALIZING THE VEHICLE OBJECT BY PASSING ARGUMENTS TO THE __INIT__() METHOD**

```
1. class Vehicle:
2.     """Class to represent any vehicle"""
3.
4.     # Define the attributes
5.     def __init__(self, manufacturer="", num_doors=0, fuel_capacity=0.0):
6.         self.manufacturer = manufacturer
7.         self.num_doors = num_doors
8.         self.fuel_capacity = fuel_capacity
9.
10.
11. # Create objects of class Vehicle
12. james_car = Vehicle()
13. her_car = Vehicle("Ford", 5, 65.5)
```

Code 1.3 shows the "__init__()" method has three parameters in addition to the parameter "self". When the object "james_car" is created in Line 12, all parameters take the default value provided in Line 5. When the object "her_car" is created in Line 13, three arguments are passed, which map to the three parameters in Line 5, excluding the first parameter, "self".

### 1.2.3 Identifying Attributes of an Object

Attributes define an object's distinctive characteristics or properties that need to be represented in software. For example, a vehicle has a manufacturer and number of doors, which are represented as variables of the vehicle class. The following steps need to be considered to identify attributes of an object that should be represented in software.

- Start by identifying the real-world object or concept to model in the software. For example, if a car is to be represented in software, think about what attributes a car has in the real world.
- List all the relevant attributes of the real-world object. These properties may include size, color, weight, speed, capacity, and other descriptive features important for the software's purpose. The more attributes, the more similar the object would be to the real-world object.
- Determine the appropriate data type for each attribute. Data types include integers, floats, strings, booleans, or custom-defined types. The choice of data type should reflect the property's nature, for example, an integer for the number of doors and a string for the car's model.
- Decide on the accessibility (public, private, and protected) and encapsulation (whether to use getter and setter methods) for each instance variable based on the design and the principle of information hiding. Public variables can be accessed directly, while private or protected variables may require accessor methods. This is discussed further in Section 1.2.5.
- In some cases, default values for the attributes need to be set for initializing the object when it is created, either in the initializer or with default values in the variable declarations.

For instance, let us consider the Vehicle class in Code 1.3. Many attributes define a vehicle. Some common attributes of most vehicles could be:

- "manufacturer": String
- "model": String
- "color": ENUM (ENUM stands for enumerated values, i.e., a list of possible values)
- "num_doors": Integer
- "fuel_capacity": Float
- "engine_state": Boolean
- "current_speed": Float

### 1.2.4 Identifying Behaviors of an Object

Behavior is an action that retrieves or changes the state of an object. It provides the means for interaction with the object and is also called a function or method. They are verbs that define the object's action. For example, a vehicle created in Code 1.3 would have methods like:

- "turn_left(angle: Float)": Boolean
  - The car turns left at a certain angle, and the method returns true or false.
- "turn_right(angle: Float)": Boolean
  - The car turns right at a certain angle, and the method returns true or false.
- "stop_car()": Boolean
  - The car stops, and the method returns true or false.
- "accelerate(speed: Float)": Boolean
  - The car increases speed by a certain value and returns true or false.
- "check_remain_distance(remain_fuel: Float, current_speed: Float)": "remaining_distance": Float
  - The method calculates the remaining distance based on the remaining fuel and current speed and returns the distance, which is a float value.

Moreover, for each attribute identified for the real-life object, the associated set (mutator) and get (accessor) methods may be included, also known as the setter/getter methods. A set method is created to provide an initial value for the attribute or update the value of an attribute, and a get method is created to get or access the value of this attribute. Object-Oriented (OO) design discourages direct access to an attribute of the class and encourages programmers to get and set the attributes through the setters/getters. Facilitating the class with setter and getter methods provides controlled access to the attributes. These attributes' implementation and validation checks would be from within these methods.

- Examples of Getter methods for the "my_car" class are:
  - "get_manufacturer()": String
  - "get_number_of_doors()": Integer
  - "get_engine_state()": Boolean
  - "get_current_speed()": Float
- Examples of Setter methods for my_car class are:
  - "set_engine_state(Boolean)": Boolean
  - "set_current_speed(float)": Boolean
  - "set_color(ENUM)": Boolean

## CODE 1.4 STRUCTURE OF A PYTHON CLASS WITH ATTRIBUTES AND BEHAVIORS TO REPRESENT A VEHICLE

```
from enum import Enum

class Color(Enum):
    """An enumerator type class that defines colors"""
    WHITE = 1
    BLACK = 2
    RED = 3


class Vehicle:
    """Class to represent any vehicle"""

    # Define the attributes
    def __init__(self):
        self.manufacturer = ""
        self.model = ""
        self.num_doors = 0
        self.color = Color.WHITE
        self.fuel_capacity = 0.0
        self.current_speed = 0.0  #Initialize a speed property with a default value of 0

    # Find below the behaviors of the object
    def turn_left(self):
        pass

    def turn_right(self):
        pass

    def stop_car(self):
        pass

    def accelerate(self):
        pass

    # The methods below represent the Setter and Getter methods

    def get_manufacturer(self):
        pass

    def get_number_of_doors(self):
        pass

    def get_engine_state(self):
        pass

    def set_engine_state(self):
        pass

    def set_current_speed(self):
        pass

    def set_color(self):
        pass


# Create an object of class Vehicle
james_car = Vehicle()
```

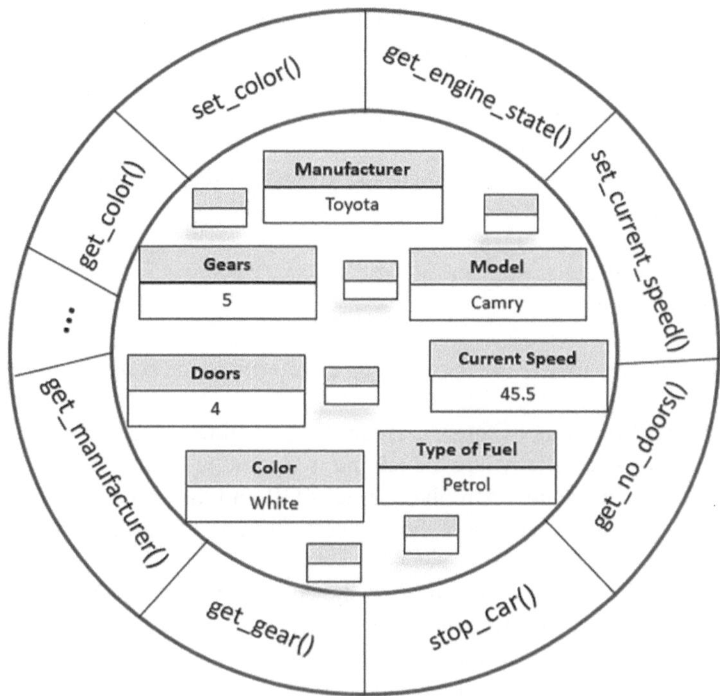


**FIGURE 1.2** Encapsulation of attributes and behaviors of a class.

### 1.2.5 Encapsulation

Encapsulation is a fundamental concept in OOP, which is the bundling of data and methods within a single unit, i.e., an object defined by its class. It allows for the abstraction of intricate systems by hiding internal details and showcasing only necessary functionalities through well-defined interfaces. Access modifiers (public, private, and protected) control access to variables, ensuring data integrity is maintained. Classes encapsulating data create a clear separation between what an object does and how it does it, making it easier to design modular, reusable, and maintainable code. This principle helps develop self-contained components that are easily manageable in software development.

For instance, in Figure 1.2, the car's attribute "current_speed" should not be accessed by an external entity without proper access rights. Therefore, access to "current_speed" is only provided through the methods "set_current_speed()" and "get_current_speed()". Encapsulation is also understood as information hiding, as this provides the required protection for the attributes to ensure that the object does not get into an error state.

Unlike earlier programming paradigms, in OOP, an object encapsulates a real-world entity's attributes, behaviors, and states in a single unit. The state of an object, represented by the value of the attributes, is hidden inside the object and is accessible via its behaviors or methods.

## 1.3 THE STATE OF AN OBJECT

An object's "state" refers to the complete set of attributes and values defining its present condition and characteristics. Within OOP, an object's state represents a snapshot of its data at any given moment. This includes the values contained within its instance variables or properties, reflecting the information the object holds. The object's state explains the behavior and interaction of the object within a program. Therefore, it is important to understand the state of an object. Amending a state's object requires a change in its attributes, which can lead to different behaviors and outcomes. Effectively managing an object's state is paramount for ensuring data integrity and achieving the desired functionality within OO systems. State changes may be triggered by either internal processes or external events, illustrating the dynamic nature of objects as they progress throughout program execution.

For example, the state of "my_car" changes depending on the values of the attributes "engine_state" and "current_speed".

- "engine_state": Boolean
- "current_speed": Float

If we consider the above attributes and their values to determine the state of "my_car", then

- "my_car" could be in a state where the engine is On and the speed is 0.
- "my_car" could be in a state where the engine is On and the speed is 40.
- "my_car" could be in a state where the engine is Off and the speed is 0.
- "my_car" cannot be in a state where the engine is Off and the speed is 40.

Therefore, the state of "my_car" is determined by the changes to the attributes. Thus, it is important to protect an object's state by protecting the values assigned to the attributes. The values of the attributes are accessed, assigned, and modified using the object's behaviors.

Consider the example shown in Figure 1.3, a class "LightBulb" with one instance of the object "smart_light". This object contains values for the attributes "is_on" and "brightness_level", which determine its current state.

A "75%" "brightness_level", for example, and if "is_on" is set to "True", this means that the bulb is actually emitting light. Modifying these attributes will change the object's state, such as adjusting the brightness level or setting "is_on" to "False", which would result in a corresponding change in its behavior, like turning off the light or changing its brightness level.

Code 1.5 shows an implementation of the example. Starting with the class initializer (Lines 2–4), it initializes a Lightbulb object with default attributes – setting the lightbulb's initial state to False, indicating it is "off", and its brightness level to 0. This establishes the initial state of the object when it is instantiated.

Several methods in the class can regulate the lightbulb's state. One method is the "turn_on()" method (Lines 7 and 8), suggesting that it's emitting light; it changes the state to True, indicating that the bulb is "on". Another method is the "turn_off()" method (Lines 10–11), turning off the light and changing the state back to false, suggesting it is "off". The class also includes the "set_brightness()" method (Lines 13–17), which is responsible for adjusting the brightness level of the lightbulb. This method takes an input "brightness_level" and checks whether it is within a feasible range of 0–100. The "__str__()" method (Lines 19–21) formats and returns a string that indicates whether the lightbulb is on or off and displays its current brightness level as a percentage, providing a human-readable format of the lightbulb's current state.

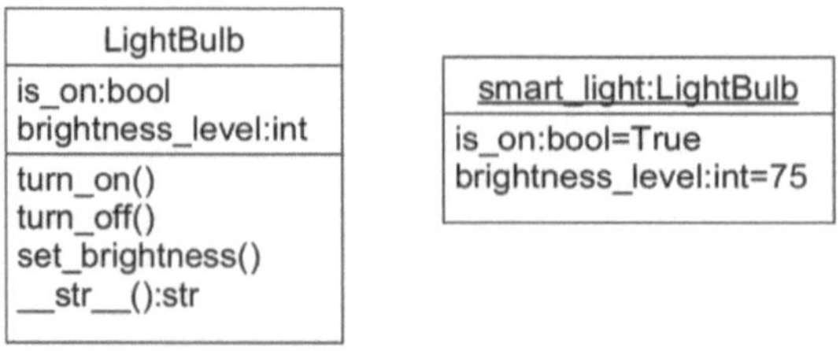

| LightBulb |
| --- |
| is_on:bool<br>brightness_level:int |
| turn_on()<br>turn_off()<br>set_brightness()<br>__str__():str |

| smart_light:LightBulb |
| --- |
| is_on:bool=True<br>brightness_level:int=75 |

**FIGURE 1.3** State of the Smart_light instance based on the Class LightBulb.

Moving on to the code's practical execution, an instance of the "LightBulb" class is instantiated (Line 25), which serves as a sample object. The code then interacts with this instance, turning on the lightbulb (Line 31), increasing its brightness to 75% (Line 32), displaying its status (Line 35), turning it off (Line 38), and lastly displaying its state (Line 41).

**CODE 1.5 LIGHTBULB CLASS WITH METHODS TO HELP SET AND UPDATE THE STATE OF AN INSTANCE**

```
1.  class LightBulb:
2.      def __init__(self, is_on=False, brightness_level=0):
3.          self.is_on = is_on
4.          self.brightness_level = brightness_level
5.
6.      # The below methods regulate the Lightbulb state.
7.      def turn_on(self):
8.          self.is_on = True
9.
10.     def turn_off(self):
11.         self.is_on = False
12.
13.     def set_brightness(self, level):
14.         if 0 <= level <= 100:
15.             self.brightness_level = level
16.         else:
17.             print("Brightness level should be between 0 and 100.")
18.
19.     def __str__(self):
20.         return ( f"Lightbulb ({'On' if self.is_on else 'Off'}), "
21.                  f"Brightness: {self.brightness_level}%")
22.
23.
24. # Create a Lightbulb object
25. my_lightbulb = LightBulb()
26.
27. # Display initial state
28. print(my_lightbulb)
29.
30. # Turn on the lightbulb and set the brightness
31. my_lightbulb.turn_on()
32. my_lightbulb.set_brightness(75)
33.
34. # Display updated state
35. print(my_lightbulb)
36.
37. # Turn off the lightbulb
38. my_lightbulb.turn_off()
39.
40. # Display final state
41. print(my_lightbulb)
```

## 1.4 ACCESSING CLASS MEMBERS

Accessing class members requires interaction with variables and methods specified within a class. Variables are responsible for storing data, whereas methods are in charge of defining particular behaviors. Variables can be instances or static. Instance variables include data unique to each object generated by the class, but static variables are shared by all class instances, providing consistent data storage. Likewise, methods might be instances or static. Instance methods allow users to access and edit object instance variables, whereas static methods are class-related but do not access instance-specific data.

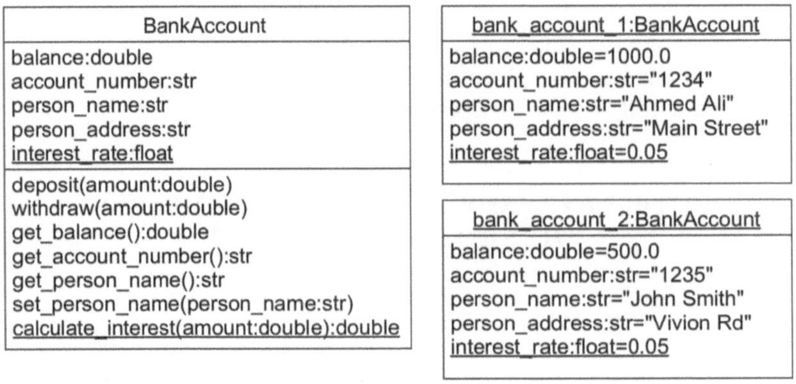

BankAccount
balance:double
account_number:str
person_name:str
person_address:str
interest_rate:float
deposit(amount:double)
withdraw(amount:double)
get_balance():double
get_account_number():str
get_person_name():str
set_person_name(person_name:str)
calculate_interest(amount:double):double

bank_account_1:BankAccount
balance:double=1000.0
account_number:str="1234"
person_name:str="Ahmed Ali"
person_address:str="Main Street"
interest_rate:float=0.05

bank_account_2:BankAccount
balance:double=500.0
account_number:str="1235"
person_name:str="John Smith"
person_address:str="Vivion Rd"
interest_rate:float=0.05

**FIGURE 1.4** The BankAccount class with two instances: bank_account_1 and bank_account_2.

Accessing class members is accomplished through the dot notation. This involves placing a dot and member name after the object's instance or the class itself.

Code 1.6 shows the "BankAccount" class; in its context, the concepts of instance and static variables and methods will be discussed, in addition to the process of accessing class members. As illustrated in Figure 1.4, the "BankAccount" class instance variables store unique data for each bank account object, such as "balance", "account_number", "person_name", and "person_address". These values can vary between different account objects. For instance, the "bank_account_1" and "bank_account_2" objects feature unique values for the instance variables of "balance", "account_number", "person_name", and "person_address". On the other hand, the static variable "interest_rate" is shared among all class instances. It holds a steady value for all bank accounts and is accessed using the class name "BankAccount.interest_rate". As shown in Code 1.6, "bank_account_1" and "bank_account_2" share the same "interest_rate" value.

Code 1.6 shows an implementation of the "BankAccount" class. The dot notation is used to access class members (variables and methods). The object instance uses the dot notation to access instance variables or instance methods. "Bank_account_1.get_account_number()", for example, returns the account number linked with "bank_account_1" (Line 45). In the same way, "bank_account_1.deposit(50)" invokes the deposit method on "bank_account_1" (Line 48).

The dot notation with the class name is used to access static variables or static methods. For instance, "BankAccount.interest_rate" retrieves the interest_rate static variable (Line 60), and "BankAccount.calculate_interest()" calls the static method calculate_interest (Line 63).

**CODE 1.6 CLASS BANKACCOUNT**

```
1. class BankAccount:
2.     interest_rate=0.05
3.     def __init__(self,balance,account_number,person_name,person_address):
4.         self.balance=balance
5.         self.account_number=account_number
6.         self.person_name=person_name
7.         self.person_address=person_address
8.
9.     # Define a method to deposit money into the account.
10.    def deposit(self,amount):
11.        if amount>=0:
12.            self.balance+=amount
13.
14.    # Define a method to withdraw money from the account.
15.    def withdraw(self,amount):
```

```
        if amount<=self.balance:
            self.balance-=amount

    # The methods below represent the Setter and Getter methods
    def get_balance(self):
        return self.balance

    def get_account_number(self):
        return self.account_number

    def get_person_name(self):
        return self.person_name

    def get_person_address(self):
        return self.person_address

    def set_person_address(self,person_address):
        self.person_address=person_address

    def calculate_interest(amount):
        return BankAccount.interest_rate * amount

# Create two instances of BankAccount
bank_account_1=BankAccount(balance=1000,account_number="1234",
                           person_name="Ahmed Ali",person_address="Main Street")
bank_account_2=BankAccount(balance=500,account_number="1235",
                            person_name="John Smith",person_address="Vivion Rd")

# The below line  will access the account number of object bank_account_1
print("Bank No: ",bank_account_1.get_account_number())
print("Depositing 50 from the account")

# The below line will access the deposit method of object bank_account_1
bank_account_1.deposit(50)

# The below line will access the balance of object bank_account_1
print("Balance: ",bank_account_1.get_balance())

# The below line  will set the address of object bank_account_1
bank_account_1.set_person_address("Union Str")

# The below line  will access the person name of object bank_account_1
print("Bank Person Name: ",bank_account_1.get_person_name())

# The below will retrieve the static variable interest_rate
print("Interest Rate: ",BankAccount.interest_rate)

# The below will call the static method calculate_interest
print("Interest: ",BankAccount.calculate_interest(bank_account_1.get_balance()))
```

## 1.5 PRIVATE AND PUBLIC ATTRIBUTES

Attributes within a class can be categorized into two types: private and public. Public attributes are accessible and can be modified from outside the class, which provides direct interaction with the data. Conversely, private attributes are encapsulated and intended to be accessed and manipulated solely within the class. This encapsulation enhances data security, prevents unintended modifications, and promotes regulated interaction with the internal workings of the class. Public attributes are denoted by their regular names, while private attributes are typically indicated with double-leading underscores (for instance, __private_attribute).

We illustrate the concept of public variables and methods with an example of a "Circle" class. Figure 1.5 shows the "Circle" class diagram. Public variables and methods are shown with a plus

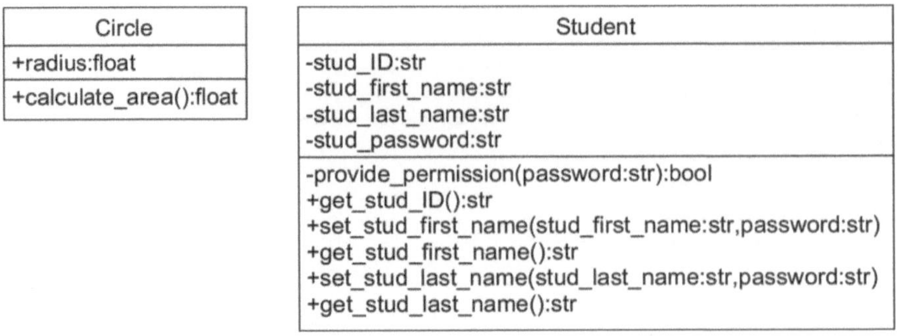

| Circle |
| --- |
| +radius:float |
| +calculate_area():float |

| Student |
| --- |
| -stud_ID:str<br>-stud_first_name:str<br>-stud_last_name:str<br>-stud_password:str |
| -provide_permission(password:str):bool<br>+get_stud_ID():str<br>+set_stud_first_name(stud_first_name:str,password:str)<br>+get_stud_first_name():str<br>+set_stud_last_name(stud_last_name:str,password:str)<br>+get_stud_last_name():str |

**FIGURE 1.5** Two class diagrams: Circle Class and Student Class with their attributes and methods.

sign in the diagram. Code 1.7 shows the implementation of the public variables and methods. On Line 5, "radius" is initialized as a public variable, accessible from outside the "Circle" class. "Radius" is identifiable as a public variable because there are no leading underscores before its name. From Lines 8 to 9, the "calculate_area ()" method is defined as a public method accessible externally. Line 16 demonstrates accessing the public radius attribute directly. Line 17 shows how to call the public "calculate_area ()" method.

**CODE 1.7 CIRCLE CLASS BASED ON THE CORRESPONDING CLASS DIAGRAM IN FIGURE 1.5**

```
1. import math
2.
3. class Circle:
4.     def __init__(self, radius):
5.         self.radius = radius
6.
7.     # The below public method will calculate the area of the circle
8.     def calculate_area(self):
9.         return math.pi * self.radius ** 2
10.
11.
12. # Create a Circle instance
13. circle1 = Circle(5)
14.
15. # Calculate and print the area and circumference
16. print("Circle Radius:", circle1.radius)
17. print("Circle Area:", circle1.calculate_area())
```

In the Student class Code 1.8, double underscores are used to define private variables, like "__stud_ID", "__stud_first_name", "__stud_last_name", and "__stud_password" (Lines 2–6). These variables can be modified and accessed only within the class methods. The "__provide_permission()" method (Lines 11 and 12) is also considered private using the double underscores. This method checks whether a given password matches a stored student's password. As it's declared private, it's designed for internal use within the class and isn't intended to be accessed from outside.

The methods in Lines 15–21, "set_stud_first_name()" and "set_stud_last_name()", support changing students' first and last names using a password. The "__provide_permission()" method validates the password before making any changes. The public "get_stud_first_name()" and "get_stud_last_name()" methods (Lines 23–27) allow retrieving students' first and last names.

An instance of the "Student" class is created (Lines 31 and 32) using the provided attributes. To access and print the initial first name of the student, the "get_stud_first_name()" method was used (Line 35). The "set_stud_first_name()" method is called to change the first name to "Jonathan" (Line 38). Since the correct password "John123" is provided, the first name is successfully updated, which is retrieved using "get_stud_first_name()" and printed (Line 41).

**CODE 1.8 CLASS STUDENT WITH PRIVATE AND PUBLIC METHODS**

```
1. class Student:
2.     def __init__(self, stud_ID, stud_first_name, stud_last_name, stud_password):
3.         self.__stud_ID = stud_ID
4.         self.__stud_first_name = stud_first_name
5.         self.__stud_last_name = stud_last_name
6.         self.__stud_password = stud_password
7.
8.     def get_stud_ID(self):
9.         return self.__stud_ID
10.
11.     def __provide_permission(self, password):
12.         return password == self.__stud_password
13.
14.     # The set and get methods for students' first and last names are defined
15.     def set_stud_first_name(self, stud_first_name, password):
16.         if self.__provide_permission(password):
17.             self.__stud_first_name = stud_first_name
18.
19.     def set_stud_last_name(self, stud_last_name, password):
20.         if self.__provide_permission(password):
21.             self.__stud_last_name = stud_last_name
22.
23.     def get_stud_first_name(self):
24.         return self.__stud_first_name
25.
26.     def get_stud_last_name(self):
27.         return self.__stud_last_name
28.
29.
30. # Create a Student instance
31. stud_1 = Student(stud_ID="S000123", stud_first_name="John",
32.                  stud_last_name="Davis", stud_password="John123")
33.
34. # Access and print the first name using the get_stud_first_name method
35. print("First Name: ", stud_1.get_stud_first_name())
36.
37. # Use set_stud_first_name to attempt changing the first name with the correct password
38. stud_1.set_stud_first_name("Jonathan", "John123")
39.
40. # Print the updated first name using get_stud_first_name
41. print("First Name: ", stud_1.get_stud_first_name())
```

## 1.6 OBJECTS AS RETURN VALUES

Programmed methods can be designed to return objects instead of simple primitive values like numbers and strings. The returned objects encapsulate related data and behaviors, promoting modularity and enhancing the code's readability. These objects can effectively represent real-world entities and facilitate interactions between different program parts, resulting in a structured approach to software design that enhances codebase maintainability and scalability.

To explain the idea, we utilize the Point class displayed in Code 1.9. The "Point" class has two attributes, "x" and "y", representing coordinates. It contains a "display()" method (Lines 6 and 7) that formats and returns the point coordinates as a string. The "create_point()" method (Lines 11 and 12) creates and returns a "Point" object using the given coordinates. The code also shows how we used the "create_point()" method to create two "Point" objects (Lines 16 and 17).

**CODE 1.9 CLASS POINT WITH A CREATE_POINT METHOD THAT RETURNS A POINT OBJECT**

```
1. class Point:
2.     def __init__(self, x, y):
3.         self.x = x
4.         self.y = y
5.
6.     def display(self):
7.         return f"({self.x}, {self.y})"
8.
9.
10. # The method below create and return a Point object using the provided coordinates
11. def create_point(x, y):
12.     return Point(x, y)
13.
14.
15. # Create points using the method
16. point1 = Point.create_point(2, 3)
17. point2 = Point.create_point(-1, 5)
18.
19. # Display the coordinates of the points
20. print("Point 1:", point1.display())
21. print("Point 2:", point2.display())
```

## 1.7 BUILT-IN ATTRIBUTES AND METHODS

Built-in variables and methods are inherent elements provided by programming languages like Python to facilitate various operations and interactions within the code. These predefined variables and methods are essential building blocks for classes. They enable functionalities such as identifying object characteristics (e.g., "__class__()" for class name), managing object string representations (e.g., "__str__()"), supporting equality checks (e.g., "__eq__()"), and comparisons (e.g., "__lt__()"). Built-in variables and methods offer standardized functionalities that can be leveraged across different classes.

Code 1.10 is an implementation of a "Person" class highlighting the usage of built-in variables and methods. The code snippet defines a "Person" class (Lines 1–4), encapsulating the attribute's "name" and "age" within its initializer method "__init_()". The "__str__()" method (Lines 7 and 8) showcases a built-in method that enables customization of the string representation of a "Person" object. By implementing this method, the display of the object can be controlled when the "__str__ ()" method is used, promoting clear and informative output. For instance, in code Line 25, when we print the "Person" object, the "print()" method invokes the built-in "__str__()" method to show the string representation of the person.

The "__eq__()" method (Lines 11 and 12) demonstrates another built-in method, allowing custom equality checks between "Person" objects. By comparing the "name" and "age" attributes, the method determines whether two instances are equal using the comparison "==" operator. In Line 26, we check if two persons are equal, which will cause Python to invoke the "__eq__()" method.

The "__lt__()" method (Lines 15 and 16) is also a built-in method allowing the customized definition of a less than comparison between two "Person" objects. In this case, we compare whether this person's age is less than the other person's. Line 27 shows how this built-in method is invoked to compare two "Person" objects.

An example of a built-in variable "__class__()" is the one shown in Line 24, allowing access to the class name, "Person".

**CODE 1.10 CLASS PERSON USING BUILT-IN METHODS AND VARIABLES**

```
1. class Person:
2.     def __init__(self, name, age):
3.         self.name = name
4.         self.age = age
5.
6.     # The below method shows the string representation of the person.
7.     def __str__(self):
8.         return f"{self.name}, {self.age} years old"
9.
10.    # The below method determines whether two instances are equal when using '=='
11.    def __eq__(self, other):
12.        return self.name == other.name and self.age == other.age
13.
14.    # The '__lt__' allows to define a custom '<' comparison between two Person objects.
15.    def __lt__(self, other):
16.        return self.name == other.name and self.age == other.age
17.
18.
19. # Create Person instances
20. person1 = Person("Alice", 30)
21. person2 = Person("Bob", 25)
22.
23. # Use built-in attributes and methods
24. print("Class Name:", person1.__class__.__name__)
25. print("String Representation:", person1)
26. print("Are they the same age?", person1 == person2)
27. print("is Alice younger?", person1 < person2)
```

## 1.8 CASE STUDY

We present a case study that exemplifies the practical application of OOP principles in a real-world scenario. Specifically, we offer an example implementation that illustrates the creation and utilization of financial transactions of transaction management with two classes: "Transaction" and "TransactionManager", as illustrated in Figure 1.6.



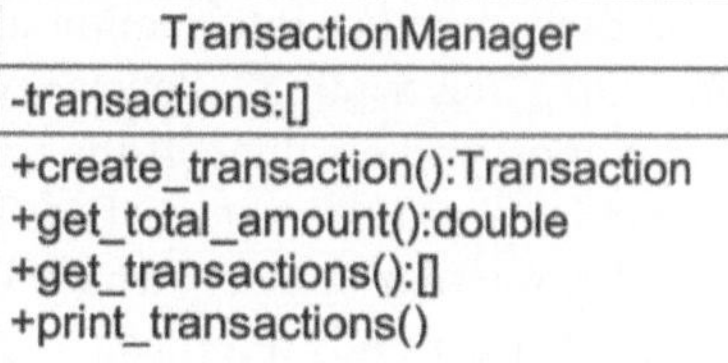


**FIGURE 1.6** Class Diagram for the Transaction class and the Transaction Manager class.

Code 1.11 shows the implementation of the transaction management system. First, we import the libraries needed in the project (Lines 1 and 2). Second, we implement two enumerator-type classes (Lines 4–14). An "Enum" class is a fixed or constant set of related values. For example, we can create an "Enum" class for a list of countries or a range of colors. In this case, we have created "Enum" classes for transaction type and currency (Lines 4–14).

Next, we show the implementation of the "Transaction" class (Lines 16–61). The class keeps track of the total number of transactions created using the "transaction_count" static variable (Line 18). We used a static variable to ensure that the number of transactions is modified at the class level. The code also includes an initializer (Lines 21–30) allowing users to initialize the transaction's private attributes: (1) "trans_id": a unique identifier. This identifier is equal to the current number of transactions. (2) "trans_type": a transaction type denoting whether it's an income or expense. (3) "trans_amount": the transaction amount (trans_amount), the currency employed in the transaction. (4) "trans_description": a transaction description that provides additional particulars. (5) "trans_currency": one of the currency types from the currency class to represent the currency of the transaction. (6) "trans_date_time": a timestamp capturing the date and time of the transaction.

The attributes are kept private within the class to ensure they cannot be modified from outside. This is crucial because transaction data is sensitive and requires protection. Further, in a real-life scenario, its information is immutable once a transaction is created. Apart from the transaction description "trans_description", the transaction attributes have getter methods but no setter methods (Lines 32–48), making them read-only. However, the "trans_description" has both a getter and a setter method (Lines 47–51).

The "__str__()" method in the "Transaction class" (Lines 54–61) creates a concise and readable representation of a transaction object when it needs to be converted to a string. This method is automatically invoked when the "__str__()" method is used or when one tries to print the object. It constructs a formatted string containing transaction details like "ID", "amount", "currency", "type", "description", and "date_time". The placeholders within the string, like {id} and {amount}, are filled in with actual values using the ".format()" method. These values are fetched from the object's attributes, such as "self.__trans_id" for the transaction ID and "self.trans_amount" for the amount. The currency and transaction type are converted from internal codes to human-readable names using the "Currency" and "TransactionType" classes.

Next, we show the implementation of the "TransactionManager" class, which manages a collection of financial transactions, allowing users to create new transactions, calculate the total transaction amount, and print out transaction details. Let's break down the class explanation by referring to the line numbers:

Lines 69–72: The "__init__()" method initializes the "TransactionManager" object by creating an empty list named "__transactions" to store the managed transactions. Chapter 2 explains the concept of lists in more detail.

Lines 75–114: The "create_transaction()" method allows users to create a new transaction. It interacts with the user to gather information about the transaction, such as its type ("expense" or "income"), "amount", "description", "currency", and "date_time". The user inputs are processed and converted into appropriate values. If the transaction type is an expense, the amount is negated. The date/time input is parsed to create a "DateTime" object. The collected information is then used to create a new "Transaction" object, which is added to the list of transactions "__transaction". Finally, the created transaction is returned.

Lines 117–123: The "get_total_amount()" method calculates and returns the total amount of all transactions managed by the "TransactionManager". It iterates through the list of transactions and accumulates their amounts.

Lines 126–128: The "print_transactions()" method prints out the details of each transaction managed by the "TransactionManager" class. It iterates through the list of transactions and uses the "print()" method to display their details.

Lines 131–137: A "TransactionManager" instance named "trans_manager" is created to manage financial transactions. Subsequently, three transactions (t1, t2, and t3) are generated using the "create_transaction()" method of the "trans_manager". Then the "print_transactions()" method displays the details of all the transactions. Finally, the code prints the total transaction amount by utilizing the "get_total_amount()" method from the "trans_manager", showing the cumulative sum of transaction amounts.

## CODE 1.11 TRANSACTION AND TRANSACTIONMANAGER CLASSES AND TWO ENUM CLASSES FOR TRANSACTIONTYPE AND CURRENCY

```
1. import datetime
2. from enum import Enum
3.
4. class TransactionType(Enum):
5.     """An enumerator type class that defines the types of transactions"""
6.     INCOME=1 #an income type defines a transaction of a gained amount of money
7.     EXPENSE=2 #an expense type defines a transaction of a spent amount of money
8.
9. class Currency(Enum):
10.     """An enumerator type class that defines the types of currencies"""
11.     USD=1 #US Dollars
12.     EUR=2 #Euro
13.     GBP=3 #Great Britain Pound
14.     AED=4 #Arab Emirati Dirham
15.
16. class Transaction:
17.     """A class that represents a financial transaction"""
18.     transaction_count=0 #a static variable that keeps track of the number of transactions
19.
20.     # Initialize the Transaction class
21.     def __init__(self,transaction_type,amount,currency,description,date_time):
22.         # Increment the number of transactions with the creation of a new transaction
22.         Transaction.transaction_count+=1
23.
24.         # Assign a unique transaction ID based on the number of transactions
25.         self.__trans_id= "T" + str(Transaction.transaction_count)
26.         self.__trans_type=transaction_type
27.         self.trans_amount=amount
28.         self.__trans_description=description
29.         self.__trans_currency=currency
30.         self.__trans_date_time=date_time
31.
32.     def get_trans_id(self):
33.         return self.__trans_id
34.
35.     def get_transaction_type(self):
36.         return self.__trans_type
37.
38.     def get_trans_amount(self):
39.         return self.trans_amount
40.
41.     def get_trans_date_time(self):
42.         return self.__trans_date_time
43.
44.     def get_trans_currency(self):
45.         return self.__trans_currency
46.
47.     def get_trans_description(self):
48.         return self.__trans_description
49.
50.     def set_trans_description(self, description):
51.         self.__trans_description=description
```

```

    # Define a string representation of the transaction
    def __str__(self):
        return ("Transaction ID: {id}, Amount: {amount} {currency}, Type: ({type}), "
                "Description: {description}, Date & Time: {date_time}"
                .format(id=self.__trans_id, amount=self.trans_amount,
                        currency=Currency(self.__trans_currency).name,
                        type=TransactionType(self.__trans_type).name,
                        description=self.__trans_description,
                        date_time=self.__trans_date_time))


class TransactionManager:
    """This class keeps track of transactions and allows users
    to create transactions, and calculate their total amount
    """

    def __init__(self):

        # Define the transactions managed by TransactionManager as an empty list
        self.__transactions=[]

    #A method that allows the creation of a transaction
    def create_transaction(self):
        trans_type_input=input("What type of transaction is it (Expense/Income)?")
        trans_type=TransactionType.EXPENSE
        if trans_type_input=="Income":
            trans_type=TransactionType.INCOME
        trans_amount=float(input("What's the amount of the transaction? "\
                                 "Please enter a number"))
        if trans_type==TransactionType.EXPENSE:
            trans_amount*=-1
        trans_description=input("What's the transaction description?")
        trans_currency_input=input("What's the transaction"\
                                   "Currency (USD/EUR/GBP/AED)?")
        trans_currency=Currency.USD
        if trans_currency_input=="EUR":
            trans_currency=Currency.EUR
        elif trans_currency_input=="GBP":
            trans_currency=Currency.GBP
        elif trans_currency_input=="AED":
            trans_currency=Currency.AED
        trans_date_time_input=input("When did the transaction happen (You can type 'now'"\
                                    "or a date/time in this format: YYYY MM DD hh mm ss)")
        trans_date_time=datetime.datetime.now()
        datetime_tokens=trans_date_time_input.split(" ")

        # If the user provides 6 numbers according to the provided format,
        # They will be used to build a date time attribute
        if len(datetime_tokens) == 6:
            trans_date_time = datetime.datetime(int(datetime_tokens[0]),
                                                int(datetime_tokens[1]),
                                                int(datetime_tokens[2]),
                                                int(datetime_tokens[3]),
                                                int(datetime_tokens[4]),
                                                int(datetime_tokens[5]))

        # Create an object of class Transaction
        transaction=Transaction(transaction_type=trans_type, amount=trans_amount,
                                currency=trans_currency, description=trans_description,
                                date_time=trans_date_time)

        # Add the newly created transaction to the list of transactions
        self.__transactions.append(transaction)
```

```
        return transaction


    def get_total_amount(self):
        total=0

        # Read the amount and then add it to the total, for each transaction
        for transaction in self.__transactions:
            total+=transaction.get_trans_amount()
        return total


    def print_transactions(self): # Print the transactions
        for transaction in self.__transactions:
            print(transaction)


trans_manager=TransactionManager() # Create a TransactionManager object

# Create three transactions defined by the user
t1=trans_manager.create_transaction()
t2=trans_manager.create_transaction()
t3=trans_manager.create_transaction()
trans_manager.print_transactions() # Print all the created transactions

# Show the total amount of transactions
print("Total Amount:{amount}".format(amount = trans_manager.get_total_amount()))
```

## 1.9 CHAPTER SUMMARY

This chapter introduces OOP, a new way of structuring programs compared to traditional methods. It starts by contrasting unstructured and structured programming with OOP. OOP focuses on objects, which combine data (attributes) and actions (behaviors) that operate on that data. This chapter delves into how these objects are created from blueprints called classes. It explores concepts like encapsulation, which restricts direct access to an object's internal data, and how objects interact. This chapter also covers accessing an object's state and its members (attributes and methods), along with the difference between private and public data. Finally, it showcases how objects can be returned from methods and explores built-in functionalities. This chapter concludes with a case study and exercises to solidify the reader's understanding of OOP concepts.

## 1.10 EXERCISES

### 1.10.1 Test Your Knowledge

1. What is the role of the constructor method in a class?
2. Explain the use of the "self" keyword in class methods.
3. How is OOP better than other programming paradigms?
4. Define the following key terms: object, object's state, object's methods, and object's attributes.
5. Why are Setter and Getter methods added in a class?
6. How does an object's behavior affect its state?
7. Why is the "__str__()" method defined in a class?
8. Explain the role of access modifiers in achieving encapsulation in OOP.
9. Explain the difference between private and public attributes.
10. Explain how returning objects instead of primitive values promotes modularity.

### 1.10.2 Multiple Choice Questions

1. When are OO programs a good choice for software development?
    a. In situations where modeling real-world objects as classes is needed.
    b. In situations requiring maintainable and scalable code.
    c. In situations where we want the code to be modular.
    d. All of the above.
2. What are the advantages of "structured" programming?
    a. Improves code readability and reusability.
    b. Increases the speed of program execution.
    c. Facilitates code functionality check.
    d. Reduces software development time.
3. What is the main difference between OOP and structured programming?
    a. Structured programming focuses on bundling data and behavior as classes, whereas OOP facilitates using global variables for easy access to values.
    b. OOP emphasizes breaking down a program into smaller methods and procedures, whereas structured programming allows continuous blocks of long, unstructured code.
    c. Structured programming supports modeling real-life objects as classes, whereas OOP is more suited for simple procedural tasks.
    d. OOP organizes attributes and methods into classes, encapsulating both an object's state and behavior, whereas structured programming provides limited encapsulation through the scope of the variable within a method.
4. Which of the following best describes a class in OOP?
    a. A template for creating objects.
    b. A method for identifying objects.
    c. A single instance of a software entity.
    d. A collection of methods.
5. What is the primary purpose of the "__init__()" method in a Python class?
    a. To create multiple instances of the class.
    b. To initialize instance variables and perform setup when an object is created.
    c. To serve as a destructor for the object.
    d. To set up the class's static variables.
6. What is encapsulation in OOP?
    a. It is a technique for storing data in a structured format.
    b. It is the process of creating multiple instances of a class.
    c. It bundles data and methods within a single unit, providing data hiding and abstraction.
    d. It is a mechanism for controlling access to the attributes and methods of a class.
7. What does an object's state refer to in OOP?
    a. The object's behavior
    b. The object's identity
    c. The complete set of attributes and values defining its present condition.
    d. The object's memory allocation.
8. How are class members accessed in Python?
    a. Using parentheses
    b. Using underscore
    c. Using dot notation
    d. Using curly braces
9. Which built-in method allows customization of the string representation of an object?
    a. "__init__"
    b. "__str__"
    c. "__eq__"
    d. "__lt__"

10. Which built-in variable in Python allows access to the class name?
    a. instance
    b. class
    c. name
    d. type

### 1.10.3 Short Answer Questions

1. Name the distinctive characteristics of an object that need to be represented in software.
2. Methods within the Python main library that facilitate operations and interactions within the code are called _________.
3. What can an object in OOP be in real-world examples?
4. How to control access to variables?
5. What methods help control access to the attributes of an object?
6. What do we call the variables that can be accessed directly?
7. What remains as a reference to the instance being constructed when creating an object while the programmer supplies values for the other arguments?
8. What defines the behavior of an object?
9. Name three pre-defined functions and state their corresponding uses.
10. What is the name of the variables shared by all class instances?

### 1.10.4 True or False Questions

1. The methods of a Class are used to modify class-level attributes and execute operations that are not specific to any particular instance.
2. In Python, class attributes with a single underscore prefix (e.g., self._variable) are considered private and should not be accessed directly from outside the class. In contrast, those without an underscore (e.g., variable) are considered public and can be accessed directly outside the class.
3. An instance variable in a class is shared among all class instances.
4. Class variables can have different values for different instances, and they are unique to each instance of a class.
5. A class can have multiple constructors with different parameter signatures.
6. Setters and getters are used in Python to control access to class attributes, providing a way to encapsulate and validate data stored in an object.
7. The "__init__" method is automatically called when a class instance is created.
8. In OOP, a class is a template used to create or model objects.

### 1.10.5 Fill in the Blanks

1. Find the appropriate code to fill into the code snippet below:

```
class DeliverySystem:
    pass
# Create an object of class DeliverySystem
Delivery1=____________
```

2. Find the appropriate code to fill into the code snippet below:

```
class DeliverySystem:
    def __init__(self,____________):
        self.delivery_number = __________
        self.delivery_status = "Pending"
```

3. Find the appropriate code to fill into the code snippet below:

```
class DeliverySystem:
  # Consider the __init__ in the previous question
  def get_delivery_number(self):
     return self.______________
  def set_delivery_status(self, new_status):
     self._delivery_status = new_status
  def get_delivery_status(self):
     return self.______________
```

### 1.10.6 Coding Problems

1. Consider the class Circle illustrated in Code 1.7 within Section 1.5. Create a new public method called "calculate_perimeter()" within the class and ensure it returns a circle's perimeter.
2. Consider the class Student in Code 1.8 within Section 1.5. Create a new method in the class called "set_student_password()". This method allows the modification of the student's password. The method will take two attributes, "current_password" and "new_password". The method will first verify the password by calling an existing method of the class. If the password is correct, then the password will be updated to the "new_password". If the "current_password" is incorrect, an error message will be displayed.

### 1.10.7 Exercises to Implement Classes and Objects

1. Create a Python class named "Product" with the following attributes: ID, name, description, cost, price, and quantity.
   a. Implement the "__init__()" method within the class to initialize attributes.
   b. For each attribute, specify the appropriate data type.
   c. Create two instances of the "Product" class and verify that all data attributes are correctly displayed.
2. Create a Python class "Patient" to manage patient records within a hospital.
   a. Each "Patient" has attributes including the patient's name (as a string), age (as an integer), phone number (as an integer), and current health status (as a string).
   b. Implement the setter and getter within the "Patient" class to facilitate updating and querying the patient's information. These methods should allow for modifying, updating, and retrieving the various attributes of the "Patient" object.
   c. Ensure the privacy of patient data by restricting direct access to their information (use appropriate access modifiers to make the attributes private).

3. Create a fitness tracking application using Python. The application should have the following classes:
   a. Create the class "User".
      - This class represents the app's users. It should have the following attributes:
        - name: a string representing the user's name.
        - age: an integer representing the user's age.
        - weight: a float representing the user's weight.
        - height: a float representing the user's height.
      - Create the appropriate "__init__()" method, as well as setters and getters methods for this class, and ensure the privacy of the user's information.
      - Create a method to calculate the BMI (BMI=weight/height$^2$).
      - Create a method to update the user's weight to a new value.
   b. Create the class "Exercise":
      - This class represents the different types of exercises. It should have the following attributes:
        - "name": a string representing the name of the exercise.
        - "target_muscle": a string representing the primary muscle targeted by the exercise.
        - "difficulty_level": an integer representing the difficulty level of the exercise (easy, moderate, and hard).
      - Create the appropriate "__init__()" method, setters, and getters methods.
   c. Add an appropriate method for both classes to provide a meaningful representation when instances are printed.
   d. Create at least two instances of the class "User" and two instances of the class "Exercise".

# 2 Python Data Structures

## 2.1 LISTS

Lists are one of the most fundamental and versatile data structures in Python. They allow the storage of a collection of items, enabling the addition, removal, or updating of items as necessary. Lists offer indexing, slicing, and a range of methods for efficient data manipulation. Figure 2.1(a) shows an example of a list that contains names of colors. The data type of stored data is "string". Each value in the list can be accessed based on its position, i.e., the index. For instance, "black" is located at index 0, while "orange" is located at index 1, and so on. It is important to note that indexes begin at 0 and not 1. Figure 2.1(b) illustrates another instance of a list that stores numbers representing student grades for a particular course, and here the data type of the values is "integer". Finally, a list containing both strings and integers is provided in Figure 2.1(c), as lists in Python can store different types of data.

### 2.1.1 Implementing Lists in Python

There are multiple techniques for creating a list in Python. The code snippet provided in Code 2.1 demonstrates how to create the lists presented in Figure 2.1 using three techniques. The first technique involves creating a list called "color_names", which contains color names such as "red", "orange", "black", and "white", by directly writing the elements within square brackets. The list is printed to the console using the "print()" statement in Line 3.

In the second technique, an empty list, "student_grades", is initialized to store student grades in Line 5 of Code 2.1. The grades "89", "55", "68", "83", and "93" are added to the list using the "append()" method in Lines 6 to 10. The "student_grades" list is then printed to the console in Line 11.

Finally, using the third technique, a new empty list named "diverse_list" is created to provide an example of creating a list with different data types. The "insert()" method is used in Code 2.1, in Lines 14–17, to add various items, including a string "Adam Smith", two integers "125923" and "125971", and another string "Ann Davis" at specific indices. The "diverse_list" is then printed to the console in Line 18, displaying the mixed data types within the list.

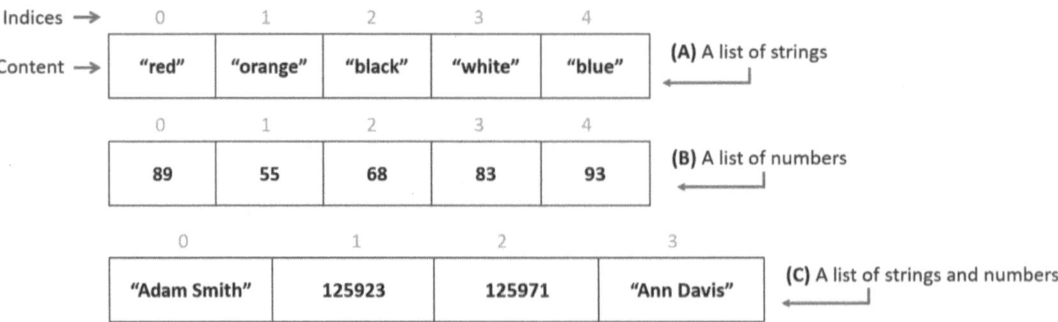


**FIGURE 2.1** Three different types of lists: (a) a list of strings, (b) a list of numbers, and (c) a list of strings and numbers.


DOI: 10.1201/9781032668321-2

It is possible to remove items from a list using various techniques. Line 21 demonstrates how to delete the last element of the "student_grades" list by calling the "pop()" method without any parameters. This method can also be given one parameter, which indicates the index of the element to be deleted, as shown in Line 22, where index "1" is specified to delete the element "55" from the "student_grades" list. Other methods that can be used to delete elements from a list are "remove()", as shown in Line 23, which takes the element to be removed as a parameter, and "del()" as shown in Line 24, which removes an element at a specific index.

**CODE 2.1 THREE TECHNIQUES FOR CREATING A LIST IN PYTHON AND REMOVING ELEMENTS FROM A LIST**

```
1. #Technique 1:
2. color_names = ["red", "orange", "black", "white", "red"] # Create a list of color names
3. print(color_names) #Print the list
4. #Technique 2:
5. student_grades = [] #Create an empty list
6. student_grades.append(89) #Add an item to the end of the list
7. student_grades.append(55)
8. student_grades.append(68)
9. student_grades.append(83)
10. student_grades.append(93)
11. print(student_grades) #Print the list
12. #Technique 3:
13. diverse_list=[] #Create an empty list with different types of data types
14. diverse_list.insert(0,"Adam Smith") #Add an item at index 0
15. diverse_list.insert(1, 125923)
16. diverse_list.insert(2, 125971)
17. diverse_list.insert(3, "Ann Davis")
18. print(diverse_list) #Print the list
19.
20. # The below code lines show the different techniques to delete an element from a list.
21. student_grades.pop() # Delete the last element of list "student_grades"
22. student_grades.pop(1) # Delete element of index 1 in "student_grades" corresponding to "55"
23. diverse_list.remove(125923) # Delete a specific element from diverse_list.
24. del diverse_list[-1] # Delete the element with index -1 : the last element in the list
```

### 2.1.2 Accessing Items with Indices

Non-negative and negative indices can be used to access items in a list. Index "0" represents the first item, index "1" represents the second item, and so on. For example, Figure 2.2 shows that "color_names[0]" allows to access the first item. Negative indices, on the other hand, start counting from the end of the sequence, with "-1" representing the last item, "-2" representing the second-to-last item, and so forth. For example, Figure 2.2 depicts that "color_names[-1]" allows to access the last item of the list.

### 2.1.3 Accessing Sublists with Index Ranges

To access sublists or slices of a list, both non-negative and negative indices can be used. For non-negative indexing, sublists can be retrieved by specifying the starting and ending indices. For instance, "color_names[0:3]" can be used to retrieve the sublist "['red', 'orange', and 'black']" from the list called "color_names". This will retrieve, as illustrated in Figure 2.3 (a), the items located from index "0" through index "2", which means that the ending index is not included in a sublist. Another example is "color_names[1:3]", which retrieves the colors at indices "1" and "2". If the starting index is not specified, as shown in Figure 2.3(b), it is assumed to be zero by default. As such, "color_names[:3]" is equivalent to "color_names[0:3]".

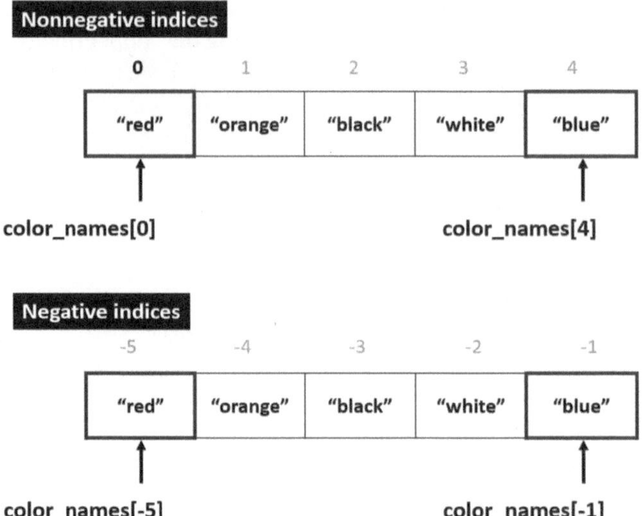


**FIGURE 2.2** Accessing items in a list using non-negative and negative indices.

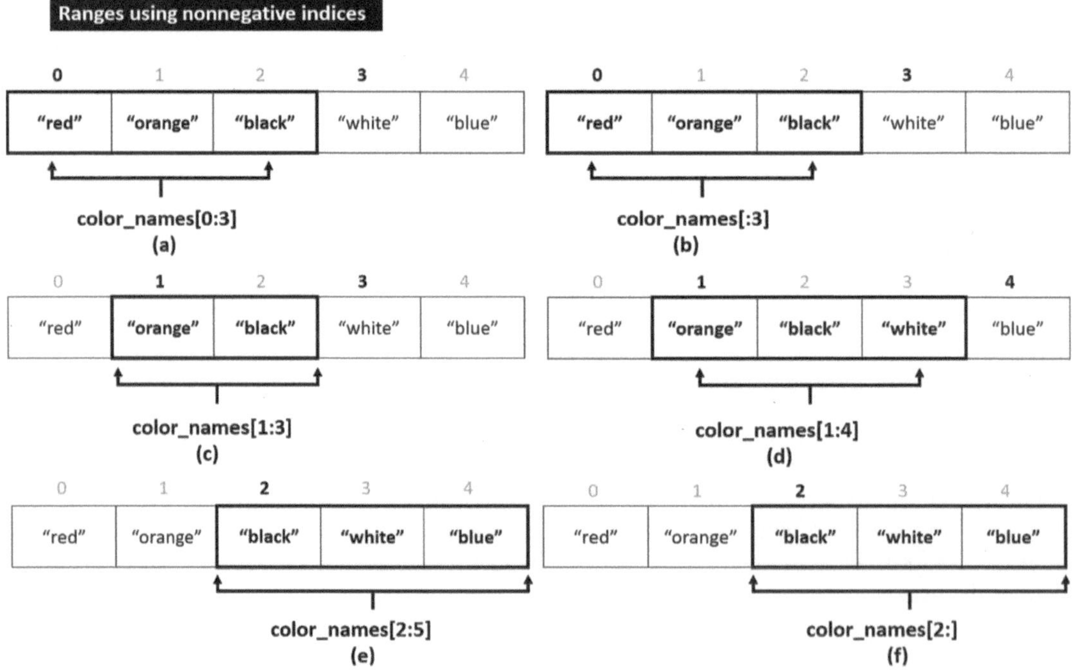


**FIGURE 2.3** An illustration of list ranges of non-negative indices. (a) color_names from index 0 to 3, (b) color_names from beginning to index 3, (c) color_names from index 1 to 3. (d) color_names from index 1 to 4. (e) color_names from index 2 to 5. (f) color_names from index 2 till the end of the list.

To ensure that the last item in the list is included in the sublist, the ending index must be equal to the size of the list. For example, "color_names[2:5]" retrieves all the items between index "2" until the last item in the list (located at index "4"), as illustrated in Figure 2.3(e). Furthermore, the last item in the list can be retrieved by not specifying the last index. As such, "color_names[2:]" is identical to "color_names[2:5]", as in Figure 2.3(f).

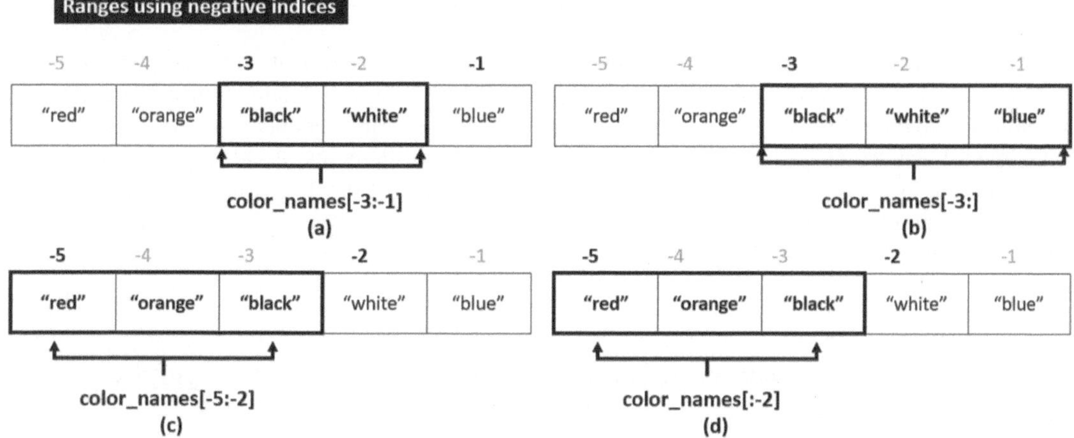


**FIGURE 2.4** An illustration of list ranges of negative indices. (a) color_names from index -3 to -1. (b) color_names from index -3 to end of the list. (c) color_names from index -5 to -2. (d) color_names from the beginning of the list till index -2

Accessing sublists in Python can also be done using negative indexing. For instance, from a list of color_names and a sublist containing the third to the second last items are to be extracted, then "color_names[-3:-1]" can be used. In Figure 2.4 (d), "color_names[:-2]" retrieves the items starting from the index "-3" and ending at the index "-5". The sublist in this range will include "red", "orange", and "black".

### 2.1.4 A List of Objects

As discussed previously, Python lists are very useful as they can hold various data types, including objects. Figure 2.5 shows a Python list that contains objects representing students. The first student at index "0" is an object that contains Ahmed Samir's details, such as his first name, last name, major (CS), and date of birth (02-05-2002). Likewise, the second student object holds Jennifer Jackson's details, including her first name, last name, major, and date of birth. This feature makes Python lists a powerful tool for managing various data types, including objects.

Code 2.1 demonstrates the implementation of the example shown in Figure 2.5. The code begins by importing the "Enum" class from the "Enum" module (Line 1). The "Enum" class creates enumerations, sets of constants representing specific values in the code. Next, a new enumeration class called "Major" is created using "Enum", with two members: "CS" for "Computer Science" and "IT" for "Information Technology" (Lines 3 and 4). The code defines a class called "Student" (Lines 5–10) with four attributes: first name, last name, major, and date of birth. It then creates four

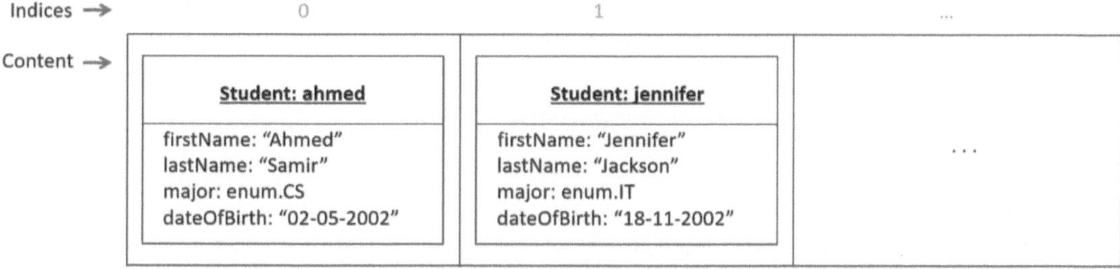


**FIGURE 2.5** An illustration of the list of student objects.

instances of the "Student" class with specific details (Lines 12–15) and adds them to a list called "student_list" (Lines 18–21).

The code then uses a "for" loop to go through each element of "student_list" and print each student's first name, last name, major (by accessing the "Enum" value), and date of birth (Lines 24–28).

Finally, two other lists are created using index slicing to separate students by major. The "cs_list" is assigned from "student_list[:2]" (Line 31), which takes the elements of index "0" and index "1" from "student_list". "it_list" is assigned from "student_list[2:]" (Line 32), which takes the elements of index "2" and "3" from "student_list". The code then prints the "cs_list" (Lines 36–40) and the "it_list" (Lines 43–48) to display the students' information grouped by major.

### CODE 2.2 CODE DEMONSTRATING THE STORAGE OF OBJECTS WITH LISTS

```
1. from enum import Enum # Import class Enum from the module enum
2. class Major(Enum): # Define a new enumeration class with 2 members CS and IT
3.     CS = "Computer Science"
4.     IT = "Information Technology"
5. class Student: # Create class "Student" with its attributes
6.     def __init__(self, first_name, last_name, major, date_of_birth):
7.         self.first_name = first_name
8.         self.last_name = last_name
9.         self.major = major
10.        self.date_of_birth = date_of_birth
11. # Create student objects
12. student_1 = Student("Ahmed", "Samir", Major.CS, "02-05-2002")
13. student_2 = Student("Jennifer", "Jackson", Major.CS, "18-11-2002")
14. student_3 = Student("John", "Smith", Major.IT, "20-05-2005")
15. student_4 = Student("Sultan", "Nasser", Major.IT, "21-12-2005")
16.
17. # Create a list to store the student objects in different techniques to show different methods.
18. student_list = [student_1] # Add student 1 to the list in one technique
19. student_list.append(student_2) # Add student_2 to the list in another technique
20. student_list.insert(2,student_3) # Add student_3 to the list in another technique
21. student_list.insert(3,student_4) # Add student_4 to the list in another technique
22.
23. # Accessing student information
24. for student in student_list:
25.     print(f"First Name: {student.first_name}")
26.     print(f"Last Name: {student.last_name}")
27.     print(f"Major: {student.major.value}")  # Access the Enum value
28.     print(f"Date of Birth: {student.date_of_birth}")
29.
30. # Separate students into CS and IT lists using index slicing
31. cs_list = student_list[:2]  # Students 1 and 2
32. it_list = student_list[2:]  # Students 3 and 4
33.
34. # Print CS students
35. print("CS Students:")
36. for student in cs_list:
37.     print(f"First Name: {student.first_name}")
38.     print(f"Last Name: {student.last_name}")
39.     print(f"Major: {student.major.value}")  # Access the Enum value
40.     print(f"Date of Birth: {student.date_of_birth}")
41.
42. # Print IT students
43. print("IT Students:")
44. for student in it_list:
45.     print(f"First Name: {student.first_name}")
46.     print(f"Last Name: {student.last_name}")
47.     print(f"Major: {student.major.value}")  # Access the Enum value
48.     print(f"Date of Birth: {student.date_of_birth}")
```

## 2.2 TUPLES

A tuple is a data structure similar to a list in that it is a collection of elements capable of holding values of different data types. Like lists, tuples can be indexed and sliced using square brackets, supporting various operations, including concatenation.

However, unlike lists, tuples are immutable, meaning their elements cannot be changed once defined, while lists are mutable and can be modified. Tuples typically store read-only sequences of values, such as coordinates or configurations. However, lists are preferred for storing collections of items that may need to be modified during the program's execution. It is also worth noting that tuples are defined using parentheses instead of square brackets.

Code 2.3 depicts an example of tuple usage to store geographic coordinates. A tuple is a good data structure type to store geographic coordinates that must be immutable. In Line 1, a tuple named "new_york_coordinates" is created with the latitude and longitude coordinates of New York: "(40.7128, -74.0060)". The coordinates are then printed with labels for clarity on Line 3.

The code defines a list called "city_coordinates_list" containing tuples for different cities, each tuple containing the city name and coordinates (Lines 5–12). A "for" loop is used to iterate through this list, extracting the city name and coordinates (Lines 14–16). The latitude and longitude values are extracted from each coordinates tuple and used to print information about each city, including its name, latitude, and longitude (Lines 17–19).

Even though geographic coordinates need to be immutable, sometimes data entry errors may happen. For example, in the code below, the GPS coordinates of the New York City location in the tuple created at Line 1 have been swapped. To fix this error, we need to convert the tuple to a list (Line 22), then swap the coordinate values (Line 23) and convert the list back to a tuple (Line 24).

### CODE 2.3 CODE DEMONSTRATING AN EXAMPLE OF TUPLE USAGE

```
1. new_york_coordinates=(-74.0060, 40.7128) # Create a tuple of New York city coordinates
2. # Print the New York city coordinate tuple using indexing
3. print("New York Coordinates: ","Lat:",new_york_coordinates[0],new_york_coordinates[1])
4. # Create a list of tuples for different cities and their corresponding coordinates.
5. city_coordinates_list = [
6.     ("New York", (40.7128, -74.0060)),
7.     ("Los Angeles", (34.0522, -118.2437)),
8.     ("Chicago", (41.8781, -87.6298)),
9.     ("San Francisco", (37.7749, -122.4194)),
10.    ("Miami", (25.7617, -80.1918)),
11.    ("Abu Dhabi", (24.4667, 54.3667))
12. ]
13. # Iterate through the list of tuples to extract the city name and coordinates from each tuple
14. for city_coordinates in city_coordinates_list:
15.     city=city_coordinates[0]
16.     coordinates=city_coordinates[1]
17.     lat=coordinates[0]
18.     lng=coordinates[1]
19.     print("City:",city," Lat:",lat," Lng:",lng) # Print the city name with its coordinates
20.
21. #How to change elements in an immutable tuple
22. new_york_list=list(new_york_coordinates) # Convert tuple to list
23. new_york_list[0], new_york_list[1]= new_york_list[1], new_york_list[0] # Swap the elements
24. new_york_coordinates = tuple(new_york_list) # Convert back to tuple
```

## 2.3 SETS

Sets are collections of unique values that are unordered and unindexed, meaning that the elements are not stored in a specific order. They are also mutable, which allows for the addition or removal of elements. Unlike lists and tuples, sets don't allow for duplicate values, making each element unique. This characteristic of sets can be used to remove duplicates in lists and tuples. For example, if a list of elements contains duplicates that need to be removed, converting it to a set will automatically remove them.

Sets also allow for efficient operations such as union, intersection, difference, and symmetric difference. The union operation creates a new set with unique values by combining elements from two sets. The intersection operation finds the shared components that two sets have in common. The difference operation helps find elements unique to one set while excluding others. Finally, the symmetric difference operation gives the elements in either of the two sets but not in their intersection. Additionally, sets are efficient for element existence checks, where fetching the existence of an element in a set is fast and has constant time regardless of the size of the set.

Code 2.4 provides an example of how to use sets in Python. In Line 1, we initialize the "colors" set with five color names. Then, we add a new element, "yellow", to the "colors" set using the "add()" method in Line 2. In Line 3, we remove "blue" from the set using the "remove()" method.

Line 5 checks if "green" is present in the "colors" set using an "if" statement. If it is, a message saying, "Yes, green is in the set", is printed.

The code then proceeds to Line 9, where we iterate through the "colors" set using a "for" loop and print each color.

In Line 12, we create a new set called "dark_colors" that contains several dark color names. Then, we perform set operations, using the "union()", "intersection()", and "difference()" methods, between the "colors" and "dark_colors" sets in Lines 15–21.

Finally, Lines 24–26 display the results of these set operations. We show all the colors, the common colors between "colors", and those not in "dark_colors".

### CODE 2.4 CODE DEMONSTRATING AN EXAMPLE OF SET USAGE

```
1. colors = {"white", "red", "blue", "green","black"} # Create a set of different Colors
2. colors.add("yellow") # Add a new element to the set
3. colors.remove("blue") # Delete an element from the set
4.
5. if "green" in colors: # Use an if statement to check if an element is in the set
6.     print("Yes, green is in the set.")
7.
8. print("Colors in the set:")
9. for color in colors:# Iterate through the elements of the set to print them
10.     print(color)
11.
12. dark_colors = {"black", "brown", "maroon"} # Create a new set with different colors
13.
14. # Create a set that combines all unique elements of the 2 other sets using "union" operation
15. all_colors = colors.union(dark_colors)
16.
17. # Create a set containing the common elements of the 2 other sets using "intersection" operation
18. common_colors = colors.intersection(dark_colors)
19
20. # Create a set that give the elements in the "colors" set but not in the "dark_colors" set
21. non_dark_colors = colors.difference(dark_colors)
22.
23. # Display the result of the 3 newly formed sets based on the 3 set operations performed
24. print("All colors:", all_colors)
25. print("Common colors:", common_colors)
26. print("Non-dark colors:", non_dark_colors)
```

## 2.4 DICTIONARIES

A dictionary is a collection of key-value pairs where keys are unique but values are not. Unlike lists and tuples, dictionaries are not index-based collections. Instead, they provide efficient key-based data retrieval. Dictionaries, similar to lists, are mutable, allowing for the addition, modification, or removal of key-value pairs. Figure 2.6 provides two examples to illustrate the concept. The first example demonstrates a dictionary with student ID as the key and the student's full name as the value. In the second example, the dictionary has the student IDs as keys and student objects as values.

Code 2.5 shows an example of managing student information using dictionaries. Line 1 begins by importing the "Enum" class from the "Enum" module to define a set of constants for the student's major field. The code defines a "Student" class with the constructor taking parameters for "first_name", "last_name", "major", and "date_of_birth". The class encapsulates this student information within its instance variables (Lines 5–8). Additionally, it defines an "Enum" called "Major" to specify three possible majors: "CS" for "Computer Science", "IT" for Information Technology, and "EE" for Electrical Engineering. This "Enum" simplifies the categorization of student majors (Lines 10–13).

Two dictionaries are created: "student_names" and "student_info". The "student_names" dictionary was created out of two lists: student IDs and student first names, which then were combined using the method "zip()" and converted into a dictionary using the method "dict()" (Lines 16–20). On the other hand, the "student_info" dictionary associates each student ID with a "Student" object, which encapsulates student information in detail (Lines 23–26).

The code demonstrates how to perform operations on dictionaries. Specifically, it adds a new entry to two dictionaries: "S0123822" with the name "Mike Johnson" in the "student_names" dictionary on Line 29, and an object of the "Student" class for "Michael Johnson" who is majoring in "Computer Science" and has a date of birth of "10-03-2001" in the "student_info" dictionary on Line 30. Furthermore, an existing entry is updated in both dictionaries. The student with the ID "S0123789" is updated to "Ahmed Samir" in "student_names" (Line 33), and the attribute "major" for "S012378" is changed to "Electrical Engineering" in student_info using update method (Line 36). The related entries are deleted from both dictionaries. "S0123790" is removed from "student_names" (Line 39), and the corresponding entry in "student_info" is deleted (Line 40).

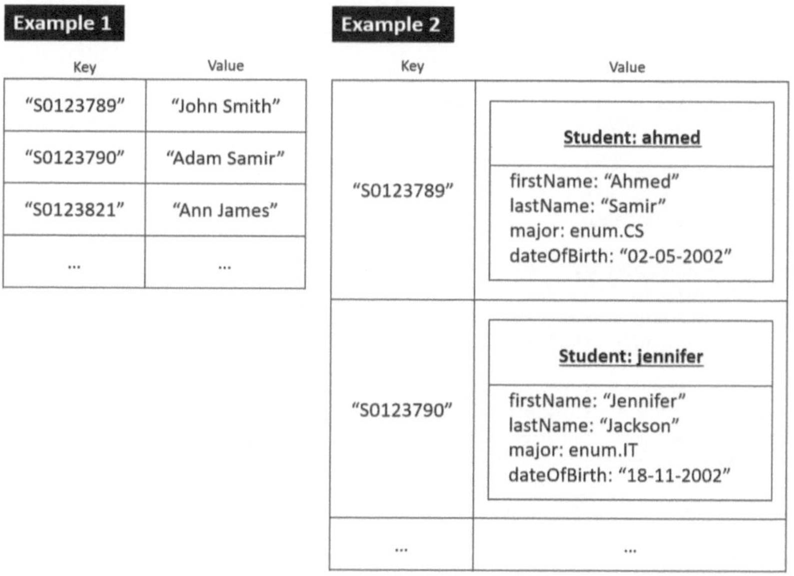


**FIGURE 2.6** An illustration of dictionary usage.

At last, the code concludes by printing the updated information in both dictionaries. The "student_id" and "name" from the first dictionary are displayed (Line 43–45), followed by student names and their detailed information, including first name, last name, major (in a human-readable form from the "Major" Enum), and date of birth from the second dictionary (Line 47–50). Lines 53–55 display the keys only of the dictionary "student_info".

### CODE 2.5 AN EXAMPLE OF USING DICTIONARIES

```
1. from enum import Enum # Import the module enum
2.
3. class Student: # Create class "Student" with its attributes
4.     def __init__(self, first_name, last_name, major, date_of_birth):
5.         self.first_name = first_name
6.         self.last_name = last_name
7.         self.major = major
8.         self.date_of_birth = date_of_birth
9.
10. class Major(Enum): # Define a new enumeration class with 3 members CS, IT and EE
11.     CS = "Computer Science"
12.     IT = "Information Technology"
13.     EE = "Electrical Engineering"
14.
15. # Create a list of student IDs and a list of first names
16. student_ids = ["S0123789", "S0123790", "S0123821"]
17. first_names = ["John", "Adam", "Ann"]
18.
19. # Create a dictionary using zip
20. student_names = dict(zip(student_ids, first_names))
21.
22  # Create "student_info" dictionary with IDs as Keys and corresponding student Object as values
23. student_info = {
24.     "S0123789": Student("Ahmed", "Samir", Major.CS, "02-05-2002"),
25.     "S0123790": Student("Jennifer", "Jackson", Major.IT, "18-11-2002")
26. }
27.
28. # Add new element to both dictionaries.
29. student_names["S0123822"] = "Mike Johnson"
30. student_info["S0123822"] = Student("Michael", "Johnson", Major.CS, "10-03-2001")
31.
32. # Change the values of a specific key "student ID" in both dictionaries.
33. student_names["S0123789"] = "Ahmed Samir"
34.
35. # Update with the use of "update" to change the major for the specific student
36. student_info.update({"S0123789": Student("Ahmed", "Samir", Major.EE, "02-05-2002")})
37.
38. # Delete one element for both dictionaries for a specific key "student ID"
39. del student_names["S0123790"]
40. del student_info["S0123790"]
41.
42. # Display the amended dictionaries
43. print("Student Names:")
44. for student_id, name in student_names.items():
45.     print(f"{student_id}: {name}")
46.
47. print("\nStudent Information:")
48. for student_id, student in student_info.items():
49.     print( f"{student_id}: {student.first_name } {student.last_name} ({student.major.value}),"\
50.            f"DOB: {student.date_of_birth}")
51.
52. # Display the keys of the student_info dictionary
53. print("Student IDs:")
54. for student_id in student_info.keys():
55.     print(student_id)
```

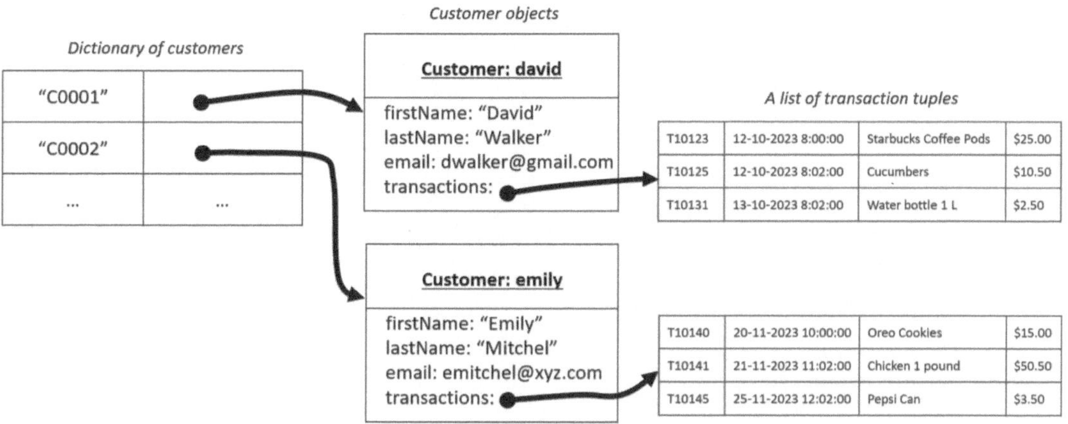


**FIGURE 2.7** An illustration of the grocery store application using dictionaries, objects, lists, and tuples.

## 2.5 CASE STUDY – GROCERY STORE APPLICATION

Figure 2.7 shows an illustration of our case study – a grocery store application. The application manages customer information by utilizing a dictionary to store customer details. The customer's unique identifier, or customer ID, serves as the key, while the corresponding value is a customer object. This object contains customer information, including their first and last name, e-mail address, and a list of transactions. The transaction history is stored as a list of read-only tuples. These transaction tuples encapsulate the transaction ID, date and time, purchased item, and transaction amount.

Code 2.6 depicts the code for the customer management system. It starts by creating a customer class (Line 1) with a constructor "__init__()" (Line 2) that takes four parameters: "first_name", "last_name", "email", and "transactions". Inside the constructor, it assigns these parameters to corresponding class attributes, which store the customer's first name, last name, email, and a list of transactions.

Next, it creates two customer objects, "customer_1" and "customer_2", with their respective information (Lines 9–18). These customers have various transactions represented as tuples within a list. The customer objects are stored in a dictionary called "customers_dictionary" (Lines 20–24), where each customer is associated with a unique customer ID.

The variable "customer_id" is created to access the specific customer's information later (Line 26). The next line of code checks if the provided customer ID exists in the customers' dictionary (Line 27). If the customer exists, it retrieves the selected customer object (Line 28) and prints details such as customer ID, first name, last name, email, and transactions (Line 30–35). The code also iterates through the "transaction" list, displaying each transaction's ID, date and time, description, and amount (Lines 37–43). If the customer ID is not in the dictionary, it prints "Customer not found" (Lines 45–46). This code allows for managing and retrieving customer information based on their unique customer ID.

### CODE 2.6 CODE FOR THE GROCERY STORE APPLICATION

```
1. class Customer: # Create a class customer with its attributes
2.     def __init__(self, first_name, last_name, email, transactions):
3.         self.first_name = first_name
4.         self.last_name = last_name
5.         self.email = email
6.         self.transactions = transactions
7.
8.
```

```
# Create two customer Objects "customer 1" and "customer 2"
customer_1 = Customer("David", "Walker", "dwalker@gmail.com",
     [("T10123", "12-10-2023 8:00:00", "Starbucks Coffee Pods", 25),
     ("T10125", "12-10-2023 8:02:00", "Cucumbers", 10.50),
     ("T10131", "13-10-2023 8:02:00", "Water bottle 1 L", 2.50)])

customer_2 = Customer("Emily", "Mitchel", "emitchel@xyz.com",
    [("T10140", "20-11-2023 10:00:00", "Oreo Cookies", 15.0),
    ("T10141", "121-11-2023 11:02:00", "Chicken 1 pound", 50.50),
    ("T10145", "25-11-2023 12:02:00", "Pepsi Can", 3.50)])

# Create a dictionary called "customers"
customers_dictionary = {
    "C0001": customer_1,
    "C0002": customer_2
 }

customer_id = "C0001" # A customer ID is selected
if customer_id in customers_dictionary: # Check if this customer is in the dictionary
    selected_customer = customers_dictionary[customer_id] # Retrieve selected student' object

    # Print the selected customer's details
    print("Customer ID:", customer_id)
    print("First Name:", selected_customer.first_name)
    print("Last Name:", selected_customer.last_name)
    print("Email:", selected_customer.email)
    print("Transactions:")

    # Iterate into the transaction to print its details.
    for transaction in selected_customer.transactions:
        print("  Transaction ID:", transaction[0])
        print("  Transaction Date and Time:", transaction[1])
        print("  Transaction Description:", transaction[2])
        print("  Transaction Amount:", transaction[3])
        print()

else: # If the customer is not in the dictionary, print the below message
    print("Customer not found.")
```

## 2.6 CHAPTER SUMMARY

This chapter explores fundamental data structures in Python, which are essential tools for organizing information within programs. It begins with lists, a versatile data structure that can store various item types in an ordered sequence. This chapter dives into creating lists, accessing elements using indexes, and extracting sublists. It then covers tuples, similar to lists but immutable (unchangeable). Sets are introduced to store unique elements and perform operations like checking for membership. Dictionaries, another key structure, are explained, allowing the linking of unique keys to associated values. To illustrate these concepts, a practical case study, a grocery store application, is presented. This chapter concludes by offering various exercises to solidify the reader's grasp of these core Python data structures.

## 2.7 EXERCISES

### 2.7.1 Test Your Knowledge

1. What is the difference between a list and a tuple?
2. List the differences between sets and dictionaries.
3. Explain what mutability is and why it is important in the context of data structures.

4. Describe the purpose and characteristics of dictionaries and provide an appropriate application for them.
5. Can the elements of a tuple be edited after it's created? Why or why not?
6. How is an element removed from a list or a dictionary? Explain with an example.
7. How are duplicates from a list removed, and which data structure is more appropriate for this task?
8. How is a new element added to the end of a list?
9. What is the significance of data structures in programming?
10. Which data structure is most appropriate for finding common elements between two data collections?

### 2.7.2 Multiple Choice Questions

1. Which data structure is a collection of pairs of keys and values?
   a. List
   b. Tuple
   c. Dictionary
   d. Set
2. How is an item added to the end of a list?
   a. Using the "append()" method
   b. Using the "insert()" method
   c. Using the "add()" method
   d. Using the "extend()" method
3. Which data structure is used to store elements unordered without duplicates?
   a. List
   b. Tuple
   c. Dictionary
   d. Set
4. Which statement accesses the last element of a list?
   a. "list.last()"
   b. "list[-1] "
   c. "list.last_element()"
   d. "list.end()"
5. Which method is used to combine two sets in Python?
   a. "combine()"
   b. "merge()"
   c. "union()"
   d. "intersect()"
6. How are elements accessed in a dictionary?
   a. By index
   b. By key
   c. By value
   d. By position
7. Which operation adds a key-value pair to a dictionary?
   a. "add()"
   b. "insert()"
   c. "append()"
   d. "update()"
8. How do you remove an item from a set?
   a. Using the "remove()" method
   b. Using the "pop()" method

   c. Using the "delete()" method
   d. Using the "discard()" method
9. What method is used to convert a list into a tuple in Python?
   a. "convert_list()"
   b. "list_to_tuple()"
   c. "tuple()"
   d. "make_tuple()"

### 2.7.3 Short Answer Questions

1. Which index number can access the last item in a list?
2. Name the dictionary methods that allow you to add, modify, or remove key-value pairs.
3. Which data structure is an immutable collection and is defined using parentheses?
4. What is the term used to describe a data structure that can be modified?
5. What is the name of the set operation that returns elements present in one set but not in the other?
6. Lists and tuples are accessed using indices. How do we access dictionaries?

### 2.7.4 True or False Questions

1. Lists and tuples in Python can only store data of a single data type.
2. In a dictionary, keys must be unique, but values do not have to be.
3. To create a list in Python, you can use square brackets with elements separated by commas.
4. You can access elements in a set using index notation, similar to lists.
5. Negative indices in a list start counting from the beginning of the list.
6. You can retrieve a Python sublist by specifying the starting and ending indices.
7. In a dictionary, data retrieval is primarily based on keys, providing an efficient way to access specific data.
8. Dictionaries are index-based collections, just like lists and tuples.
9. Tuples and lists share a common characteristic of having mutable elements.
10. By slicing and providing the start and end indices, you may extract certain parts of a tuple, generating new tuples from the original.

### 2.7.5 Fill in the Blanks

1. Fill in the appropriate code in the blanks.

```
# Create an empty list
my_list = []

# Add elements to the list
my_list.______(0,2)
my_list.______(1,"Hello")

# Print the list
print(my_list)
```

2. Fill in the appropriate code in the blanks.

```
list1 = ["Georges", "Ahmad", "Jinesh", "Afra", "Anna"]
list2 = [1, "Ahmad",2, "Anna"]
# Convert the lists to sets and take their union
union_result = list(_____(list1).union(_____(list2))
print(union_result)
```

3. Fill in the appropriate code in the blanks.

```
class DeliverySystem: # Create the DeliverySystem class
    def __init__(self, delivery_number, delivery_status="Pending"):
        self.delivery_number = delivery_number
        self.delivery_status = delivery_status
# Create objects of the DeliverySystem class
delivery1 = DeliverySystem("56973", "In Transit")
delivery2 = DeliverySystem("56974", "Delivered")
# Create lists of objects and delivery numbers
delivery_objects = [delivery1, delivery2]
delivery_numbers = [delivery1.delivery_number, delivery2.delivery_number]
# Create a dictionary by combining the 2 lists above and using the zip method.
delivery_dict = ______(zip(___________, delivery_objects))
```

### 2.7.6 Coding Problems

1. You are tasked to create two dictionaries, one to map colors to personality traits and another to map colors to mood indicators.
   a. "color_traits" dictionary:
      - Create a dictionary named "color_traits" that takes colors as keys and a list of at least two traits per color as values (e.g., the black color is associated with the traits: mystery and power).
      - Implement a method to prompt the user to select their favorite color, then display the associated personality traits for that color, suggesting they are the user's personality traits.
   b. "color_moods" dictionary:
      - Create a dictionary named "color_moods" that takes colors as keys and mood indicators as values.
      - Ensure that the method implemented in Part A also considers the selected color to display the corresponding mood indicator for the user.
2. In Chapter 1, a Python class to manage patient records was created. The class now needs to include details of patients' chronic diseases.
   a. The attributes were name, age, phone number, and current health status. Change the current health status to a "list" of chronic diseases.
   b. Develop a method for updating the list whenever a patient is diagnosed with a new chronic disease or recovers from an existing one.
   c. Implement a method for identifying patients with the same chronic disease and maintaining a count of such occurrences.

d. Create at least two instances of the class "Patient" with at least two chronic diseases. Ensure the "__str__" method is included to provide a meaningful representation when instances are printed.

### 2.7.7 Data Structures in Classes Problems

1. Consider the classes "Major" and "Student" in Code 2.2.
   a. Generate at least five additional instances of the class "Major" to represent various majors students can choose. Include all these major instances in a tuple.
   b. Identify five job roles and positions individuals can pursue upon graduating with the major instances. Proceed as follows:
      - Develop a dictionary where major instances serve as keys, and the corresponding values are sets of associated job roles for each major (e.g., a computer science graduate can be a data scientist, a cyber security analyst, or an IT consultant).
      - Create a method to identify common job roles between different majors, revealing the majors that will help hold these roles.
      - Create a method prompting students to select a major and display the potential job roles achievable upon graduation.
   c. Provide user interactions as follows:
      - Request the user to choose a major and display the associated job roles.
      - Prompt the user to select another major, showcasing the job roles linked to it.
      - Compare the job roles between the first and second majors. If there are no shared job roles, inform the user that these two majors lead to distinct career paths.
2. Develop an Object-Oriented program (OOP) to create an Emirati food ordering system. The program will showcase a menu of Emirati dishes with corresponding prices, allow users to select items from the menu, generate an invoice detailing the selected items and their prices, and calculate the total amount payable. Use OOP concepts (classes, attributes, and methods) to provide the following functionalities for the system:
   a. The program will display a menu of Emirati dishes with prices for user selection.
   b. Users will be prompted to choose items from the menu, and the program will facilitate order placement.
   c. Upon order completion, the program will generate an invoice detailing the selected items and their prices, providing the total amount payable.
   d. Users can request concise descriptions of the ingredients for each food item on the menu.
   e. Ensure the privacy of attributes by applying encapsulation.

# 3 Exception Handling

## 3.1 INTRODUCTION

An exception is an unwarranted error that occurs during program execution. In most cases, it happens when an external input is handled within a program. If the external input is not verified and handled within the program, the program raises an error and abruptly stops. This situation is disruptive for the user trying to accomplish a task. To ensure the graceful execution of the program without throwing errors, such exceptions need to be caught, and a useful message needs to be displayed to the user.

**CODE 3.1 PROGRAM TO DIVIDE TWO NUMBERS**

```
1. # Input: Get two numbers from the user
2. num1 = int(input("Enter the first number: "))
3. num2 = int(input("Enter the second number: "))
4.
5. # Process: Divide the two numbers
6. result = num1 / num2
7.
8. # Output: Display the result
9. print(f"Dividing {num1} and {num2} gives us: {result}")
```

**BOX 3.1 OUTPUT FOR CODE 3.1**

**Output**

```
Enter the first number: 23
Enter the second number: 2
Dividing 23 and 2 gives us: 11.5
Process finished with exit code 0
```

Consider the program shown in Code 3.1. The program works perfectly; if the correct input is given for "num1" and "num2", for example, "23" and "2", respectively, it will give a result of "11.5", as shown in Box 3.1.

However, the program depends on external inputs given by the user and converts the two input values to an integer data type. If the user chooses to give input that is not an integer, the program will throw an error. Similarly, if the user chooses to give "0" for the second input, "num 2", the program will throw an error in Line 6 because division by zero is invalid. If an exception occurs on a particular line in the code, the program will not continue to the next line, and instead, the program will terminate while the exception's traceback is displayed in the terminal. The traceback provides useful information regarding the exception, i.e., the name of the exception and the line number that caused the exception.

DOI: 10.1201/9781032668321-3

- When the user provides a non-integer input like the word "Hello", a "ValueError" is raised, as shown in Output 1, Box 3.2. The "ValueError" is raised because an unsupported value is assigned to an integer data type.
- Since division by "0" is not supported, Python raises the "ZeroDivisionError", as shown in Output 2, Box 3.2.

When such errors occur in computer programs, programmers commonly use the expressions *"An exception was raised"* or *"An exception was thrown"*.

### BOX 3.2 RAISING EXCEPTIONS WITH INCORRECT INPUT

**Output 1**

```
Process finished with exit code 1
Enter the first number: Hello
Traceback (most recent call last):
  File "D:\Users\pr1.py", line 2, in <module>
    num1 = int(input("Enter the first number: "))
           ^^^^^^^^^^^^^^^^^^^^^^^^^^^^^^^^^^^^^^^
ValueError: invalid literal for int() with base 10: 'Hello'

Process finished with exit code 1
```

**Output 2**

```
Enter the first number: 50
Enter the second number: 0
Traceback (most recent call last):
  File "D:\Users\pr1.py", line 6, in <module>
    result = num1 / num2
             ~~~~~^~~~~~
ZeroDivisionError: division by zero
```

"ValueError" and "ZeroDivisionError" are built-in classes defined in Python to describe the respective errors. Similarly, other errors are defined by Python as exception classes. Some of the commonly seen errors are:

1. "IndexError": Raised when an index of a sequence (list, tuple, or string) does not exist.
2. "TypeError": Raised when there is a conflict between data types.
3. "NameError": Raised when a variable does not exist.
4. "FileNotFoundError": Raised when a file or directory is requested but doesn't exist.

All exception classes, such as the ones mentioned above, are extensions of the parent class, "BaseException". As a result, user-defined exception classes can be created by inheriting from the "BaseException" class or any other built-in exceptions.

Another interesting construct that programmers can utilize is the ability to explicitly trigger an exception by using the "raise" statement. The "raise" statement creates an instance of the specified exception class. The basic syntax of the "raise" statement is:

"raise ExceptionClassName"

For example:

"raise ValueError"

Optionally, the exception class name may include arguments to provide a custom error message. For example:

"raise ValueError('A string value is given instead of an integer')"

To illustrate the advantage of raising an exception, consider the following example in Code 3.2. A programmer explicitly raises an exception in this code if the user enters a negative number for their age.

**CODE 3.2 RAISING AN EXCEPTION**

```
age=int(input("Enter your age: "))
if age < 0:
    raise ValueError("Age cannot be negative")
```

Exceptions or errors are either explicitly triggered by the programmer or actual errors that occur during program execution. They must be handled in both cases so that programs do not end abruptly. Instead, programs must gracefully exit, providing meaningful feedback to the users. Programmers handle smooth program execution with the use of "try-except-else-finally" blocks. These blocks work together to handle the exceptions raised in a program. There are several ways to use these blocks, but the basic version is the "try/except" statement.

## 3.2 THE "try/except" STATEMENT

Python uses the "try/except" statement to handle exceptions that may occur in a program. The "try/except" statements must include the "try" clause, the "try" block, the "except" clause, and the "except" block.

As shown in Figure 3.1, Line 1 has the "try" clause followed by a colon ":" and then a block of indented Python statements (Lines 2 and 3) that constitute the "try" block. The "try" block is followed by the "except" clause with a colon ":" and then a block of indented Python statements (Lines 5 and 6) that constitute the "except" block. Any statement in the "try" block may raise an exception, and if so, the remaining lines in the "try" block are not executed. However, the program does not end abruptly. Instead, the exception triggers the execution of the "except" clause. As a result, all the

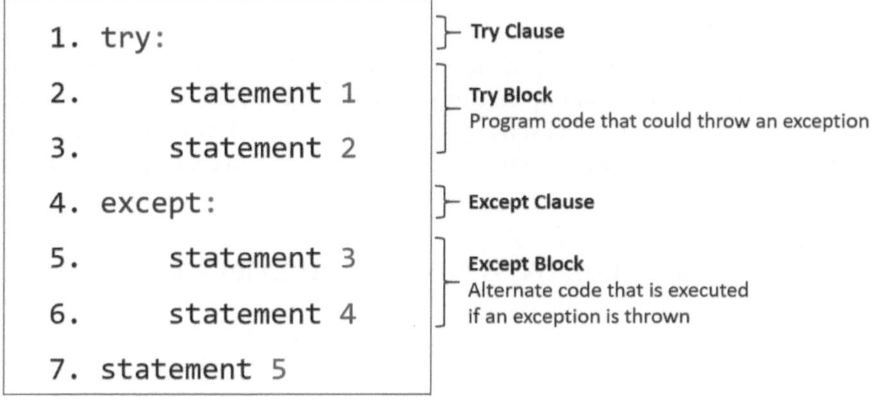


**FIGURE 3.1** Try/Except statements.

statements in the “except” block are executed, and the compiler continues to execute the remaining statements of the program (Line 7).

**CODE 3.3 HANDLING EXCEPTIONS WITH THE TRY/EXCEPT STATEMENT**

```
1. try:
2.     # Input: Get two numbers from the user
3.     num1 = int(input("Enter the first number: "))
4.     num2 = int(input("Enter the second number: "))
5.
6.     # Process: Divide the two numbers
7.     result = num1 / num2
8.
9.     # Output: Display the result
10.    print(f"Dividing {num1} and {num2} gives us: {result} ")
11. except:
12.    print("Something went wrong.")
```

**BOX 3.3 OUTPUT OF CODE 3.3**

**Output**

```
Enter the first number: 5
Enter the second number: 0
Something went wrong.

Process finished with exit code 0
```

Consider the example in Code 3.3, where Line 1 has the “try” clause, and Lines 2–10 constitute the “try” block. Code Line 11 has the “except” clause, and Line 12 is the “except” block. Let us consider that the user enters zero for the second input, as shown in Box 3.3, leading to a “ZeroDivisionError” to be raised by Line 7. The rest of the “try” block is skipped, and the control goes to Line 11, the “except” clause. The “except” block prints the generic statement as an output and gracefully exits the program. Contrast the output with the traceback that was produced earlier in Box 3.2; in that case, the “try/except” statement was not used to handle the error.

### 3.2.1 Multiple “except” Blocks

The “except” block in Code 3.3 is a generic except block that handles all types of exceptions. However, a “try” block can have multiple “except” blocks for specific exceptions. In such cases, when there are multiple “except” blocks, the generic “except” block will be written at the end and will be executed only if the specific “except” blocks cannot handle the exception that was raised.

**CODE 3.4 PROGRAM WITH POSSIBLE MULTIPLE EXCEPTIONS**

```
# Create a list of numbers
nums = [2, 4, 6, 8, 10]
print(nums)
# Input: Get a position in the list
position = int(input("Choose the position in the list: "))
# Input: Get denominator for division
```

```
7. denominator = int(input("Choose a number to divide your choice with: "))
8. # Process: Divide the value from the list with the denominator
9. result = nums[position] / denominator
10. # Output: Display output
11. print(f"Your choice {nums[position]} divided by {denominator} is: ", result)
```

Code 3.4 requires two integer inputs from the user on Lines 5 and 7. The first input is expected to be an integer between "0" and "4" to indicate a position in the list on Line 2. An input that is outside that range would raise an "IndexError". The second input is expected to be an integer but not a 0, as this would raise a "ZeroDivisonError". If any input given by the user is not an integer, then a "ValueError" is raised. The different possible outputs are shown in Box 3.4.

## BOX 3.4 POSSIBLE OUTPUTS FOR CODES 3.4 AND 3.1

**Output 1 – No Errors**

```
[2, 4, 6, 8, 10]
Choose the position in the list: 3
Choose a number to divide your choice with: 2
Your choice 8 divided by 2 is:  4.0

Process finished with exit code 0
```

**Output 2 – IndexError**

```
[2, 4, 6, 8, 10]
Choose the position in the list: 7
Choose a number to divide your choice with: 2
Traceback (most recent call last):
  File "D:\Users\pr1.py", line 9, in <module>
    result = nums[position] / denominator
             ~~~~^^^^^^^^^^
IndexError: list index out of range

Process finished with exit code 1
```

**Output 3 – ZeroDivisionError**

```
[2, 4, 6, 8, 10]
Choose the position in the list: 3
Choose a number to divide your choice with: 0
Traceback (most recent call last):
  File "D:\Users\pr1.py", line 9, in <module>
    result = nums[position] / denominator
             ~~~~~~~~~~~~~~~^~~~~~~~~~~~
ZeroDivisionError: division by zero

Process finished with exit code 1
```

**Output 4 – ValueError**

```
[2, 4, 6, 8, 10]
Choose the position in the list: Hello
Traceback (most recent call last):
  File "D:\Users\pr1.py", line 5, in <module>
    position = int(input("Choose the position in the list: "))
               ^^^^^^^^^^^^^^^^^^^^^^^^^^^^^^^^^^^^^^^^^^^^^^^
ValueError: invalid literal for int() with base 10: 'Hello'

Process finished with exit code 1
```

Now, Code 3.5 illustrates how two of the exceptions, "IndexError" and "ZeroDivisionError", are handled in separate except blocks (Lines 13 and 15), and the generic "except" block (Line 17), which would be the last one, handles any other exception like "ValueError" that may be raised. Every "try" block must be followed by at least one "except" clause and its respective "except" block. The possible outputs are shown in Box 3.5.

## CODE 3.5 HANDLING EXCEPTIONS WITH MULTIPLE EXCEPT BLOCKS

```
1. try:
2.     # Create a list of numbers
3.     nums = [2, 4, 6, 8, 10]
4.     print(nums)
5.     # Input: Get a position in the list
6.     position = int(input("Choose the position in the list: "))
7.     # Input: Get denominator for division
8.     denominator = int(input("Choose a number to divide your choice with: "))
9.     # Process: Divide the value from the list with the denominator
10.    result = nums[position] / denominator
11.    # Output: Display result
12.    print(f"Your choice {nums[position]} divided by {denominator} is: ", result)
13. except IndexError:
14.    print("Position should be between 0 and 4 only.")
15. except ZeroDivisionError:
16.    print("Denominator cannot be 0.")
17. except:
18.    print("Something went wrong.")
```

## BOX 3.5 POSSIBLE OUTPUTS FOR CODE 3.5

**Output 1 – IndexError handling**

```
[2, 4, 6, 8, 10]
Choose the position in the list: 8
Choose a number to divide your choice with: 2
Position should be between 0 and 4 only.
```

**Output 2 – ZeroDivisionError handling**

```
[2, 4, 6, 8, 10]
Choose the position in the list: 2
Choose a number to divide your choice with: 0
Denominator cannot be 0.
```

**Output 3 – Any other error or ValueError handling**

```
[2, 4, 6, 8, 10]
Choose the position in the list: hello
Something went wrong.
```

It is common to have a "try" block followed by several "except" clauses and their respective "except" blocks. If exceptions need to be handled collectively, then the exceptions can be grouped in a tuple with a single "except" clause. For instance,

"except (exception 1, exception 2, …) as variable_name:"

The "as" clause is optional. If the program needs a reference to the object of the raised exception class, then this can be done by referencing the "variable_name" in the "as" clause.

For example, consider Code 3.6. The "except" clause in Line 10 handles a range of exceptions that might occur while working with files. For example, the "FileNotFoundError" occurs if the filename does not exist in the respective directory, as shown in Output 1 of Box 3.6. A "PermissionError" occurs if the file has restricted access and an attempt was made to read or write to the file, as shown in Output 2 of Box 3.6. An "OSError" is a general error class that can be raised due to various concerns such as operating system, hard disk, or network issues. The "except" clause in Line 10 assigns the type of exception raised to a variable named "ex". Suppose we need to access the information related to the specific type of error. In that case, we can use the variable "ex", as shown in Line 14, which prints the default error message for the specific error class, helping us identify the type of error raised.

**CODE 3.6 HANDLING A RANGE OF EXCEPTIONS IN ONE EXCEPT CLAUSE**

```
1.  try:
2.      # Open a file
3.      file = open("data1.txt", "r+")
4.      # Read from a file
5.      content = file.read()
6.      # Print everything in the file
7.      print("File content:", content)
8.
9.  # Catch file-related errors
10. except (FileNotFoundError, PermissionError, OSError) as ex:
11.     # ]Display User-defined error message
12.     print("An error occurred during file operations:")
13.     # Prints the message provided by the built-in error class
14.     print(ex)
```

**BOX 3.6 POSSIBLE OUTPUTS FOR CODE 3.6**

**Output 1 – FileNotFoundError handling**

```
An error occurred during file operations:
[Errno 2] No such file or directory: 'data1.txt'
```

**Output 2 – PermissionError handling**

```
An error occurred during file operations:
[Errno 13] Permission denied: 'data1.txt'
```

### 3.2.2 The Code in the "try" Block

Which set of code statements should be placed in the "try" block? Most programs that require explicit access to external resources, such as files, network connections, database connections, or user input, have potential exceptions that may be raised when processing those resources. For this reason, most code statements that access external resources should be scaffolded within a "try" block.

Rather than using separate "try" statements around every statement that could raise an exception, it is better to have a "try" block to contain a significant section of a program where several statements can raise exceptions. However, for proper granularity, each "try" statement should enclose a section of code small enough that, when an exception occurs, the specific context can be identified, errors are easily isolated, and the "except" blocks can process specific exceptions.

Exceptions may also surface via statements in a "try" block that are calls to methods or methods called directly or indirectly via the Python interpreter as it executes the code. If the method that is called raises an exception, then the "try" block surrounding the method call must handle the exception(s) thrown by the method. It is essential to know what exceptions are raised (if any) by the method or method and understand reasons why such exceptions may occur before using the methods. Code 3.7 illustrates an example where an exception is raised within the method call. The built-in "power()" method is used, which calculates "x" to the power of "y". The program begins with initializing "x" and "y" in Lines 2 and 3. A "try" clause has been added in Line 4 to handle an exception caused by the "power()" method. The "power()" method throws a "ValueError" in case the user enters a decimal value for the exponent. As a result, the "except" block at Line 9 catches a "ValueError" and displays an appropriate error message to the user, as shown in Box 3.7.

### CODE 3.7 EXCEPTION HANDLING FOR METHODS

```
1. import math
2. x=-1.7
3. y=0.3
4. try:
5.     # Use the built-in power method
6.     result=math.pow(x,y)
7.     print(result)
8. #This except clause will catch if a decimal value is entered as an exponent
9. except ValueError as ve:
10.    print("The exponent cannot be a decimal value")
```

### BOX 3.7 OUTPUT FOR CODE 3.7

**Output**

```
The exponent cannot be a decimal value
```

To ensure classes handle exceptions properly, we need to add "try-and-catch" blocks to class methods. To illustrate exception handling for classes, consider a "HealthApp" class that takes the "weight" and "height" to calculate the body mass index (BMI) of a person in Code 3.8. The "set_weight()" method of the "HealthApp" class has its own "try-and-catch" blocks to register an error if the "weight" is negative (Lines 7–13). Handling the exception within the class method helps programmers identify the code causing the error. If a negative value is provided for "weight", the "set_weight()" method identifies the "ValueError" and prints the class name and the exception "e" (Line 13). Similarly, the "set_height()" method throws and catches an exception if the "height" provided is negative (Lines 15–21). Moreover, the "calculate_bmi()" method ensures that "weight" and "height" are provided before a calculation is performed. Otherwise, the method raises an error to be thrown back to the main program. This ensures that the main program identifies that the BMI was not calculated successfully within the method. Box 3.8 shows the result of the program if negative values for "weight" and "height" were entered.

## CODE 3.8 HANDLING EXCEPTIONS IN A CLASS

```
import math
class HealthApp: # Create a class called HealthApp
    def __init__(self):
        self.weight = None
        self.height = None

    def set_weight(self, weight): #This method will ensure weight is a positive number
        try: # Handle the exception when a weight is negative
            if weight <= 0:
                raise ValueError("Weight must be greater than 0.")
            self.weight = weight
        except ValueError as e:
            print(f"Error Occured In:{self.__class__.__name__} Class, Type of Error: {e}")

    def set_height(self, height):
        try: # Handle the exception when a height is negative
            if height <= 0:
                raise ValueError("Height must be greater than 0.")
            self.height = height
        except ValueError as e:
            print(f"Error Occured In:{self.__class__.__name__} Class, Type of Error: {e}")

    def calculate_bmi(self):
        try: # Handle the exception when a weight and height are not provided
            if self.weight is None or self.height is None:
                raise ValueError("Weight and height must be recorded to calculate BMI.")
            return self.weight / math.pow(self.height, 2)
        except ValueError as e:
            print(f"Error Occured In:{self.__class__.__name__} Class, Type of Error: {e}")
            raise e


#Instance of the HealthApp class is created
health_app = HealthApp()
try:
    weight = float(input("Enter your weight (in kg): "))
    health_app.set_weight(weight)

    height = float(input("Enter your height (in meters): "))
    health_app.set_height(height)

    bmi = health_app.calculate_bmi()
    print(f"Your BMI is: {bmi:.2f}")

except ValueError:
        print("Error taking Input")
```

## BOX 3.8 OUTPUT FOR CODE 3.8

**Output**

```
Enter your weight (in kg): -76
Error Occurred In:HealthApp Class, Type of Error: Weight must be greater than 0.
Enter your height (in meters): -68
Error Occurred In:HealthApp Class, Type of Error: Height must be greater than 0.
Error Occurred In:HealthApp Class, Type of Error: Weight and height must be recorded to calculate BMI.
Error taking input.
```

## 3.3 THE "else" CLAUSE

The "try/except" statement may include an optional "else" clause, which appears after all the "except" blocks. The statements in the "else" block are executed after the statements in the "try" block only if no exceptions were raised. If an exception is raised, the "else" suite is skipped. Code 3.5 with the addition of the "else" block is shown in Code 3.9. Line 17 of Code 3.9 includes the "else" clause, executed only when no exceptions are raised, as shown in Box 3.9, where we used a correct index and a denominator different than "0". In this case, the "else" block (Lines 18 and 19) prints out the expected output of a successful division.

### CODE 3.9 THE USE OF THE ELSE BLOCK FOR HANDLING EXCEPTIONS

```
1. try:
2.     # Create a list of numbers
3.     nums = [2, 4, 6, 8, 10]
4.     print(nums)
5.     # Input: Get a position in the list
6.     position = int(input("Choose the position in the list: "))
7.     # Input: Get denominator for division
8.     denominator = int(input("Choose a number to divide your choice with: "))
9.     # Process: Divide the value from the list with the denominator
10.    result = nums[position] / denominator
11. except IndexError as ex:
12.    print("Position should be between 0 and 4 only.", ex)
13. except ZeroDivisionError as ex:
14.    print("Denominator cannot be 0.", ex)
15. except:
16.    print("Something went wrong.")
17. else: # This clause will execute only when there are no exceptions raised
18.    # Output: Display result
19.    print(f"Your choice {nums[position]} divided by {denominator} is: ", result)
```

### BOX 3.9 OUTPUT FOR CODE 3.9

**Output**

```
[2, 4, 6, 8, 10]
Choose the position in the list: 3
Choose a number to divide your choice with: 3
Your choice 8 divided by 3 is:  2.6666666666666665
```

## 3.4 THE "finally" CLAUSE

The "try/except" statement may have an optional "finally" clause, which must appear after all the "except" clauses and after the "else" clause if it is used. The statements under the "finally" clause are known as the "finally" block. The statements in the "finally" block always execute after the "try" block and after any "except" blocks have been executed. Therefore, the statements in the "finally" block execute irrespective of whether an exception occurs. The "finally" block usually creates backup error logs or performs cleanup operations, such as closing files or other resources. Lines 20 and 21 in Code 3.10 depict the "finally" clause and block. The output "Bye" is printed whether an exception was raised or not, as shown in Outputs 1 and 2 in Box 3.10.

### CODE 3.10 THE USE OF THE FINALLY BLOCK FOR HANDLING EXCEPTION

```
try:
    # Create a list of numbers
    nums = [2, 4, 6, 8, 10]
    print(nums)
    # Input: Get position in list
    position = int(input("Choose the position in the list: "))
    # Input: Get denominator for division
    denominator = int(input("Choose a number to divide your choice with: "))
    # Process: Divide the value from list with the denominator
    result = nums[position] / denominator
except IndexError as ex:
    print("Position should be between 0 and 4 only.", ex)
except ZeroDivisionError as ex:
    print("Denominator cannot be 0.", ex)
except:
    print("Something went wrong.")
else:
    # Output: Display output
    print(f"Your choice {nums[position]} divided by {denominator} is: ", result)
finally: # This clause will run whether an exception occurred or not
    print("Bye")
```

### BOX 3.10 OUTPUT FOR CODE 3.10

**Output 1 – "finally" clause output with an error**

```
[2, 4, 6, 8, 10]
Choose the position in the list: 3
Choose a number to divide your choice with: 0
Denominator cannot be 0. division by zero
Bye
```

**Output 2 – "finally" clause output without error**

```
[2, 4, 6, 8, 10]
Choose the position in the list: 2
Choose a number to divide your choice with: 2
Your choice 6 divided by 2 is:  3.0
Bye
```

## 3.5 THE "assert" STATEMENT

Assertions are Boolean expressions that check if a condition is true or false. If the result of the condition is true, the program continues to the next line of code. However, if the result is false, the program stops and throws an error. Python has a built-in "assert" statement to use the assertion condition in a program. The "assert" statement has a conditional expression, which is supposed to be always *true*. If the condition is *false*, the "assert" statement raises an "AssertionError". The syntax of the "assert" statement is:

"assert <condition>, <exception message>"

The "assert" statement may include the optional exception message. If the condition is not satisfied, i.e., if it is not *true*, an "AssertionError" along with the error message is raised.

### CODE 3.11 THE USE OF ASSERT STATEMENT

```
1. # Create a method to square a negative number
2. def square_neg(val):
3.     assert val < 0 # Raise an AssertionError if the value is positive
4.     return val ** 2
5.
6. try:
7.     # Input: Get a user input
8.     negativenumber = int(input("Enter a negative number: "))
9.     # Call method to square the negative number
10.    result = square_neg(negativenumber)
11. except AssertionError:
12.    print("We need a negative value.")
13. except:
14.    print("Something went wrong.")
15. else:
16.    print(result, "is the square of your number")
17. finally:
18.    print("Bye")
```

### BOX 3.11 OUTPUT FOR CODE 3.11

**Output**

```
Enter a negative number: 9
We need a negative value.
Bye
```

In Code 3.11, the “assert” statement in Line 3 checks if the value of a number “val” passed to the square_neg() method is negative. If the condition “val < 0” evaluates to *true*, then the code continues, and the method returns the square of the negative number and does not raise an exception. However, if the condition evaluates to *false*, i.e., if the value passed to the method is a positive number such as the output shown in Box 3.11 where ”9” is entered by the user, then an exception is raised. The “assert” statement always raises the exception “AssertionError”. The exception is handled in the “except” block (Lines 11 and 12). All other exceptions are handled in the generic “except” block (Lines 13 and 14). If no exceptions are raised, the “else” block (Lines 15 and 16) is executed, and the result is printed. Next, the “finally” block (Lines 17 and 18) is always executed, and therefore, the string “Bye” is printed out.

### CODE 3.12 THE ASSERT STATEMENT WITH THE OPTIONAL EXCEPTION MESSAGE

```
# Set age limit for a job application
try:
    age = int(input("Enter your age: "))
    assert (age > 20 and age < 60), age
    print("You can apply for the job!")
except AssertionError as ex:
    print("Age", ex, "is not suitable for this job.")
except:
    print("Something went wrong.")
```

**BOX 3.12 OUT FOR CODE 3.12**

**Output**

```
Enter your age: 12
Age 12 is not suitable for this job.
Process finished with exit code 0
```

The "assert" statement in Code 3.12 includes the optional exception message in Line 4 following the "assert" condition and a comma, which is the "age" variable in this code. Here, the "assert" condition "age > 20 and age < 60" will be evaluated. If the "age" given by the user is between "20" and "60", the program continues without throwing an exception. However, if the condition evaluates to *false*, the "AssertionError" is thrown and handled in Line 6, and the exception message "age" is passed to the variable "ex" as shown in the output, Box 3.12. The output shows the assertion message when the condition evaluates to *false*.

## 3.6 HIERARCHY OF EXCEPTIONS

Python has a rich suite of built-in exception classes. The hierarchy of exception classes is shown in Box 3.13, where all exceptions are extended from the parent class "BaseException". Typically, "BaseException" is never raised on its own and is ideally inherited by other extended exception classes that can be raised. The "Exception" class is the most inherited exception type of the "BaseException" class. In addition, all exception classes that are errors are subclasses of the "Exception" class.

Most subclasses under the "Exception" class have a suffix of "Error". Python does not differentiate between fatal and catchable errors (those handled with "try/except" statements). Python recognizes all exceptions that would essentially crash a program as errors and, therefore, the suffix "Error". Several exceptions represent warning categories. This categorization is useful to be able to filter out groups of warnings.

**BOX 3.13 HIERARCHY OF EXCEPTION CLASSES IN PYTHON**

- BaseException
  - GeneratorExit
  - KeyboardInterrupt
  - SystemExit
  - Exception
    - ArithmeticError
      - FloatingPointError
      - OverflowError
      - ZeroDivisionError
    - AssertionError
    - AttributeError
    - BufferError
    - EOFError
    - ImportError
      - ModuleNotFoundError
    - LookupError

        - IndexError
        - KeyError
    - MemoryError
    - NameError
        - UnboundLocalError
    - OSError
        - BlockingIOError
        - ChildProcessError
        - ConnectionError
            - BrokenPipeError
            - ConnectionAbortedError
            - ConnectionRefusedError
            - ConnectionResetError
        - FileExistsError
        - FileNotFoundError
        - InterruptedError
        - IsADirectoryError
        - NotADirectoryError
        - PermissionError
        - ProcessLookupError
        - TimeoutError
- ReferenceError
- RuntimeError
    - NotImplementedError
    - RecursionError
- StopIteration
- StopAsyncIteration
- SyntaxError
    - IndentationError
        - TabError
- SystemError
- TypeError
- ValueError
    - UnicodeError
        - UnicodeDecodeError
        - UnicodeEncodeError
        - UnicodeTranslateError
- Warning
    - BytesWarning
    - DeprecationWarning
    - FutureWarning
    - ImportWarning
    - PendingDeprecationWarning
    - ResourceWarning
    - RuntimeWarning
    - SyntaxWarning
    - UnicodeWarning

## 3.7 CASE STUDY

The case study presented in the first chapter applied the object-oriented approach to creating bank transactions and maintaining an account balance. The same example is presented here, with the exception handling included.

### CODE 3.13 EXCEPTION HANDLING FOR USER INPUTS FOR TRANSACTIONS

```
import datetime
from enum import Enum


class TransactionType(Enum):
    """An enumerator type class that defines the types of transactions"""
    INCOME = 1  # an income type defines a transaction of a gained amount of money
    EXPENSE = 2  # an expense type defines a transaction of a spent amount of money


class Currency(Enum):
    """An enumerator type class that defines the types of currencies"""
    USD = 1  # US Dollars
    EUR = 2  # Euro
    GBP = 3  # Great Britain Pound
    AED = 4  # Arab Emirati Dirham


class Transaction:
    """A class that represents a financial transaction"""
    # a static variable that keeps track of the number of transactions
    transaction_count = 0

    # Initialize the Transaction class
    def __init__(self, transaction_type, amount, currency, description, date_time):
        # Increment the number of transactions with the creation of a new transaction
        Transaction.transaction_count += 1

        # Assign a unique transaction ID based on the number of transactions
        self.__trans_id = "T" + str(Transaction.transaction_count)
        self.__trans_type = transaction_type
        self.trans_amount = amount
        self.__trans_description = description
        self.__trans_currency = currency
        self.__trans_date_time = date_time

    def get_trans_id(self):
        return self.__trans_id

    def get_transaction_type(self):
        return self.__trans_type

    def get_trans_amount(self):
        return self.trans_amount

    def get_trans_date_time(self):
        return self.__trans_date_time

    def get_trans_currency(self):
        return self.__trans_currency

    def get_trans_description(self):
        return self.__trans_description

```

```
    def set_trans_description(self, description):
        self.__trans_description = description

    # Define a string representation of the transaction
    def __str__(self):
        return (
            'Transaction ID: {id}, Amount: {amount} {currency}, '
            'Type: ({type}), Description: {description}, '
            'Date & Time: {date_time}'.format(
                id=self.__trans_id,
                amount=self.trans_amount,
                currency=Currency(self.__trans_currency).name,
                type=TransactionType(self.__trans_type).name,
                description=self.__trans_description,
                date_time=self.__trans_date_time))


class TransactionManager:
    """
        This class keeps track of transactions and allows users
        to create transactions, and calculate their total amount
    """

    def __init__(self):
        # Transactions managed by TransactionManager as a list
        self.__transactions = []

    # A method that allows the creation of a transaction
    # User Input is collected for each transaction
    def create_transaction(self):
        # Transaction type
        trans_type_input = input("Transaction type (Expense/Income)? ")
        assert trans_type_input in ["Expense", "Income"], "Invalid Input"
        trans_type = TransactionType.EXPENSE  # Defaulting the type to Expense
        if trans_type_input == "Income":
            trans_type = TransactionType.INCOME
        trans_amount = float(input("Transaction amount? "))
        if trans_type == TransactionType.EXPENSE:
            trans_amount *= -1

        # Transaction description
        trans_description = input("Transaction description: ")

        # Transaction currency
        trans_currency_input = input("Currency of Transaction (USD/EUR/GBP/AED)? ")
        assert trans_currency_input in ["USD", "EUR", "GBP", "AED"], "Invalid Input"
        if trans_currency_input == "USD":
            trans_currency = Currency.USD
        elif trans_currency_input == "EUR":
            trans_currency = Currency.EUR
        elif trans_currency_input == "GBP":
            trans_currency = Currency.GBP
        else:
            trans_currency = Currency.AED

        # Transaction date & time
        trans_date_time_input = input("Transaction date - 'now' or YYYY-MM-DD hh:mm:ss: ")
        if trans_date_time_input == "now":
            trans_date_time = datetime.datetime.now()
        else:
            trans_date_time = datetime.datetime.strptime(
                                trans_date_time_input,
                                "%Y-%m-%d %H:%M:%S")

```

```
119.         # Create an object of class Transaction
120.         transaction = Transaction(transaction_type=trans_type,
121.                                   amount=trans_amount,
122.                                   currency=trans_currency,
123.                                   description=trans_description,
124.                                   date_time=trans_date_time)
125.
126.         # Add the newly created transaction to the list of transactions
127.         self.__transactions.append(transaction)
128.         return transaction
129.
130.     def get_total_amount(self):
131.         total = 0
132.         # For each transaction,  read the amount and then add it to the total
133.         for transaction in self.__transactions:
134.             total += transaction.get_trans_amount()
135.         return total
136.
137.     # Print the transactions
138.     def print_transactions(self):
139.         for transaction in self.__transactions:
140.             print(transaction)
141.
142.
143. # Create a Transaction Manager object
144. trans_manager = TransactionManager()
145. # Create two transactions defined by the user
146. # Handle all possible exceptions
147. for counter in range(2):
148.     while True:
149.         try: # Handle exceptions that occur within create_transaction() method
150.             t1 = trans_manager.create_transaction()
151.         except ValueError as ve:
152.             print(f"Ensure proper input is provided. {ve}")
153.         except AssertionError as ae:
154.             print(f"Ensure proper input is provided. {ae}")
155.         except Exception as e: # Exception is the parent of non-fatal exceptions
156.             print(f"An unexpected error occurred: {e}")
157.         else: # This block will run only if no exception is raised.
158.             # Print all the created transactions
159.             trans_manager.print_transactions()
160.             # Display the total amount of transactions
161.             print("Balance Amount: {amount}".format(
162.                             amount=trans_manager.get_total_amount()))
163.             break
164.
```

In Code 3.13, in Line 87, the "assert" statement is used to ensure that the user's input is restricted to two values, either "Expense" or "Income". If the input is not one of the two string values, then an "AssertionError" is raised. Similarly, in Line 100, the "assert" statement ensures that the currency type is restricted to "USD", "EUR", "GBP", and "AED". Again, an "AssertionError" will be raised if the user does not adhere to these values.

Note that the above lines of code are where the program receives input from an external source, i.e., a user input. These "assert" statements are in the method named "create_transaction()" of the "TransactionManager" class. The call to this method is seen in Line 150. Therefore, the call to the method is part of the "try" block. Any exception that occurs within the method "create_transaction()", is handled in this "try" block. The specific exceptions handled are "ValueError" and "AssertionError", and any other exceptions will be handled in the generic exception block Lines 155 and 156. If no exceptions are raised, the "else" block from Lines 157 to 163 provides continuity for the program.

## 3.8 CHAPTER SUMMARY

This chapter explains managing unexpected events and errors in a program. It covers the fundamental concepts of try, except, else, and finally blocks, allowing programmers to handle exceptions and maintain program stability gracefully. Moreover, it covered raising exceptions and assertions, allowing for precise management within the current try block, or propagating the exception to the calling code for appropriate handling. Moreover, a general outline of exception class hierarchy was introduced, and the application of some exception classes was explained with practical examples. Therefore, this chapter provides comprehensive support for writing robust code and ensuring reliable programming execution.

## 3.9 EXERCISES

### 3.9.1 Test Your Knowledge

1. Explain what happens when an exception occurs in Python.
2. What is the purpose of the "try" and "except" blocks?
3. Explain the significance and purpose of the "finally" block in a "try/except" statement.
4. What is the difference between a "TypeError" and a "ValueError"?
5. Describe the functionality of the "else" clause within a "try/except" statement.
6. With an example, explain the "assert" statement.
7. With an example, explain the "raise" statement.

### 3.9.2 Multiple Choice Questions

1. What is an exception?
   a. A warning that there is something that might go wrong.
   b. An error that occurs during program execution.
   c. A way to prevent errors from occurring.
   d. A way to recover from errors.
2. What purpose does the "assert" statement serve in Python?
   a. It defines a variable and assigns a value to it
   b. It prints a message to the console
   c. It checks if a condition is true to continue program execution.
   d. It terminates the program regardless of the condition's result.
3. What is the "raise" statement used for?
   a. To create an instance of an exception class
   b. To catch an exception
   c. To prevent an exception from occurring
   d. To handle an exception
4. What is the purpose of an "except" block in "try/except" statements?
   a. To protect code from occurred errors.
   b. To catch errors that occur in the "try" block.
   c. To execute code regardless of a raised exception.
   d. To stop program execution if an exception occurs.
5. How does one handle multiple exceptions in a single "try/except" block?
   a. By using the "raise" statement.
   b. By using multiple "except" blocks.
   c. It is not possible.
   d. By using the "else" block.

6. Which of the following statements about the "finally" block in a try/except statement is true?
   a. The "finally" block executes only if an exception occurs.
   b. The "finally" block executes before the "try" block.
   c. The "finally" block is optional and comes after all the other except clauses.
   d. The "finally" block is optional and can be placed before the except clauses.
7. Which of the following is the most accurate description of Python's "BaseException" class?
   a. It is a concrete class that can be instantiated directly.
   b. Exception classes in Python cannot be customized or extended to create user-defined exceptions.
   c. It is a built-in exception class that is raised when a generic error occurs.
   d. It is the root class of all exception classes in Python.
8. What happens if an exception is raised within a "try" block that contains an "else" clause in Python?
   a. The "else" block is executed regardless of whether an exception is raised or not.
   b. The "else" block is skipped if an exception is raised.
   c. The "else" block is executed before handling the exception.
   d. The "else" block raises an additional exception.

### 3.9.3 Short Answer Questions

1. Name the two optional clauses that appear after the "try" and "except" clauses.
2. What is the use of the "raise" statement?
3. What block is executed immediately after a "try" block when there is no exception?
4. What error is raised when there is an attempt to divide a number by zero?
5. What clause is always executed, regardless of whether an exception occurs?
6. What happens when an object of the incorrect type is passed to a method?
7. What error occurs if a file does not exist in the suggested location in a program?
8. When handling an exception, can the "else block" be written above the "except block"?
9. Name the parent class from which all exceptions in Python are inherited.

### 3.9.4 True or False Questions

1. All exceptions in Python are built-in classes.
2. The "except" block is used to catch exceptions.
3. A "PermissionError" is a general error class that can be raised due to various concerns, such as operating system, hard disk, or network issues.
4. The statements in the "finally" block of a "try/except" statement execute only if an exception is raised.
5. A "NameError" occurs when a variable that has not been defined is used.
6. An "IndexError" occurs only when trying to access elements of a sequence using an index within the valid range of indices for that sequence.
7. The "assert" statement in Python raises an error only if the condition provided is true.
8. The "Exception" class in Python cannot be inherited to create custom user-defined exceptions.
9. The "else" clause in a "try/except" statement is executed only if an exception is not raised within the "try" block.
10. We can have multiple "try" clauses and multiple "except" clauses in an exception syntax.

### 3.9.5 Fill in the Blanks

1. Add the appropriate code snippet to fill in the blanks in Code 3.14:

**CODE 3.14 FILL IN THE BLANKS WITH APPROPRIATE CODE SNIPPETS**

```
____:
    # Code that might raise an exception
    result = 10 / 0  # This line might raise a ZeroDivisionError
______ ZeroDivisionError:
    # Handling the specific ZeroDivisionError exception
    print("Error: Division by zero is not allowed")
```

2. Add the appropriate code snippet to fill in the blanks in Code 3.15:

**CODE 3.15 FILL IN THE BLANKS WITH APPROPRIATE CODE SNIPPETS**

```
class StringToInteger:
    def __init__(self, value):
        try:
            self.number = int(value)  # Convert 'value' to an integer
        except ____________:
            print("Error: Please provide a valid integer value.")


try:
    obj = StringToInteger("Hello")  # Create an instance with a non-integer value
except:
    pass  # Handle exception for creating an instance with an invalid value
```

3. Add the appropriate code snippet to fill in the blanks in Code 3.16:

**CODE 3.16 FILL IN THE BLANKS WITH APPROPRIATE CODE SNIPPETS**

```
class Product:
    def __init__(self, name, price):
        self.name = name
        self.price = price

    def set_price(self, new_price):
        _________ new_price >= 0, "Price should be a non-negative value"
        self.price = new_price


# Create an instance of the Product class
product = Product("Phone", 500)

# Set a negative price
product.set_price(-100)  # This will cause an AssertionError with the specified message
```

### 3.9.6 Coding Problems

1. In Chapters 1 and 2, a Python class was developed to manage patient records. This class had a method for updating the list of chronic diseases whenever a patient is diagnosed with a new chronic disease or recovers from an existing one.
   a. Modify the program to include a "try/except" block, which ensures that the new disease added by the user is of type "string". An "AssertionError" must be raised if a non-string value for the disease is provided.
   b. Add a "finally" block to print a message indicating the completion of the update operation.
2. In Chapter 2, classes "Student" and "Major" were implemented in Code 2.2. Update this code as follows:
   a. Filter students by their birth year:
      - Introduce a method to filter students based on a specific birth year entered by the user, returning a new list of these students along with their count.
      - Implement a "try/except" block to handle a potential "TypeError" if the entered year type is incorrect.
   b. Split students by major:
      - Develop a method to split the students' lists into different lists based on their majors.
      - Include a "try/except" block to handle a potential "IndexError" that may arise during the iteration over the "student_list".

### 3.9.7 Exception Handling Problems

1. Consider the "BankAccount" class from Code 1.6 in Chapter 1. Perform the following suggested modifications:
   a. Rewrite the constructor to include a "try/except" block for initializing the "account_number". This block should raise a "ValueError" to handle any negative entry of the "account_number".
   b. Enhance the "withdraw()" method by incorporating an "assert" statement to ensure two conditions:
      - The "amount" to be withdrawn is positive.
      - Verify that the "amount" to be withdrawn is less than the available "balance".
   c. Implement a "try" block within the "deposit()" method to execute the deposit operations:
      - If the deposit "amount" is greater than or equal to zero, proceed with the deposit and display a success message in an "else" block.
      - In case of a negative "deposit" amount, raise a "ValueError", handle it in an "exception" block, and print an appropriate error message.
      - Introduce a "finally" block to consistently display the current "balance", regardless of the deposit's success or failure. This ensures the balance's visibility after each deposit attempt.
2. Consider the Emirati food ordering system that you were asked to develop in Problem 2 of Section "Data structures in classes Problems":
   a. Modify the user selection method:
      - Implement a "try/except" block to handle user input validation within the appropriate range or existing in the menu.
      - Raise a "ValueError" if the input does not match and prompt the user to re-enter a valid choice.

b. Modify the display food item method:
   - Introduce a "try/except/else" block to manage the retrieval of descriptions for selected food items.
   - Raise a "NameError" if the description is unavailable, catch the error, and inform the user accordingly.
   - Display the description to the user in the "else" block upon successful retrieval.

# 4 Fundamentals of Object-Oriented Analysis

## 4.1 STRUCTURED VS. OBJECT-ORIENTED ANALYSIS

The software development process demands a deep understanding and analysis of the requirements before the software is implemented. Over time, various approaches have been employed to comprehend and define these requirements. The conventional method is structured analysis, which is still in use. However, a more recent and evolved method of analysis is Object-Oriented analysis (OOA). In this section, the two methods will be compared, shedding light on their differences and guiding the selection of the most appropriate approach for software analysis and design needs. This contrast will provide insight into the evolution of software development methodologies, providing a deeper understanding of the field.

Structured analysis, a well-documented process model, is a graphical description of a software system. It primarily focuses on functional decomposition and data modeling. The system's functions are depicted as a series of interconnected steps to illustrate the system's overall functional requirements. This technique is process-centered, where a data flow diagram illustrates the data that flows in and out of the sub-systems that make up the system. Within each function, business rules transform input data to generate output. Data modeling, typically done using ER (entity relationship) diagrams, is used to describe the data structures and relationships that are used by the functions and sub-systems. Understanding the role of structured analysis in software development is crucial for comprehending its contrast with OOA and choosing the most suitable approach for your software analysis and design needs.

Even in modern projects, structured analysis is still employed as an effective choice for requirement analysis, particularly in simple projects with well-defined requirements. As structured analysis has a well-documented process, it can be used to ensure that all of the required requirements are met and that the system is compliant with all applicable regulations. For this reason, it is favored in industries with strict regulatory and compliance requirements and projects where clear and thorough documentation is essential, such as in government contracts. Historical successes and related education in software development programs and hybrid approaches that blend structured analysis with modern methodologies further underline its enduring relevance in the field.

OOA is a more recent approach to software development that meticulously examines and defines software requirements. Unlike structured analysis, which treats data and processes as distinct elements, OOA unifies the data and the processes that operate on the data to form independent units called "objects". This methodology emphasizes creating a comprehensive system encompassing collaborating objects. OOA divides a software system into interconnected objects that communicate and cooperate to achieve specific goals. These objects encapsulate data (attributes or properties) and methods (functions or behaviors) that manipulate the data. The primary objective is to model real-world entities as objects within the software system and define their relationships and interactions.

By encapsulating data and the methods within objects, OOA ensures that data is only accessible and modified through controlled methods, preventing unauthorized access and maintaining data integrity and protection. Providing data integrity at this level is a unique feature of OOA, also termed "Information Hiding". Encapsulation promotes modularity by dividing the system into independent units, each with well-defined responsibilities and data. This modularity simplifies code organization, enhances reusability, and facilitates easier maintenance and testing. Additionally, OOA emphasizes concepts like inheritance, allowing objects to inherit properties and behaviors from other related

DOI: 10.1201/9781032668321-4

objects, effectively reusing code, and reducing redundancy. This is particularly useful for modeling real-world entities with shared characteristics.

Another important Object-Oriented (OO) concept is polymorphism, which enables different objects to be treated uniformly based on a common interface. Polymorphism enables flexible and dynamic behavior within the system. It is often implemented through abstract classes or interfaces, which define objects' common structure and behavior while allowing subclasses to provide their implementations.

Compared to structured analysis, OOA offers a more intuitive, flexible, and real-world approach to analyzing software requirements. It allows users to easily map real-world entities to software objects, which provides a better representation of complex real-world systems and promotes modularity, extensibility, and maintainability in software design. OOA's emphasis on modeling real-world entities as objects aligns well with modern software development practices, making it a preferred choice for projects with dynamic or evolving requirements.

## 4.2 THE UNIFIED MODELING LANGUAGE (UML)

Unified modeling language (UML) is a visual modeling language that can capture both dynamic behavior and static structure of systems. In software development, UML provides a fundamental and indispensable framework that functions as a general-purpose and flexible visual modeling language used to define, visualize, build, and record the elements of a software system. The main goal of UML is to visually capture important decisions, functionalities, and insights about systems that need to be built. UML is a set of tools for understanding, creating, exploring, defining, preserving, and organizing data and functions related to a system. UML combines previous modeling experiences with the best practices. Unlike a programming language, UML uses symbols and notations to show the different parts of a system and how they interact. The main features of UML diagrams are as follows:

- **Standard**: UML is a widely accepted standard, enabling developers working on different platforms to understand what is required.
- **Visual**: UML uses simple diagrams to represent the structure and behavior of a system, making it easier for developers to understand complex systems.
- **Versatile**: UML can model different systems, not just software, and therefore, it can also be used for business process modeling and other purposes.

UML is a powerful tool for software developers, streamlining system development and promoting team collaboration. Its formal language ensures precise modeling, facilitating effective stakeholder communication. UML's concise notation simplifies learning and usage while ensuring comprehensive coverage of significant system aspects. Its scalability adapts seamlessly to projects of varying sizes without unnecessary complexity. Additionally, UML's evolution incorporates best practices from the OO community, enhancing adaptability to emerging technologies. Its standardization ensures seamless transformability and interoperability across diverse platforms and systems.

UML diagrams can be broadly categorized into two main types:

1. **Structural Diagrams**: These diagrams depict the static aspects of a system, focusing on the components that make up the system and the relationships between them. Here are some common structural diagrams:
   - **Class Diagram**: Shows classes, their attributes (properties or data), methods (operations or behaviors), and relationships like inheritance and association.
   - **Component Diagram**: Illustrates the structural relationships between software components and their dependencies.

- **Deployment Diagram**: Shows the physical deployment of a system on hardware components, like servers and databases.
- **Object Diagram**: Represents a specific instance of a class diagram, showcasing objects and their links at a particular point in time.
- **Package Diagram**: Depicts how parts of a system (packages) are organized and their dependencies on each other.
- **Composite Structure Diagram**: Shows the internal structure of a complex component and its constituent parts.

2. **Behavioral Diagrams**: These diagrams represent the dynamic aspects of a system, focusing on how the components interact and behave over time. Here are some common behavioral diagrams:
   - **Use Case Diagram**: Illustrates the interaction between actors (users) and the system, depicting functionalities delivered through use cases.
   - **Activity Diagram**: Shows the flow of activities within a system, including sequential, branching, or concurrent steps.
   - **State Diagram**: Illustrates the different states of a system or object and the events that cause transitions between those states.
   - **Message Sequence Diagram**: Focuses on the message flow between objects in a specific scenario, visualizing the sequence of interactions.
   - **Communication Diagram**: Similar to message sequence diagrams, it uses links with roles instead of objects, offering a more general view of interactions.
   - **Interaction Overview Diagram**: Provides a high-level view of interactions within a system, showing them as frames.
   - **Timing Diagram**: Visualizes the time-sensitive behavior of a system, focusing on message exchange and its timing constraints.

UML has organizational features that facilitate the grouping of models into easier-to-handle packages. This partitioning helps software teams understand and manage package dependencies and version model units in complicated development environments. Additionally, it has techniques that facilitate the representation of implementation choices and the componentization of runtime elements.

UML is an important tool for managing the inherent complexity of system analysis and design. The many components required for anything from desktop applications to complex enterprise-scale systems require a structured way of monitoring and understanding their functions and connections. The core of modeling is managing the complexity involved in software design.

By generating an abstraction of the system, UML enables developers to concentrate on and clarify the essential elements of the system's design. It serves as a condensed representation that facilitates a more rapid assessment and comprehension of the system's feasibility. An essential component of system modeling is the use of descriptive language. In this sense, UML functions as a language with an exact but abstract notation. Because of its accuracy, machines can read this data, making it easier to interpret, execute, and convert data between systems.

UML is a foundation for interactive visual modeling tools integrating report writers and code generators. While it does support most current OO development methods, it may not be the most suitable option for some specific fields, such as rule-based artificial intelligence or Graphical User Interface (GUI) design. Additionally, UML is not intended to be used for modeling continuous systems like those found in engineering and physics because it is designed for discrete systems such as software, firmware, or digital logic.

In essence, UML empowers software developers to effectively manage system complexities, comprehend system design, and facilitate seamless communication within development teams, making it an indispensable tool in software and systems development. The comprehensive overview delineates UML's purpose, structure, application, and significance, highlighting its pivotal role in the developer's collection.

## 4.3 UML USE CASE DIAGRAMS

UML use case diagrams provide a visual representation depicting the interactions between the system's users, known as "actors", and the system over time. They model the system's functional requirements and emphasize what it will do. They are designed to clarify the system's behavior from the user's perspective without representing how to implement this behavior. A use case diagram model contains four components that are briefly described here.

1. **Actors**: The users of the system.
2. **Use Cases**: These define what the actors can do in the system.
3. **Associations**: These relationships indicate a dependency between the actor and the use case or between use cases.
4. **System Boundary**: This is a visual representation, shown as a box, that defines the limits of our system. Everything inside the box represents the functionalities (use cases) that make up the system.

### 4.3.1 Actors

Actors are the system's users, denoting the external entities that must engage with the system. They represent roles held by individuals, things, or other systems that must be engaged by exchanging messages and providing inputs to one or more system functionalities. A stick figure represents an actor in a use case diagram, as shown in Figure 4.1a. The actor is placed outside the system's boundary and named with a short, clear, and descriptive functional role in the system, such as "Customer" (Figure 4.1b) in an online shopping system or a "Patient" (Figure 4.1c) in a medical appointment system. A primary actor can initiate one or more tasks in the system.

The first step in creating a use case diagram for a system is to identify its actors. Identifying actors involves examining who initiates, interacts with, maintains, sends, or receives information or participates in predetermined system functions. For instance, in a hospital administration system, doctors, nurses, pharmacists, system administrators, and patients are potential actors performing various roles and engaging with the system differently.

### 4.3.2 Use Case

A use case is represented by an ellipse, as shown in Figure 4.2a, to portray how a system responds when users (actors) interact with it from outside the system. It showcases how the system behaves to meet the user requirements by exchanging messages with actors who initiate these use cases. A use case hides the system's internal complexity and is consistently presented from the actor's viewpoint. A use case name must be concise and unique, represented by a string of short active verb phrases that describe a specific behavior or action of the system. Examples of use cases in an online shopping system are "Place Order" (Figure 4.2b) or "Pay for Order" (Figure 4.2c).

**FIGURE 4.1** Examples of actors in a use case diagram. (a) The stick man notation to represent an actor represented in use case diagrams, (b) an example of an actor "Customer", (c) another example of an actor "Patient".

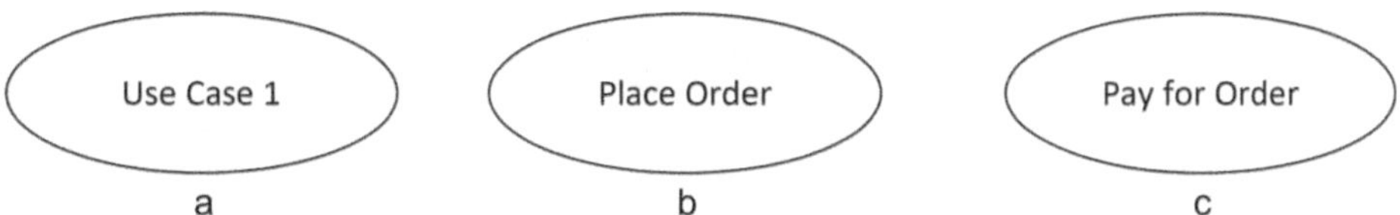


**FIGURE 4.2** Ellipses for presenting use cases. (a) An ellipse shape represents a use case, (b) an example of a use case "Place Order", (c) another example of a use case "Pay for Order".

Use cases are important in system development as they demonstrate measurable outcomes for users or external systems. They help design the system's behavior and specify finer components. Some use cases are specialized versions of others and can be broken down into simpler use cases, creating relationships and interactions.

Use cases may have relationships between them. A use case can include the functionality of one or more other use cases, called an "includes" relationship, where the included use case is always performed as part of the use case derived from. In an online shopping system example, the "Place Order" use case can be completed only if the customer's information is verified, so an included use case can be "Verify Customer Information", as shown in Figure 4.3. Notice that the "includes" relationship is represented by a dashed arrow labeled <<includes>> pointing toward the included use case.

The second inter-use-case relationship is the "extends" relationship. A use case can also be defined as extending the behavior of another core use case. It is considered an additional behavior not dependent on the base use case, representing the fundamental sequence of actions and behaviors that need to occur. For instance, in the online shopping system example, "Add Special Handling" use case <<extends>> "Pay for Order", as shown in Figure 4.4, where special handling could involve extra care in packaging, expedited shipping, or any additional service that can be extended with the

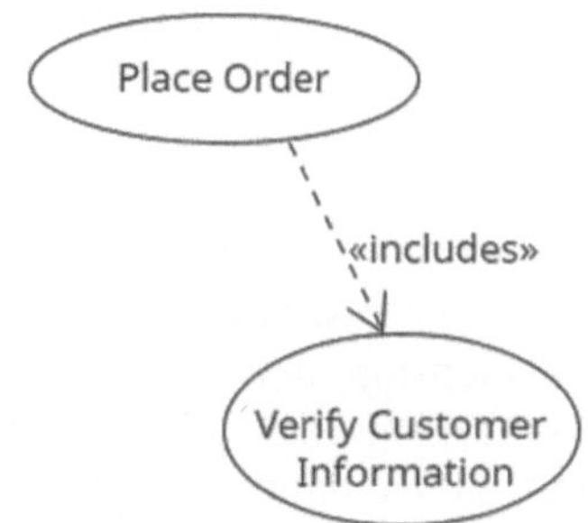


**FIGURE 4.3** An example illustrating an included use case.

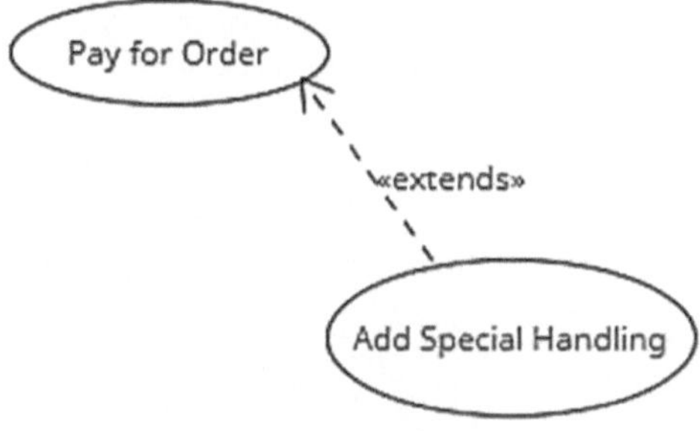


**FIGURE 4.4** An example of an extended use case.

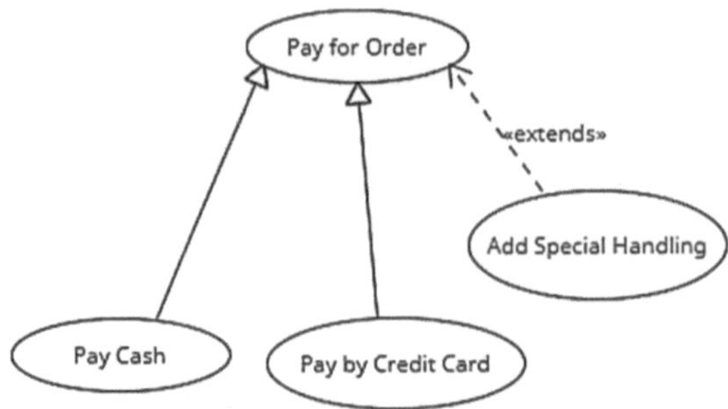


**FIGURE 4.5** Pay Cash and Pay by Credit Card Use Cases generalize from Pay for Order.

order for the customer. This extension is optional and doesn't affect the core process of paying for the order. Notice that the "extends" relationship is also represented by a dashed arrow labeled with <<extends>> that points toward the use case being extended.

Additionally, a use case can have multiple child use cases, known as use case "generalization". Use case generalization portrays a relationship where a use case inherits the behavior and functionalities of another use case. This relationship mirrors the concept of inheritance in OO programming. A child use case inherits properties, functionalities, and behavior from a parent use case but can also have additional specific functionalities. For instance, Figure 4.5 shows some use cases from an online shopping system where various payment methods, such as credit cards or cash, are used. There may be child use cases such as "Pay by Credit Card" and "Pay Cash" in this case. These child use cases would have some common functionalities to make payments, which can be generalized under the parent use case "Pay for Order". Notice that the illustration of use case generalization follows a standard generalization format, represented by a line extending from the child use case to the parent use case, featuring a larger triangular arrowhead on the parent end.

### 4.3.3 Associations

Use cases may depend on actors or other use cases. Such relationships are called associations and are represented by lines connecting actors and use cases. They do not specify the nature or frequency of interaction, only that there is a relationship in some capacity. These associations in a use case diagram help visualize the interactions between actors and use cases.

Figure 4.6 illustrates an example of an online shopping system, which includes an actor "Customer" associated with the previously discussed use cases "Place Order" and "Pay for Order".

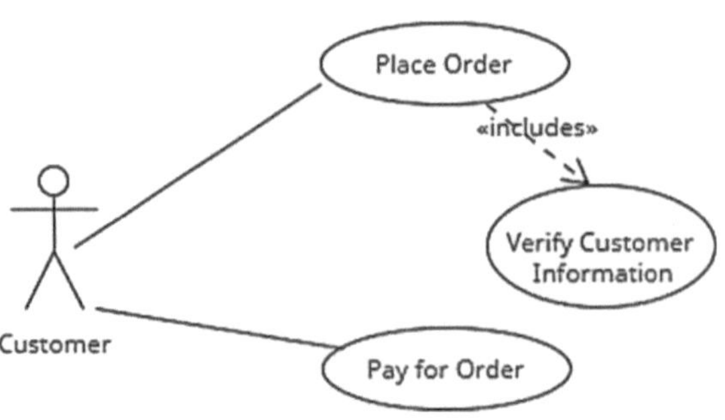


**FIGURE 4.6** An example of an actor engaged in use cases.

### 4.3.4 System Boundary

Determining a system's limitations or boundaries is an essential step in creating a use case diagram. This entails defining precisely what exists outside the system's boundaries and what is part of it. While this may seem apparent, it is a crucial component that specifies the scope of the system.

The definition of the system boundary significantly impacts the functional (what the system does) and sometimes non-functional (how well it performs) requirements. Ambiguous or incomplete specifications frequently cause project failures. As seen in Figure 4.7, the system boundary is a box with the system's name written in it. The different use cases and all their variants are displayed inside this barrier, while actors are depicted outside it. This limit may not always be obvious when modeling use cases. However, the system boundary gradually gets more defined and obvious as more actors and use cases are identified.

The simple example in Figure 4.7 shows one actor – a customer who places and pays for orders (Place Order and Pay for Order). The information is verified (Verify Customer Information), with an option to add packaging details (Add Special Handling). Payment options diversify via Pay by Credit Card and Pay Cash, generalizing under Pay for Order. Many more actors and use cases would be included in an actual real-world system for Online Ordering.

Defining use cases with actors, associations, and the system boundary aids in determining a system's scope. The system boundary becomes clearer as one gains a deeper comprehension of the system's functionalities and users, which helps to define the system's requirements and scope more precisely.

The benefits of use case diagrams become very relevant when analyzing requirements for a system where the domain specifics are unclear to the analyst or the analyst is new to the domain. Use cases enable the analyst to ask essential questions that would help remove ambiguities and clarify requirements before stepping into the system's design phase.

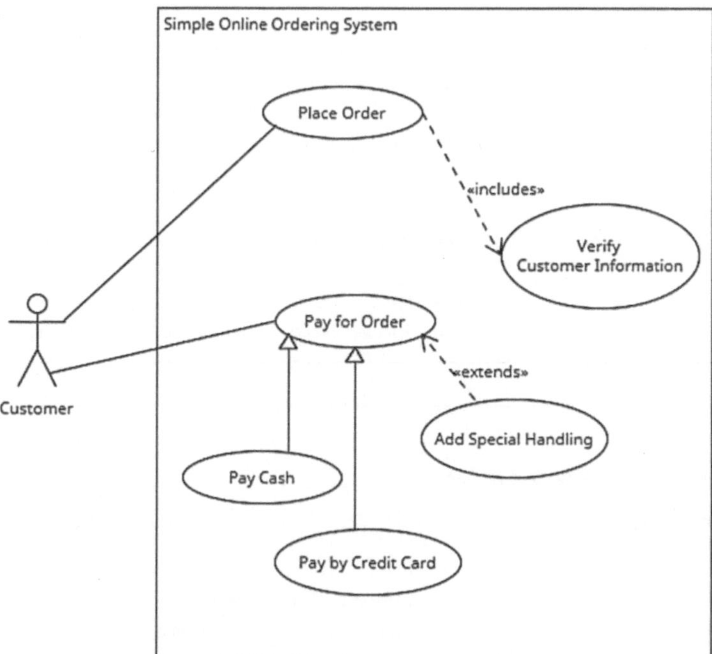


**FIGURE 4.7** An example of the system boundary in a use case diagram.

## 4.4 UML USE CASE DESCRIPTION

While visualizing functional requirements for developers, use case diagrams lack details to understand system behavior and user requirements. The best way to note all important information is by using a text-based description for each use case.

It is important to note that various formats exist to create a use case description. A commonly used style is presented here, covering essential aspects to capture while analyzing requirements. It typically starts with the name of the use case, stating the actor or actors involved with the use case, the preconditions of the use case, the trigger, the main success scenario, and the exceptional scenarios. A use case description will also indicate when it includes or extends relationships. Table 4.1 shows the components in this description table, what the detail means, and why it is useful. While components like preconditions, triggers, and post-conditions can be added to a use case to provide more context, including only essential information is crucial. Lengthy and overly detailed descriptions often hinder comprehension and defeat their purpose. Instead, focus on clarity and readability.

The level of detail in a use case should align with the associated risks and requirements. Early-stage use cases may only need basic details, with more comprehensive information added before implementation.

Let's consider our example in Section 4.3 about the online shopping system, where the corresponding use case diagram is shown in Figure 4.7. Table 4.2 shows the use case description for the use case "Place Order" and indicates that "Verify Customer Information" is included within this use case. Table 4.3 shows the use case description for "Pay for Order".

Table 4.4 shows the use case description for the use case "Add Special Handling" extended from "Pay for Order". Notice that the level of detail has changed based on the requirements of this use case.

**TABLE 4.1**
**Common use case description format**

| Use Case Description Components | Use and Purpose |
|---|---|
| Use Case Name | *Write the name of the described use case.* |
| Actor(s) | *List the actor(s) involved in the use case with an indication of whether or not a primary actor triggers the use case.* |
| Preconditions | *The event needs to be done before the execution of the use case.* |
| Trigger | *The event that initiates the execution of the use case* |
| Main Success Scenario | *The step-by-step description* |
| Post Conditions | *The condition that the system should perform after the successful completion of the use case* |
| Exceptions | *Describe how the system handles exceptional conditions, errors, or unexpected events during the execution of the use case.* |

**TABLE 4.2**
**Use Case Description for Place Order**

| Use Case Description Components | Use and Purpose |
|---|---|
| Use Case Name | Place Order |
| Actor(s) | Customer (Primary actor) |
| Preconditions | The customer is authenticated. |
| Trigger | The customer initiates an Order. |
| Notes | The "Verify Customer Information" is included within this use case. |

(*Continued*)

**TABLE 4.2 (Continued)**

| Use Case Description Components | Use and Purpose |
|---|---|
| Main Success Scenario | 1. The system asks the customer to provide his information.<br>2. The customer provides the requested information.<br>3. The system checks the information provided by the customer.<br>4. The system verifies the information.<br>5. The customer selects items.<br>6. The customer adds items to the cart.<br>7. The customer confirms orders. |
| Post Conditions | The order is placed successfully. |
| Exceptions | 1. The customer cancels an order.<br>2. Invalid order details. |

## TABLE 4.3
## Use Case Description for Pay for Order

| Use Case Description Components | Use and Purpose |
|---|---|
| Use Case Name | Pay for Order |
| Actor(s) | Customer (Primary actor) |
| Preconditions | The customer has a confirmed order. |
| Trigger | The customer selects payment. |
| Main Success Scenario | 1. The customer selects a payment method.<br>2. The system verifies and processes payment.<br>3. The system confirms the payment.<br>4. The system generates a receipt. |
| Post Conditions | 1. The order is paid.<br>2. Payment details are recorded. |
| Exceptions | 1. The payment fails while trying to verify and process payment.<br>2. Insufficient fund. |

## TABLE 4.4
## Use Case Description for "Add Special Handling"

| Use Case Description Components | Use and Purpose |
|---|---|
| Use Case Name | Add Special Handling |
| Notes | The extension point of this use case happens after selecting a payment method in "Pay for Order", which gives an optional feature. |
| Precondition | 1. The order has been placed.<br>2. The order is in the process of payment.<br>3. The customer wants to request special handling. |
| Main Success Scenario | 1. The customer ticks the checkbox to select the "Add Special Handling" option for the order.<br>2. The customer writes an optional comment for the Special Handling.<br>3. The system adds the customer's request to the order. |
| Post Conditions | The Customer's payment for the order is completed. |

**TABLE 4.5**
**Use Case Description for "Pay by Credit Card"**

| Use Case Description Components | Use and Purpose |
|---|---|
| Use Case Name | Pay by Credit Card |
| Actor(s) | Customer (Primary actor) |
| Preconditions | The order is pending payment. |
| Notes | Extends "Pay for Order" |
| Trigger | The customer chooses a credit card. |
| Main Success Scenario | 1. The customer selects payment method: select credit card.<br>2. The customer enters credit card information.<br>3. The system verifies and processes payment.<br>4. The system confirms the payment.<br>5. The system generates a receipt. |
| Post Conditions | 1. The order is paid via credit card.<br>2. Payment details are recorded. |
| Exceptions | 1. The payment fails: Incorrect card details.<br>2. Insufficient fund. |

Table 4.5 shows the use case description for the use case "Pay by Credit Card", which is a specific instance use case derived from the base use case "Pay for Order". The "Pay by Credit Card" use case focuses specifically on the payment via credit card method, inheriting the main success scenario and exceptions and many points from the general "Pay for Order" use case.

The use case description examples that have been supplied above highlight their importance in comprehending system behaviors. They provide a thorough understanding of how the system operates by going into detail about several topics, including the actors engaged, triggers, exception handling, and more. These tables are a useful tool for understanding system functionalities in software development because they strike a balance between necessary detail and simplicity to provide clarity without overwhelming comprehension.

## 4.5 CASE STUDY

In this case study, a grocery store records customers' financial transactions at the store with a self-checkout system. The user purchases items at the grocery store, and the system keeps a record of the purchases as financial transactions. The grocery store manager can view these transactions and determine profit and loss. It is assumed that the transaction management system has the following requirements:

- The user can purchase an item at the store by scanning an item with details such as the description and price.
- The user can select to purchase the scanned items and pay using a credit card.
- The system will use an external credit card service to authenticate the credit card and process the payment with the bank.
- Before making a payment, customers can apply for a discount by scanning a loyalty card.
- After the user pays for the items, a new financial transaction is created in the system with details such as transaction type, amount, description, currency, and date/time. If the transaction is unsuccessful, the system will alert the user about the error and request the correct information.
- The store manager can view daily, monthly, or yearly transactions made at the store.
- The manager can also select a transaction to view its details.

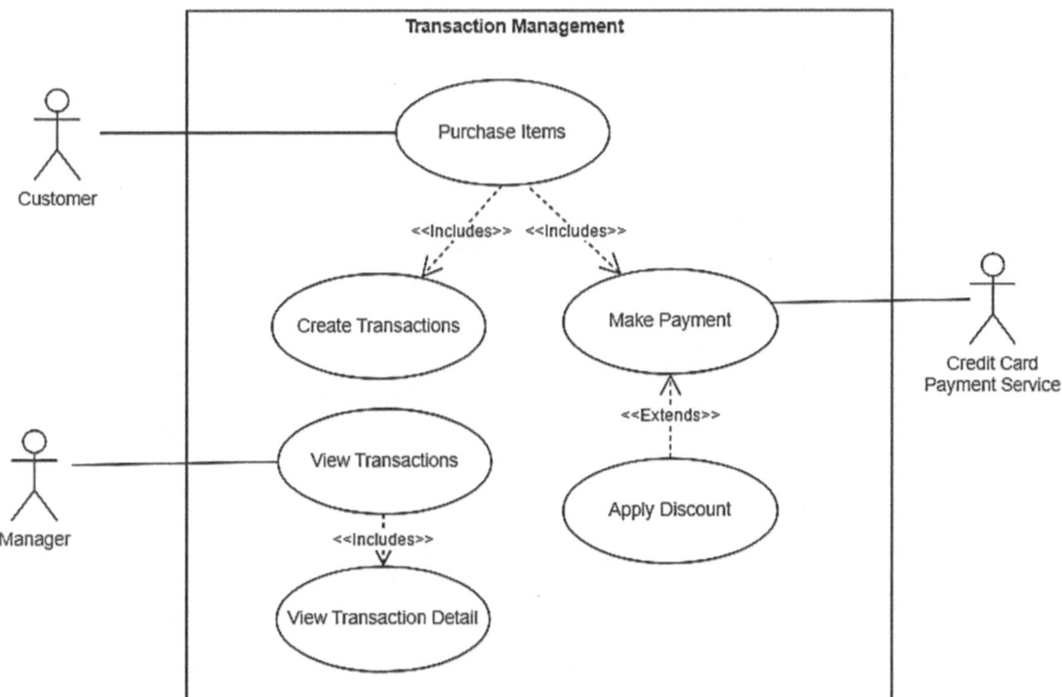


**FIGURE 4.8** The use cases involved in the financial transaction management system.

Starting with identifying the actors in this system, we can see that the actors are the "User" and "Manager" in its working scenario.

Next, identify the use cases by investigating how the users interact with the system. What operations are performed? Based on the provided scenario, the user purchases items, resulting in a new transaction in the system. These two operations can be linked to use cases, such that "Purchase Items" can include "Create Transactions". When purchasing items, a customer may select to apply for the discount, which is not a compulsory step but rather an extended use case, "Apply Discount". Similarly, the purchase items use case will require the customer to "Make Payment", which is a compulsory step and needs to be included with the "Purchase Items" use case. The system also needs to verify a credit card and withdraw the payment, which can be handled by an external credit card payment service and can be drawn as an external stakeholder. Another important functionality is that a user can apply a discount to the total amount before payment. As a result, "Apply Discount" is an extended functionality to the purchase use case. Now, let's identify the operations the manager can perform, such as "View Transactions", created as another use case of the system. This view of transactions can allow a manager to select one transaction and view its details, which can be designed as a separate functionality, "View Transaction Detail". Based on the above discussion, this system has six use cases illustrated in Figure 4.8.

As discussed in Section 4.4, the use case diagram alone does not provide enough details to let the developers understand the specifications of each use case and a detailed understanding of the system behavior. Therefore, the use case descriptions for each of the use case diagrams must be created.

Table 4.6 shows the use case description of the "Purchase Items" use case. Table 4.7 shows the use case description of the "Make Payment" use case.

Table 4.8 shows the use case description of "Create Transactions". Table 4.9 shows the use case description of "Apply Discount".

Table 4.10 shows the use case description of "View Transactions". Table 4.11 shows the use case description of "View Transactions".

**TABLE 4.6**
**Use Case Description for Purchase Items**

| Use Case Description Components | Use and Purpose |
|---|---|
| Use Case Name | Purchase items |
| Actor(s) | User (Primary actor) |
| Preconditions | The user has vegetables weighed with labeled prices. |
| Trigger | The user wants to purchase one or more items. |
| Main Success Scenario | 1. The system prompts the user to scan items.<br>2. The user scans items one by one.<br>3. The system displays item description and price as a list.<br>4. The system calculates the price of the scanned items to display the total.<br>5. The user chooses to pay for the scanned items. <Invoke Make Payment use case><br>6. If the payment is successful, the system creates a transaction. <Invoke Create Transactions use case><br>7. The system generates a receipt. |
| Post Conditions | A system resets and prompts the user to scan items. |
| Exceptions | 1. If the item is not scanned properly, an error message is displayed. The system asks the user to re-scan.<br>2. The user chooses not to pay for an item and deletes it from the list of scanned items. |

**TABLE 4.7**
**Use Case Description for Make Payment**

| Use Case Description Components | Use and Purpose |
|---|---|
| Use Case Name | Make payment |
| Actor(s) | User (Primary actor) |
| Preconditions | The user has scanned at least one item |
| Trigger | The user chooses to pay for scanned items. |
| Main Success Scenario | 1. The system prompts the user to scan the credit card.<br>2. The user scans a credit card for payment.<br>3. The system highlights the total amount and requests for confirmation.<br>4. The user confirms the payment.<br>5. The system sends the credit card information to the payment service gateway with the total purchase amount.<br>6. The system receives a confirmation of payment from the gateway. |
| Post Conditions | The system displays a payment success message to the user. |
| Exceptions | 1. If an invalid credit card is scanned, an error message is displayed.<br>2. The user cancels the payment. |

**TABLE 4.8**
**Use Case Description for Create Transactions**

| Use Case Description Components | Use and Purpose |
|---|---|
| Use Case Name | Create transactions |
| Precondition | The credit card payment is successful |
| Main Success Scenario | 1. The system retrieves the list of all purchased items.<br>2. The system creates a transaction for each item purchased with details such as transaction type, amount, description, currency, and date/time.<br>3. The system also calculates the total amount paid. |
| Post Conditions | A receipt with all the transaction details is generated. |

**TABLE 4.9**
**Use Case Description for Apply Discount**

| Use Case Description Components | Use and Purpose |
|---|---|
| Use Case Name | Apply discount |
| Actor(s) | User (Primary actor) |
| Preconditions | The user has scanned at least one item and is ready for payment. |
| Trigger | The extension point of this use case happens if the user opts to use a loyalty card in the Make Payment use case. |
| Main Success Scenario | 1. The system prompts the user to scan the loyalty card.<br>2. The user scans the loyalty card.<br>3. The system applies the discount on the items. |
| Post Conditions | A list of items with discounted prices is generated. |
| Exceptions | 1. The loyalty card is invalid, and an error message is displayed. |

**TABLE 4.10**
**Use Case Description for View Transactions**

| Use Case Description Components | Use and Purpose |
|---|---|
| Use Case Name | View transactions |
| Actor(s) | Manager (Primary actor) |
| Trigger | The manager wants to view the transactions |
| Precondition | The manager is authenticated before using the system |
| Main Success Scenario | 1. The system retrieves all the transactions.<br>2. The system displays the most recent transactions to the manager.<br>3. The manager can select a date and time to filter transactions.<br>4. The system retrieves the transactions based on the selected date and time and displays them with the total amount.<br>5. If the manager selects a transaction to view its details. <Invoke View Transaction Detail use case> |
| Post Conditions | The manager sign-outs. |

**TABLE 4.11**
**Use Case Description for View Transactions**

| Use Case Description Components | Use and Purpose |
|---|---|
| Use Case Name | View transaction detail |
| Actor(s) | Manager (Primary actor) |
| Trigger | The manager wants to view a transaction |
| Precondition | The manager has selected a transaction |
| Main Success Scenario | 6. The system retrieves the selected transaction details.<br>7. The system displays the details.<br>8. The system prompts the manager to go back to the main view.<br>9. The user closes the detail screen. |

This case study described a grocery store's financial transactions system, including information on its actors, use cases, and descriptions. It offered a clear idea of the system's functionality and user interactions with its comprehensive use case descriptions and relationships.

## 4.6 CHAPTER SUMMARY

This chapter explores OOA, a fundamental approach for analyzing and designing software systems. It contrasts traditional structured analysis with OOA, highlighting the benefits of modeling systems using objects and their interactions. This chapter introduces the UML and specifically focuses on UML use case diagrams, a tool for OOA. Use case diagrams depict the functionality of a system from the perspective of its users, known as actors. These diagrams illustrate the different use cases, which represent functionalities offered by the system and how actors interact with them. Associations between actors and use cases and the system boundary that defines its scope are also explained. Moving beyond the basic structure, this chapter explores UML use case descriptions. These detailed narratives provide a deeper understanding of each use case, outlining its steps, potential alternate flows, and exceptions. Studying this chapter provides a foundation in OOA focusing on UML use case diagrams and descriptions. This knowledge allows for precisely modeling and communicating software functionalities from a user-centered perspective.

## 4.7 EXERCISES

### 4.7.1 Test Your Knowledge

1. Compare the advantages of OOA to structured analysis in promoting modularity, scalability, and maintainability in software design.
2. In which scenarios or industries might structured analysis still be preferred over OOA in modern software development projects?
3. What is the primary focus of structured analysis in software development?
4. What key concepts are emphasized in OOA for modeling real-world entities within a software system?
5. Explain the purposes of using UML in software development.
6. Describe the different parts that form a use case diagram.
7. Describe the purpose of the "system boundary" in a use case diagram.
8. Explain the significance of including essential information while avoiding overly detailed descriptions.
9. How do use case descriptions help in clarifying system functionality?

### 4.7.2 Multiple Choice Questions

1. Which methodology treats data and processes as distinct elements in software analysis?
   a. Structured analysis
   b. Object Oriented Analysis
   c. Both structured and OO analysis
   d. None of the above
2. Which methodology emphasizes modeling system processes as a series of connected steps?
   a. Object Oriented Analysis
   b. Structured analysis
   c. Agile development
   d. Waterfall model
3. Which software development projects are likely to benefit more from structured analysis?
   a. Projects with evolving requirements.
   b. Projects requiring complex system representations.
   c. Projects with well-defined and static requirements.
   d. Projects emphasizing object encapsulation.
4. Which view of the system does the UML's dynamic view primarily focus on?
   a. Interaction with external users
   b. Structural elements
   c. Constraints among elements
   d. Relationships between objects
5. How does UML assist in handling complexity in system design?
   a. By encouraging continuous system modeling
   b. By omitting abstract notation
   c. By generating complex diagrams for better understanding
   d. By providing a structured way to monitor and understand functions and connections
6. Which use case relationship shows an additional behavior that is not always a part of the base use case?
   a. Include
   b. Extend
   c. Generalization
   d. Interaction
7. What does the "Include" relationship between use cases mean?
   a. A mandatory part of the behavior of the base use case
   b. An optional addition to the base use case
   c. A relationship representing inheritance
   d. Unrelated use case behaviors
8. How is an actor represented in a use case diagram?
   a. An oval shape inside the system's boundary
   b. A stick figure outside the system's boundary
   c. An oval figure enclosed in a box
   d. An arrow pointing to a use case
9. What is the primary goal of a use case description?
   a. To specify the actors.
   b. To specify every aspect of user interaction.
   c. To outline the interaction between actors and system functions.
   d. To provide a detailed system.

10. Which components are commonly included in a use case description table?
    a. The system's boundary
    b. Sequential description of the whole system.
    c. Project target
    d. Main success scenario

### 4.7.3 Short Answer Questions

1. What is polymorphism, and how does it benefit the analysis phase?
2. Compare Object-Oriented or Structured Analysis in industries with strict regulatory requirements that require thorough documentation.
3. What are the benefits of encapsulation and information hiding?
4. Despite being an earlier methodology to model software requirements, structured analysis is still used widely. Why?
5. Why is a visual modeling language like UML important for software analysis?
6. In UML, which diagrams describe the system's elements, relationships, and constraints within the static view?
7. How does a use case diagram help the programmers?
8. How are actors in use case diagrams identified?
9. What states the event that initiates the execution of the use case?
10. What details are captured in a use case description table?

### 4.7.4 True or False Questions

1. Encapsulation in OOA ensures that data within objects can be accessed and modified directly, promoting flexibility in data manipulation.
2. OOA doesn't emphasize modularity in software design and typically results in monolithic systems.
3. Structured and OOA helps in understanding software requirements.
4. UML is limited to specific application domains and lifecycle stages.
5. UML's organizational features help manage package dependencies and versioning model units.
6. UML is designed specifically for modeling continuous systems in engineering and physics.
7. When a use case inherits from another, the child use case cannot add additional functionalities.
8. Actors in a Use Case Diagram represent internal system functionalities.
9. An actor can only initiate a single task in the system and cannot engage in multiple functionalities.
10. Lengthy and detailed use case descriptions enhance comprehension and effectively fulfill their purpose.

### 4.7.5 Exercises on Creating Use Case Diagrams

1. A university wants to provide a personalized learning platform. The platform should adapt to individual student needs and learning styles, offering relevant resources and activities. Construct the use case diagram and the use case descriptions for the details below.
    - The professors manage the learning materials and activities on the system, personalize content for different students, monitor their progress, and analyze data to identify areas of improvement in learning objectives.
    - Students view and access their personalized materials, complete exercises and assessments, and track progress and performance through personalized reports.

2. A football club wants to create a match management mobile app for coaches, players, and fans. The app is designed to provide various functionalities tailored to the needs of each type of user. Construct the use case diagram and the use case descriptions for the details below. Ensure "includes", "extends", and generalization relationships are included where necessary.
   - A coach can use the app to view planned matches and opponent profiles, plan training sessions, assign roles, analyze match statistics and player performance, and review fans' ratings.
   - A player can access the training schedule, view assigned roles, track personal performance statistics, and review match footage and coaches' feedback.
   - A fan can view match scores and updates, access player profiles and match statistics, and rate players' performance.
3. A rapidly expanding supermarket chain in Sharjah, UAE, has grown to include six branches and a large warehouse. However, the lack of an online grocery ordering and delivery system is causing significant losses. In response, the supermarket management has developed a mobile application to facilitate online orders through a user-friendly app for its customers. Construct the use case diagram and the use case descriptions for the details below. Ensure "includes", "extends", and generalization relationships are included where necessary. The envisioned mobile app will have the following functionalities:
   - Customers can register to the app, browse available items, add items to the cart, modify quantities, and make payments.
   - Supermarket staff can receive a paid order, prepare the order, and indicate that the order is ready for delivery.
   - Delivery staff receives notifications for assigned deliveries, indicates the order is out for delivery when picking up the order, and updates the order status when delayed or delivered.

### 4.7.6 Build Use Case Diagrams/Descriptions

1. Identify the different use cases for the given scenario and create the corresponding use case descriptions. Ensure that the necessary details and relationships for each use case description table are included to help developers understand the system's functionalities. Details of a parking system scenario are given below.
   - The driver approaches the Parking Lot and interacts with the system to Enter the Parking Lot. The system checks availability and issues tickets. The driver finds an available space using the system's guidance.
   - Upon returning, the driver initiates "Pay for Parking" at a payment kiosk. The system processes and validates the payment. The driver proceeds to the exit gate. The system verifies the payment and raises the gate, completing the Exit Parking Lot use case.
   - You are also provided with the following use cases' relations:
     - "Enter Parking space" (includes "Check Availability" and "Request Ticket")
     - "Pay for Parking" (extends "Exit Parking Lot")
     - "Find Available Space" (inherits into "Find Regular Space" and "Find Handicapped Space")
     - "Exit Parking Lot" (includes "Validate Payment")
2. For the given scenario, construct the use case diagram for the pharmacy ordering app with detailed use case descriptions for the corresponding use cases.
   A well-established chain of pharmacies spanning the UAE is developing a dedicated mobile app ordering system where customers can seamlessly place their orders using mobile devices. The user journey involves registration on the app, entering personal details,

browsing items, adding them to a virtual cart, specifying quantities, making payments, and scheduling deliveries.

- User Registration and Profile Management:
  - This use case extends to include "Edit Profile Details".
  - The "Register" scenario includes names, contacts, and addresses.
- Order Placement and Cart Management:
  - Includes the sub-use cases "Add to Cart" and "Remove from Cart".
  - The user adds desired items to the cart and can remove them if necessary.
- Payment Processing and Checkout:
  - The user selects a payment method, confirms payment, and checks out.
  - Extends to scenarios like "Pay with Credit Card" and "Pay with Cash".
- Delivery Scheduling:
  - Allows the user to set the preferred date and time for delivery.
  - No inclusion or extension relationships are present in this straightforward process.

3. For the given scenario, construct the use case diagram with detailed use case descriptions for the corresponding use cases.
   A tourist company requires a system that manages bookings for desert safaris in Dubai, UAE. The system includes various activities and customer preferences during the safaris. Below are specific requirements for the system:
   - Display Safari Packages:
     - The system should display the packages available, including duration, activities, and prices.
     - The user should be able to view detailed descriptions of each package.
   - Book a Safari:
     - The user should be able to select a package, specify the number of participants, choose a preferred date and time, and provide contact and payment information.
     - The system should check availability and confirm bookings.
   - Generate Vouchers:
     - The system should create printable vouchers or e-tickets for confirmed bookings containing essential details for the safari.

# 5 Fundamentals of Object-Oriented Design

## 5.1 UML CLASS DIAGRAMS

Unified modeling language (UML) class diagrams are an essential component of object-oriented (OO) modeling. They provide a unique visualization of the various components of a software application and offer a concise overview of the software design.

A class diagram is a graphical representation of a software's structure. Each class is presented as a rectangle with three distinct compartments: class name, attributes, and methods. The relationships between classes are represented by different types of lines and connectors. Class diagrams are pivotal in understanding, defining, and communicating a system's structure and behavior. They are widely used in software development for planning and documenting system design, making them an indispensable tool for any software engineer.

An example of a class diagram is shown in Figure 5.1, depicting three classes: Student, Section, and Course. The "Student" class contains attributes like first name, last name, date of birth, and gender. It also contains a few methods like enroll_course(), drop_course(), and a few getter methods. The "Course" class contains its attributes course_ID, course_title, description, and a few getter methods. The "Section" class contains its attributes section_ID, classroom, weekly schedule, and a few getter and setter methods. The lines between classes represent relationships. This real-world

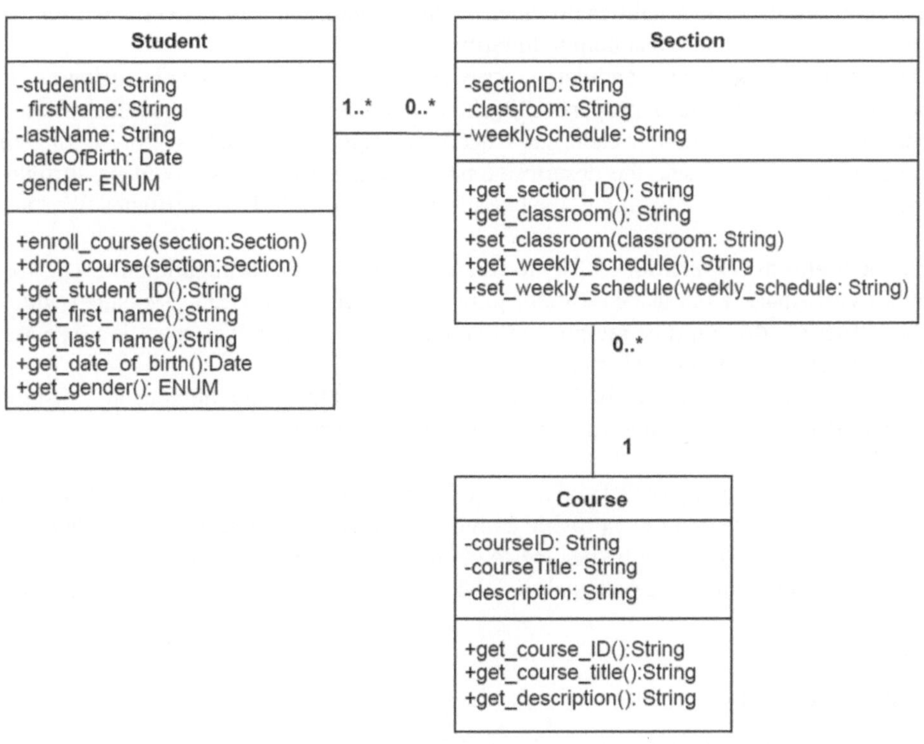


**FIGURE 5.1** An example of a UML class diagram.

DOI: 10.1201/9781032668321-5

example demonstrates the practical application of class diagrams, inspiring and motivating software engineers to use them in their own projects.

For example, course and section are related. The numeric annotations on the lines represent the multiplicities of the relationships. 1 means one instance, 0..* means zero or many, and 1..* means one or many. As such, a course has zero or many sections, while a section is related to one course. A student can enroll in zero or many sections, while a section can have one or more students.

### 5.1.1 From Use Case to Class Diagrams

The UML use case diagram provides a high-level view during the analysis phase, focusing on actors (users or external systems) and the functionalities they interact with (use cases). By studying these use cases, we move to the next step, i.e., the design phase, which uses the UML class diagram.

This analysis of use cases lays the foundation for the UML class diagram, which translates the functionalities into the system's building blocks. The nouns in use case diagrams and descriptions transform into candidate classes, while verbs suggest methods (actions) these classes can perform. The class attributes are derived from the information each class needs to store based on use case details. This transition is not obvious, as some nouns become class attributes instead of classes, and not all actors translate directly to classes. However, by systematically analyzing use cases, we extract the core elements – classes, their attributes, and the operations they can perform – forming the foundation for the system's design represented in the UML class diagram.

The following steps will help to identify classes, attributes, and methods for a UML class diagram based on a UML Use Case diagram:

1. **Analyze Actors and Use Cases:** The actors in the use cases represent the users or external systems that interact with the system. Identify nouns and actor names in the use case descriptions. These are strong candidates for becoming classes in the class diagram. The use cases are the functionalities provided by the system. Identify the verbs within the use case descriptions. These can help to identify potential methods.
2. **Define Attributes:** Once candidate classes are identified, identify their attributes. These are also called the class's properties or characteristics. Refer to use case descriptions to understand the information each class needs to store. While nouns representing major concepts become classes, the descriptive nouns related to a class would be attributes of the identified class. For example, the class "Customer" might have attributes like "name", "email", and "address".
3. **Define Methods (Operations):** Methods represent the actions or functionalities performed by a class. Analyze the use case descriptions and identify the verbs associated with each class. These verbs could be methods of that class. Focus on actions specific to a class and how it interacts with other classes.
4. **Refine and Iterate:** Review the identified classes, attributes, and methods and ensure they reflect the functionalities described in the use case diagram.

It is important to note that all nouns from use cases do not become classes. Some might be attributes of other classes. Use case diagrams provide a high-level view of the requirements. Therefore, additional details or user stories may be needed to elicit and fine-tune the required details of the class diagram. This would require a deeper analysis of the requirements, considering various stakeholders of the system.

### 5.1.2 Visual Representation of a Class

The rectangle with three compartments in Figure 5.2 (A) represents a single class. The first compartment displays the class name. By convention, the class name is a noun, singular, centralized, and

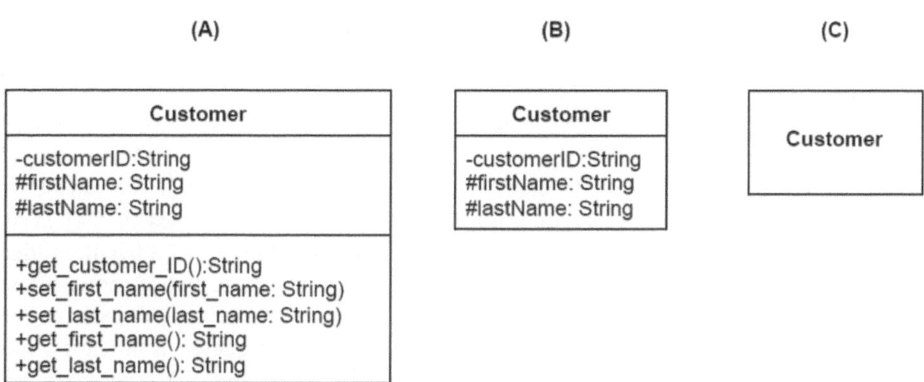


**FIGURE 5.2** Visual representation of a Class Diagram. (A) A class diagram with class name, attributes and methods. (B) A class diagram with class name, and attributes. (C) A class diagram with only the class name.

starts with a capital letter. The second compartment shows the attributes of the class with access modifiers: plus (+), minus (–), or hash (#). The third compartment defines the methods of the class. However, as Figure 5.2 (B) and (C) shows, it is also possible to represent classes with only the class name and attributes or just the class name. More compact versions of classes provide brevity to documentation.

The access modifiers are defined as follows:

- A minus (–) means the attribute or method is private and only accessible within the class.
- A plus (+) means public attributes or methods accessible inside and outside the class (by other classes).
- A hash (#) means protected, which means it is accessible within the class and the child classes (classes inheriting the attributes and methods).

## 5.2 CLASS RELATIONSHIP – ASSOCIATIONS

Relationships in a UML class diagram are represented by connections or lines drawn between classes called associations. The description on the line describes the relationship between the classes. For example, Figure 5.3 shows that a customer can place multiple requests. The arrow pointing from

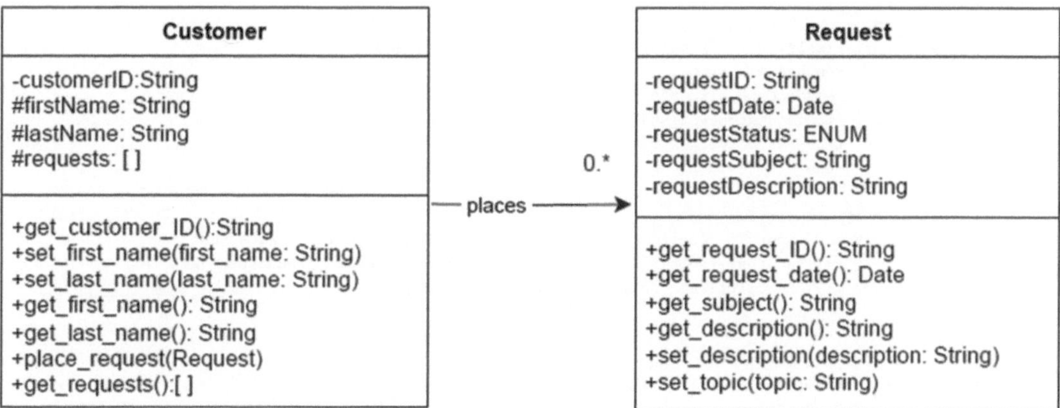


**FIGURE 5.3** Association called "Customer **places** Requests".

the class "Customer" to the class "Request" means that customers keep track of requests. Since the arrow does not point to the other side, other side, which indicates a Unary Association where a customer keeps track of requests and not the other way around.

Code 5.1 illustrates the Python classes for the "Customer" and "Request" classes. The "Customer" class has some essential attributes outlined in the UML class diagram Figure 5.3, customer_ID, first name, and last name initialized in the constructor (Lines 2–6). Similarly, the attributes of the "Request" class outlined in the diagram are initialized in the constructor (Lines 19–24). However, to implement the relationship that a "Customer" places requests, observe an attribute added to the "Customer" class constructor (Line 6), a list to store customers' requests. To add requests to this list, a method called "place_request()" is created, which receives a "Request" type object as a parameter and adds it to the list (Lines 11 and 12). To retrieve "Customer" requests, a method called "get_requests()" is created, which returns the list of requests (Lines 14 and 15). The "display_info()" method (Lines 26–29) is used to print the details of the "Request", such as in Lines 43–45. The remaining code creates a "Customer" type object, "customer1" (Line 33), and two "Request" type objects, "request1" and "request2" (Lines 35–38). The objects of the "Request" class are added to the "Customer" using the "place_request()" method (Lines 41 and 42). The list of "Customer" requests is retrieved using the "get_requests()" method of the "Customer" class. The "for" loop retrieves the "request" list, iterates through each request (Lines 44), and prints its details on Line 45. The "display_info()" method is an example of a public method of the class "Request" as it was accessible from outside its class and was called and used by the class "Customer".

**CODE 5.1 CUSTOMER PLACES A REQUEST**

```
1. class Customer:
2.     def __init__(self,id,fname,lname):
3.         self.__customerID=id
4.         self.__firstName=fname
5.         self.__lastName=lname
6.         self.requests = []  # A list to store requests made by the customer
7.
8.     # Getters and setters for the class may be included here
9.
10.    # Method to add a request placed by the customer
11.    def place_request(self, request):
12.        self.requests.append(request)
13.    # Method to return all requests added to the customer class
14.    def get_requests(self):
15.        return self.requests
16.
17.
18. class Request:
19.    def __init__(self, r_id, r_date, r_status, r_subject, r_description):
20.        self.__requestID = r_id
21.        self.__requestDate = r_date
22.        self.__requestStatus = r_status
23.        self.__requestSubject = r_subject
24.        self.__requestDescription = r_description
25.
26.    def display_info(self):
27.         print(f"Request ID: {self.__requestID}|| Request Date: {self.__requestDate}"
28.               f"|| Status: {self.__requestStatus}|| Subject: {self.__requestSubject}"
29.               f"|| Description: {self.__requestDescription}")
30.
31.
32. # Create a Customer Object
33. customer1 = Customer(1, "Ismail", "Khalifa")
34. # Create two Request objects
35. request1 = Request(101,"2023-06-24","Open","Technical Support",
```

```
                      "Having trouble with my computer")
request2 = Request(102,"2023-06-25","Open","Return/Exchange",
                      "Does not meet the required specifications")

# Add requests to the customer
customer1.place_request(request1)
customer1.place_request(request2)
print(f"Requests placed by the customer:")
for request in customer1.get_requests():
    request.display_info()
```

### 5.2.1 Binary Association

An association can be unidirectional or bidirectional. A binary association indicates that two classes are associated with each other. In a unidirectional association, one class references another class, as illustrated in the example in Figure 5.3, where the "Customer" objects have references to "Request" objects, and the association is shown with an arrow pointing from the "Customer" class to the "Request" class.

In a bidirectional association or binary association, both classes reference each other's objects. As shown in Figure 5.4, the "Employee" class keeps track of multiple tasks that the employee is working on, and the "Task" class tracks the employee working on it. Although the figure shows arrowheads on both sides of the line, such arrowheads are optional in a binary relationship. However, arrowheads can be used to provide clarity on whether it is a unidirectional or a bidirectional relationship.

The bidirectional relationship between the "Employee" and "Task" classes is implemented in Code 5.2. The code sample contains an "Employee" class with a constructor (Lines 2–6) initializing some essential attributes outlined in the UML class diagram (Figure 5.4). Additionally, a list attribute named "tasks_assigned" is added to the constructor (Line 6) to ensure that the "Employee" keeps track of all the tasks assigned to him. Similarly, the "Task" class has some essential attributes initialized in the constructor (Lines 29–3). An attribute named "assigned_employee" is added to store the responsible employee (Line 33). A public method, "assign_task()", is added to the "Employee" class (Lines 11–15). This method adds a "Task" object as a parameter and adds it to the "tasks_assigned" list (Line 13) and sends the reference of the "Employee" itself to the "Task" object to save the employee in the "Task" class by calling the "assign_to_employee()" public method (Line 15). The dual operation within the "assign_task()" method maintains data integrity by updating attributes in both classes simultaneously, ensuring correct references are stored by both classes.

The program creates two "Employee" objects (Lines 49 and 50) and three "Task" objects (Lines 53–55). The "employee1" is assigned to a task, and the task is assigned to the employee by calling the "assign_task()" method (Lines 58–59). Line 62 displays information about "employee1" and all the assigned tasks. In Line 65, "task3" is displayed, which is not assigned to any "employee".

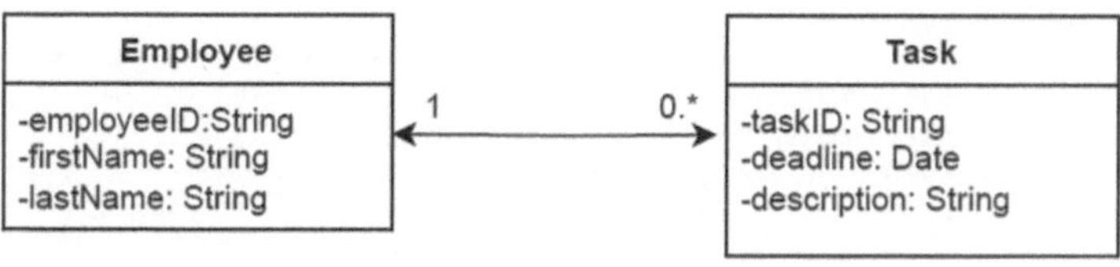


**FIGURE 5.4** Bidirectional association between Employee and Task.

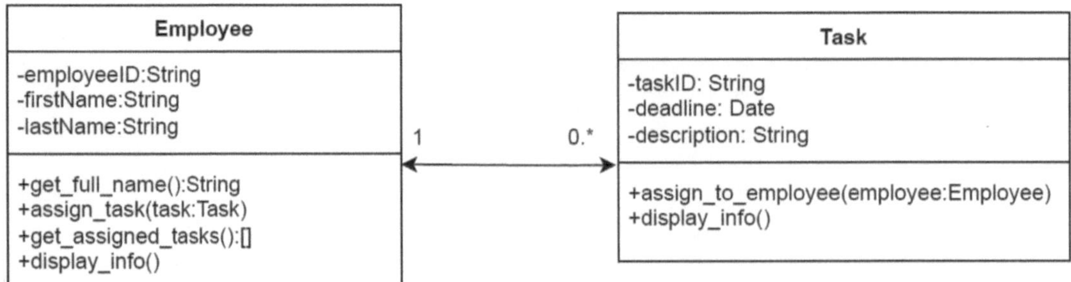


**FIGURE 5.5** A class diagram depicting the attributes and methods of the Employee and Task classes.

### CODE 5.2 EMPLOYEE AND TASK WITH A BINARY RELATIONSHIP

```
class Employee:
    def __init__(self, emp_id, first_name, last_name):
        self.__employeeID = emp_id
        self.__firstName = first_name
        self.__lastName = last_name
        self.__tasks_assigned = []  # List to store tasks assigned to the employee

    def get_full_name(self):
        return f"{self.__firstName} {self.__lastName}"

    def assign_task(self, task):
        # Add the task to the employee task list
        self.__tasks_assigned.append(task)
        # Update the task attribute __assigned_employee
        task.assign_to_employee(self)

    def get_assigned_tasks(self):
       return self.__tasks_assigned

    def display_info(self):
        print(f"Employee ID: {self.__employeeID} || Name: {self.get_full_name()}"
              f"\nAssigned Tasks:")
        for task in self.__tasks_assigned:
            task.display_info()
        print("_"*100)


class Task:
    def __init__(self, task_id, deadline, description):
        self.__taskID = task_id
        self.__deadline = deadline
        self.__description = description
        self.__assigned_employee = None  # Employee to whom the task is assigned

    def assign_to_employee(self, employee):
        self.__assigned_employee = employee

    def display_info(self):
        assigned="Not assigned"
         # Add the name of the employee if a task has been assigned to an Employee
        if self.__assigned_employee!=None:
            assigned=self.__assigned_employee.get_full_name()
        print(f"Task ID: {self.__taskID} || Deadline: {self.__deadline} ||"
              f"Description: {self.__description} || Assigned to: {assigned}")
```

```
47. # Example Usage:
48. # Create employees
49. employee1 = Employee(1, "Andrew", "Baker")
50. employee2 = Employee(2, "Sarah", "Williams")
51.
52. # Create tasks
53. task1 = Task(101, "2024-01-15", "Code Login Screen")
54. task2 = Task(102, "2024-01-20", "Code Home Screen")
55. task3 = Task(103, "2024-01-25", "Code Adding Screen")
56.
57. # Assign tasks to employees
58. employee1.assign_task(task1)
59. employee1.assign_task(task2)
60.
61. # Display an Employee's information
62. employee1.display_info()
63.
64. # Display Task3 assigned to no Employee
65. task3.display_info()
```

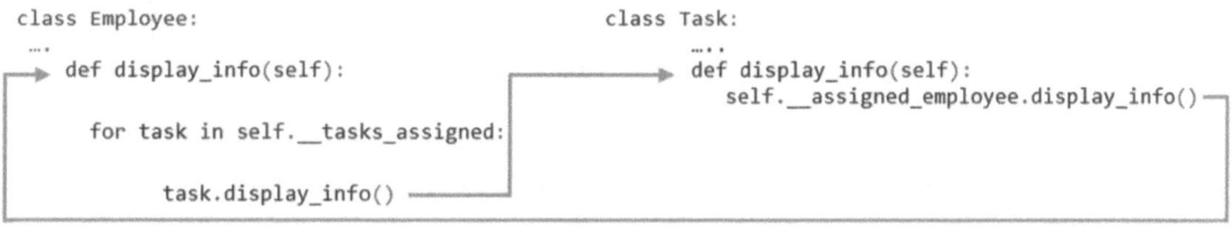

```
class Employee:                                  class Task:
  ....                                             ....
  def display_info(self):                          def display_info(self):
                                                      self.__assigned_employee.display_info()
      for task in self.__tasks_assigned:

          task.display_info()
```

**FIGURE 5.6** Cyclic dependency.

However, attempting to save information twice creates redundancy, leading to a potential cyclic dependency. Cyclic dependency occurs when two or more modules or classes depend on each other directly or indirectly, forming a loop in their dependencies and causing the program to run in a cycle. For example, in Code 5.2, the "Employee" class requires a task to work on, and a "Task" requires an "Employee" to be assigned to. When displaying employee information (Lines 20–24), each assigned "Task" is looped through and displayed (Line 23). While displaying "Task "information (Lines 38–44), the "assigned_employee" is accessed, and the employee display method is called at Line 42, resulting in a potential circular dependency. Figure 5.6 illustrates the circular dependency created if both display methods call each other. To avoid this, the task's display method does not directly call the employee's display method but rather the "get_full_name()" method (Line 42).

In OO programming, binary associations must avoid cyclic dependencies. Cyclic dependencies potentially cause memory leaks or infinite loops during object creation or destruction. A common approach to resolving bi-directional relationships is to use whole-part relationships, discussed further in Section 5.2.4. In rare cases where two or more objects have a temporary or independent association, the binary association may be used while ensuring that the code does not have cyclic dependencies. For example, for a system that manages student registrations in a university, the "Student" class can have a binary association with a "Course" class for enrollment purposes. The objects exist independently, and their lifecycles aren't strictly tied.

### 5.2.2 Self Association

The self association is another example of a unary association, as shown in Figure 5.7. In this example, tasks are associated with other tasks. In particular, a task can have zero or multiple subtasks.

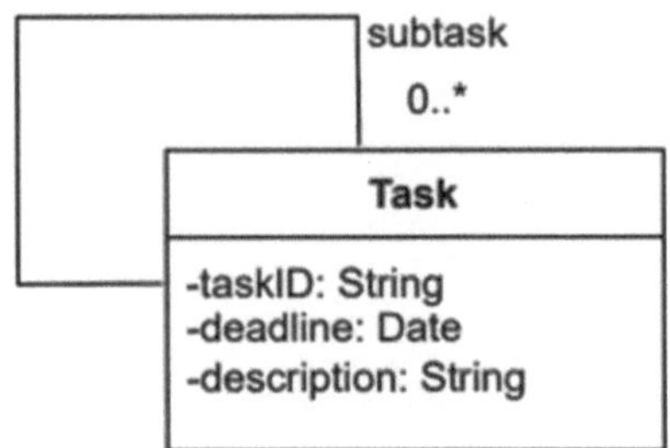


**FIGURE 5.7** An example of a self-association.

To illustrate the implementation of this self-association, consider Code 5.3. In this code sample, the class "Task" initializes its attributes in the constructor (Lines 2–6). Since the class "Task" has a relationship with itself to store objects of itself as "subtasks", a list is added to the constructor (Line 6). The "add_subtask()" method (Lines 9–11) enables the class to receive a "Task" object and add it to the "subtasks" list. The "display_info()" method (Lines 13–19) displays task information, including each subtask, by iterating through each task in the subtask list and calling the "display_info()" method on each object (Lines 18 and 19). Three "Task" objects are created (Lines 23–27). The first task, named "maintask", adds two "subtasks" using the "add_subtask()" method (Lines 30 and 31). Line 34 prints the "maintask", which includes the two subtasks.

## CODE 5.3 CLASS TASK WITH SELF ASSOCIATION

```
1. class Task: # Task Class that represents a main class with subtasks.
2.     def __init__(self, task_id, deadline, description):
3.         self.__taskID = task_id
4.         self.__deadline = deadline
5.         self.__description = description
6.         self.__subtasks = []  # List to store subtasks
7.
8.     # Add a subtask to the list
9.     def add_subtask(self, task):
10.        # Add a subtask to the list
11.        self.__subtasks.append(task)
12.
13.    def display_info(self):
14.        print(f"Task ID: {self.__taskID} || Deadline: {self.__deadline} ||"
15.              f"Description: {self.__description}")
16.        if self.__subtasks: # Check if there are subtasks
17.            print("Subtasks:")
18.            for task in self.__subtasks: # Display information for each subtask
19.                task.display_info()
20.
21.
22. # Create a main task
23. maintask = Task(100, "2024-01-10", "Website Design for SOM")
24.
25. # Create subtasks
26. subtask1 = Task(101, "2024-01-15", "Homepage development")
27. subtask2 = Task(102, "2024-01-20", "Payment Request development")
28.
29. # Add subtasks to the main task
30. maintask.add_subtask(subtask1)
31. maintask.add_subtask(subtask2)
32.
33. # Display details for the main task, including subtasks
34. maintask.display_info()
```

**TABLE 5.1**
**Examples of Cardinalities**

| Cardinality | Description |
|---|---|
| 0..1 | Zero or one instance |
| 0..5 | Zero to five instances |
| 1..* | One or more instances |
| * | Many instances |
| 1 | Exactly one instance |
| 6..8 | Exactly six to eight instances |

### 5.2.3 Cardinality

Cardinalities are a crucial aspect of UML class diagrams as they facilitate the definition of the relationships between different classes in a model. They specify the number of instances of one class that can be associated with one instance of another class. In UML class diagrams, cardinalities are typically denoted with numeric and symbolic annotations placed on top of the lines in class diagrams. These symbols can include exact numbers, a range represented by ".." or "*" that denotes "many". Table 5.1 provides various examples of cardinalities in UML class diagrams.

The diagram depicted in Figure 5.8 displays various types of cardinalities. In Figure 5.8 (A), we can see that a "Car" is linked to four instances of the "Wheel" class. Conversely, the "Wheel" class may be related to either zero or one instance of the "Car" class. In Figure 5.8 (B), we have three classes, "Passenger", "Ticket", and "Flight". The cardinality reveals that a passenger can buy any number of tickets. A ticket is associated with one flight, which can have zero or many tickets. Figure 5.8(C) shows two classes: "Flight" and "Airport". A flight is related to two airport instances – the airport of origin and the destination airport – while an airport can manage many flights (departing or landing).

### 5.2.4 Whole-Part Associations

Whole-part associations in UML class diagrams are essential for representing relationships where one entity, known as the whole, comprises several other entities, referred to as parts. These associations are categorized either as aggregation or composition.

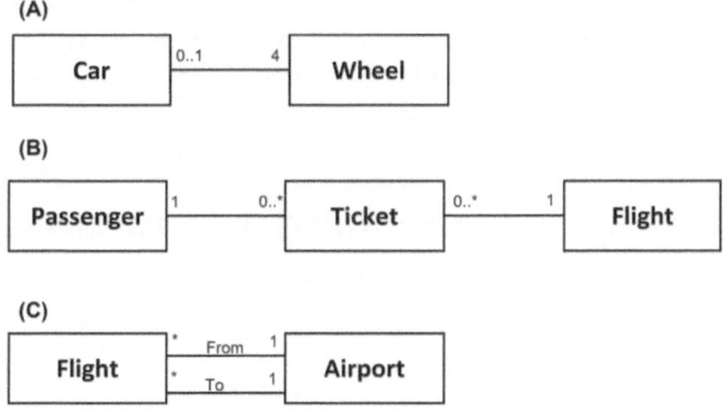


**FIGURE 5.8** Examples of different types of cardinalities. (A) Cardinality reads: a car can have 4 wheels and a wheel belongs to 0 or 1 car. (B) Cardinality reads: A passenger can buy 0 or more tickets and a ticket is for for one flight, and a flight can have 0 or more tickets and a ticket belongs to one person. (C) Cardinality reads: A flight can fly to and from an Airport, and an airport can have many flights with the airport as destination and the source.

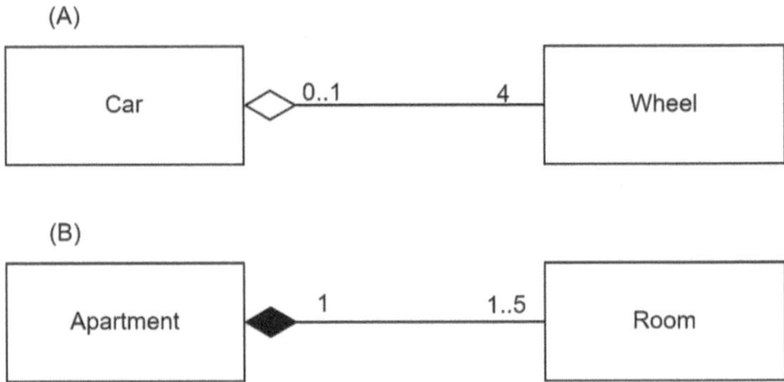


**FIGURE 5.9** Examples of whole-part relationships. (A) Aggregation whole-part relationship where the Car is the whole and the Wheel is the part. (B) Composition whole-part relationship where the Apartment is the whole and the Room is the part.

Aggregation is a weaker form of the whole-part relationship symbolized by a line with a hollow diamond arrowhead. It implies that the parts can exist independently of the whole. For instance, the relationship between a car and its wheels, where a wheel can exist separately from the car, can be represented as an example of aggregation (Figure 5.9 (A)).

On the other hand, composition is a stronger, more restrictive form of the whole-part relationship. It is denoted by a filled diamond, indicating that the parts cannot exist independently of the whole. This is shown in the relationship between an apartment and its rooms, where a room is integral to the apartment and does not have an independent existence (Figure 5.9 (B)).

Code 5.4 defines two classes, "Wheel" and "Car" shown in Figure 5.9 (A). This code section illustrates the aggregation relationship, where "Wheel" instances are created independently and associated with a "Car" instance. From Lines 1 to 3, the "Wheel" class is defined with its "__init__ ()" method. The method initializes each instance of the "Wheel" class with a "position" attribute. This attribute allows each "Wheel" to have a specified position like "Front Left" or "Rear Right". Lines 6–14 define the "Car" class. This class has its constructor that accepts a list of "Wheel" objects as an argument. In the constructor, there's a condition check (Line 8) that raises a "ValueError" if the number of wheels passed is not exactly four. This ensures that a "Car" object always has four wheels. The list of "wheels" is then assigned to the "wheels" attribute of the "Car" instance (Line 10). Additionally, the "Car" class includes a method, "list_wheels()" (Lines 12–14), which iterates over each "Wheel" instance in the "wheels" attribute and prints its position.

The last part of the code (Lines 18–21) demonstrates how these classes can be used. A list named "wheels" is created with four "Wheel" instances, each initialized with a distinct position. This list is then passed to create a "Car" instance named "my_sedan". Finally, "my_sedan.list_wheels()" calls the method in the "Car" class to print the positions of all four wheels.

**CODE 5.4 AN EXAMPLE OF AGGREGATION IMPLEMENTATION**

```
1. class Wheel: # Class that represents individual wheels
2.     def __init__(self, position):
3.         self.position = position # Initialize wheel position
4.
5.
6. class Car: # Class that represents a Car with four wheels
7.     def __init__(self, wheels):
8.         if len(wheels) != 4: # Check if the car has exactly 4 wheels
9.             raise ValueError("A sedan must have exactly four wheels.")
10.        self.wheels = wheels
```

```
    def list_wheels(self): # Display the position of each wheel in the Sedan
        for wheel in self.wheels:
            print(f"Wheel Position: {wheel.position}")


# Example of using these classes
wheels = [Wheel("Front Left"), Wheel("Front Right"),
          Wheel("Rear Left"), Wheel("Rear Right")]
my_sedan = Car(wheels)
my_sedan.list_wheels()
```

Code 5.5 shows a composition relationship to model an apartment and its rooms, as shown in Figure 5.9 (B). This code effectively models the composition relationship where the rooms are part of the apartment (the instances are created in the Apartment class) and do not have an independent existence outside of it. The "Room" class, defined at the beginning, is a simple class with a constructor that initializes each "Room" instance with a "name" (Lines 1–3). Following this, the "Apartment" class is defined to manage the relationship between the apartment and its rooms (Lines 6–14). Within the "Apartment" constructor, an empty list of "rooms" is created, and then up to five "Room" instances are generated, each labeled sequentially (e.g., "Room 1" and "Room 2"), and added to this list (Lines 7–10). This illustrates the composition aspect, where the apartment creates and manages the rooms. The "list_rooms()" method in the "Apartment" class iterates over the list of "Room" instances, printing out the name of each room, thus providing a way to list all rooms in the apartment (Lines 12–14). The code creates an instance of "Apartment" named "my_apartment" to use these classes, which automatically creates its associated rooms (Line 18). The "list_rooms()" method is then called for this instance to display the names of the rooms (Line 19).

**CODE 5.5 AN EXAMPLE OF A COMPOSITION IMPLEMENTATION**

```
1. class Room: # Class to represent individual rooms
2.     def __init__(self, name):
3.         self.name = name # Initialize room name
4.
5.
6. class Apartment: # Class to represent an apartment consisting of multiple rooms
7.     def __init__(self):
8.         self.rooms = [] # A list to store rooms
9.         for i in range(5): # Create 5 rooms and add them to the list
10.            self.rooms.append(Room(f"Room {i + 1}"))
11.
12.    def list_rooms(self): # display all rooms in the apartment
13.        for room in self.rooms:
14.            print(f"Room Name: {room.name}")
15.
16.
17. # Example of using these classes
18. my_apartment = Apartment()
19. my_apartment.list_rooms()
```

## 5.3 CLASS RELATIONSHIP – INHERITANCE

In UML class diagrams, inheritance refers to the ability of a child class or subclass to inherit the attributes and methods of its parent or superclass. Only public and protected members are inherited, while private members are not. To illustrate this relationship, consider different types of accounts in a bank. For instance, a bank may have savings, deposits, or current accounts. While all types of

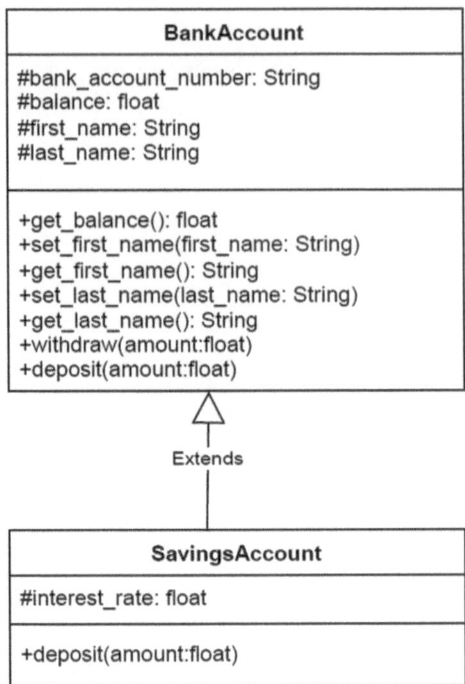


**FIGURE 5.10** An example of inheritance.

accounts would have some common members (attributes and methods), each of the account types would have some characteristics that are unique to them. Therefore, a general "BankAccount" class could be identified as the parent class with all the shared attributes and methods, and the specific types of accounts would have their unique members.

If we take the example of a "SavingsAccount" that inherits from a "BankAccount" (Figure 5.10), it can access the protected attributes (marked by the # in Figure 5.10) such as "first_name", "last_name", "balance", and "account_number", as well as the public methods like "get_balance()", "withdraw()", and "deposit()" of its parent class.

Inheritance facilitates code reuse and reduces redundancy; it also allows for customization in the child class. In the case of "SavingsAccount", it can add unique attributes, such as interest_rate, and override inherited methods to alter their behavior. For instance, in the "SavingsAccount" class, the "deposit()" method may be modified to apply an interest rate to the balance before adding the deposit amount. This feature, known as "*method overriding*", is an essential aspect of OO programming. It enables the "SavingsAccount" class to use the same method name as those described in the "BankAccount" class but with different code execution, thereby providing behavior specific to a savings account.

In UML class diagrams, inheritance is depicted with an arrow pointing from the child class to the parent class (Figure 5.10). This graphical representation offers a clear view of the inheritance structure, emphasizing the inherited attributes and methods and any overridden methods in the child class. Only unique attributes and methods and overridden methods must be shown in the child class. Additionally, the child has access to all public and protected members of the parent class.

Code 5.6 shows an example of how to implement the inheritance relationship illustrated in Figure 5.10. The "SavingsAccount" class inherits from "BankAccount" and is defined from Lines 26 to 36. The name of the parent class is included in parentheses next to the name of the child class in Line 26. The "__init__()" method of the child class (Lines 27–29) extends the initialization of "BankAccount" by adding an "interest_rate" attribute. It also calls the "__init__()" method of

"BankAccount" using the "super()" method. This ensures that "SavingsAccount" contains all the attributes of "BankAccount" in addition to its own attribute, i.e., "interest_rate".

The "deposit()" method in SavingsAccount (Lines 32–36) is an example of method overriding. It changes the behavior of the "deposit()" method inherited from "BankAccount". When depositing money, the "interest" is first calculated based on the deposit "amount" and the "interest_rate" (Line 34). Then, it adds the "amount" and the calculated "interest" to the "balance" (Line 35). This is different from the "deposit()" method in the "BankAccount" class, which only adds the deposit "amount" to the "balance". Finally, the code from Lines 40 to 46 demonstrates how to use these classes. An instance of "BankAccount" is created and used to deposit and withdraw money. Then, an instance of "SavingsAccount" is created to show how the overridden "deposit()" method works differently from that in "BankAccount" by applying an "interest_rate" to the deposit "amount". The "withdraw()" method (Lines 16–22) behaves the same in both classes since it was not overridden in "SavingsAccount".

### CODE 5.6 AN EXAMPLE OF POLYMORPHISM AND METHOD OVERRIDING IN PYTHON

```
1. class BankAccount: # Class to represent a Bank Account
2.     def __init__(self, first_name, last_name, account_number, balance=0):
3.         self._first_name = first_name
4.         self._last_name = last_name
5.         self._account_number = account_number
6.         self._balance = balance
7.
8.     def get_balance(self):
9.         return self._balance
10.
11.     def deposit(self, amount):
12.         if amount > 0:
13.             self._balance += amount
14.             print(f"Deposited: {amount}. New Balance: {self._balance}")
15.
16.     def withdraw(self, amount):
17.     # This method withdraws a specified amount if sufficient balance is available.
18.         if 0 < amount <= self._balance:
19.             self._balance -= amount
20.             print(f"Withdrawn: {amount}. New Balance: {self._balance}")
21.         else:
22.             print("Insufficient balance for the withdrawal.")
23.
24.
25. # Class to represent a savings account, inheriting from BankAccount.
26. class SavingsAccount(BankAccount):
27.     def __init__(self, first_name, last_name, account_number, interest_rate, balance=0):
28.         super().__init__(first_name, last_name, account_number, balance)
29.         self._interest_rate = interest_rate
30.
31.     # This method overrides the deposit method to include interest calculation
32.     def deposit(self, amount):
33.         if amount > 0:
34.             interest = amount * self._interest_rate
35.             self._balance += amount + interest
36.             print(f"Deposited: {amount}. Interest: {interest}. New Balance: {self._balance}")
37.
38.
39. # Example Usage
40. account = BankAccount("John", "Doe", "123456789", 1000)
41. account.deposit(200)
42. account.withdraw(500)
43.
44. savings_account = SavingsAccount("Jane", "Doe", "987654321", 0.05, 1000)
45. savings_account.deposit(200)
46. savings_account.withdraw(500)
```

### 5.3.1 Generalization and Specialization

As explained above, inheritance in OO programming allows for creating new classes based on existing ones. These existing classes serve as blueprints, providing the new classes with inherited attributes and methods. Inheritance can be understood through two complementary concepts: generalization and specialization.

Generalization and specialization are opposite sides of the same coin in OO programming inheritance, but both are useful for creating a well-organized hierarchy of classes in your OO program.

Generalization is a bottom-up approach in which common characteristics of specific classes are identified and a more general parent class is created. The general class inherits its features from the specific classes.

Specialization, on the other hand, is a top-down approach. Here, the focus is on an existing general class, and more specific child classes are created. The child classes present specific features of the general class. They can also have their own unique attributes and methods specific to their specialized function.

## 5.4 POLYMORPHISM

Polymorphism is derived from the Greek words "poly", meaning many, and "morph", meaning "*form*" or "*method*". In OO programming, polymorphism is the concept of creating generic interfaces and methods that can take different forms based on adapted implementations in derived classes. By applying polymorphism, programmers design code that can adapt to diverse object types, which leads to maintainable and scalable software.

For example, in Section 5.3, the "SavingsAccount" class (child class) has a version of the "deposit()" method, adapted from the "deposit()" method in the "BankAccount" (parent class). Though both methods have the same name, the "deposit()" method in the child class behaves differently from the "deposit()" method in the parent class. This provides a solid example of polymorphism. As such, polymorphism can be achieved by using method overloading or operator overloading.

### 5.4.1 Method Overloading

In Python, method overloading is achieved by allowing default values for parameters and facilitating variable-length arguments.

In Code 5.7, the class "Polygon" (Lines 1–24) exemplifies the concept of polymorphism through method overloading in calculating the polygon area. The "calculate_area()" method (Lines 4–24) dynamically adapts its behavior based on the number of "sides" provided. For instance, it computes the "area" for both triangles and quadrilaterals in case of "3" or "4" arguments and prints messages in case of zero or no arguments. This is an example of how Python achieves method overloading through variable-length arguments.

The default value for the parameter "sides" in the "__init__()" function is "None". The instance "polygon1" is created in Line 28 with no arguments. Then, the instance "polygon2" is created with a list [5, 6, 2, and 3] as an argument. This would not work if a default value was not assigned to the parameter "sides". This is an example of how Python achieves method overloading by allowing default values.

Then, the "calculate_area()" method of the instance "polgon1" is called on Line 32, and no arguments are provided. The method prints a message indicating the number of "sides" is not provided. On Line 33, the "calculate_area()" method of the instance "polgon1" is called with three arguments. It interprets these as the "sides" of a triangle and calculates and prints the triangle's "area", defined in Lines 7–14. On Line 34, the "calculate_area()" method of the instance "polgon1" is called with four arguments. It interprets these as the "sides" of a quadrilateral and prints the quadrilateral's "area", defined in Lines 16–22. For any other number of arguments, as indicated in Lines 23 and 24, it prints a message indicating that the number of "sides" is unsupported.

**CODE 5.7 EXAMPLE OF METHOD OVERLOADING**

```
class Polygon: # Class to represent a Polygon
    def __init__(self, sides=None):
        self.sides=sides
    def calculate_area(self, *args):
        if len(args) == 0:
            print("Number of sides for the polygon are not provided.")
        elif len(args) == 3:
            # A triangle with three sides a, b, and c
            self.sides = args
            a,b,c=self.sides
            # Calculate the Area of a triangle using the parameter
            s = (a + b + c) / 2
            area = (s * (s - a) * (s - b) * (s - c)) ** 0.5
            print(f"Area of the Triangle: {area}")
        elif len(args) == 4:
            # A quadrilateral with four sides a, b, c, and d
            self.sides = args
            a, b, c, d = self.sides
            # Calculate the area for a general quadrilateral based on the parameter
            s = (a + b + c + d) / 2
            area = (s - a) * (s - b) * (s - c) * (s - d)
            print(f"Area of the Quadrilateral: {area}")
        else:
            print("Unsupported number of sides for calculation.")


# Examples of method overloading with and without parameters
polygon1 = Polygon()
polygon2 = Polygon([5,6,2,3])

# Examples of method overloading with arguments of variable length
polygon1.calculate_area()            # Calculate the area if sides are not given.
polygon1.calculate_area(3, 4, 5)      # Calculate the area for a triangle
polygon1.calculate_area(1, 2, 3, 4)   # Calculate the area for a quadrilateral
```

### 5.4.2 Operator Overloading

Operator overloading is another form of polymorphism exhibited in OO programming. Python allows the customization of operator functionality for user-defined objects. Appropriate implementations of operations such as addition, subtraction, and comparison operators of objects are included within a class to override the action of the default operator.

Consider Code 5.8, which provides a "SchoolTeam" class to encapsulate the information and behavior of teams that play inter-school matches (Lines 1–17). The constructor initializes two attributes: the team's "name" and the "points" gained during a match (Lines 2–4).

Though it is not explicitly specified, all classes in Python inherit from a parent class called "object". In Code 5.8, the "__add__()", "__gt__()", and "__str__()" methods are inherited from the parent class "object". By overriding the "__add__()" method (Lines 7–9), we are allowing two objects of the "SchoolTeam" class to be added using a plus operator, as shown in Line 27. The "__add__()" method requires two "SchoolTeam" type objects as parameters: "self" and "other_team". In Line 8, the two teams' points are accessed and added to create a total. A new "SchoolTeam" is created and initialized with the added names and total points of both teams and returned after the addition (Line 9). To display how this method will operate, consider code Line 27. Where the "swim_team_yasmina" is added to the "soccer_team_Yasmina". The plus operator (+) between these "SchoolTeam" objects invokes the "__add__()" method. The new instance of the "SchoolTeam" class is returned in Line 9 of the "__add__()" method. This new instance is stored in the variable "totalPerformance_Al_Yasmina" in Line 27.

Overriding the "__gt__()" method (Lines 12 and 13) allows the two "SchoolTeam" type objects to be compared using the greater than (>) operator, as shown in Line 31. Similar to the "__add__()" method, the "__gt__()" method takes two "SchoolTeam" type objects as parameters, "self" and "other_team". The method returns "True" if the current object's points are greater than the "other_team" or "False" if the points are lower than the "other_team". To understand how to use the greater-than operator, skip to Line 31, where "totalPerformance_Al_Yasmina" teams are compared with "totalPerformance_Al_Raha", and based on the team with greater points, the "print()" statement at Line 32 or 34 is printed.

Code 5.8 also overrides the "__str__()" method (Lines 16 and 17), which can be invoked by printing the object, such as in Line 32, where the object is placed within the "print()" method. Instead of creating a "display_info()" method for the classes, as an alternative, the "__str__()" method can be overridden to display object details. The "__str__()" method returns the object's details concatenated as a string, the format required by the "print()" method to display information to the console.

**CODE 5.8 EXAMPLE OF OPERATOR OVERLOADING**

```
 1. class SchoolTeam:
 2.     def __init__(self, name, points):
 3.         self.name = name
 4.         self.points = points
 5.
 6.     # Overload addition operator
 7.     def __add__(self, other_team):
 8.             new_points = self.points + other_team.points
 9.             return SchoolTeam(f"{self.name} & {other_team.name}", new_points)
10.
11.     # Overload greater than operator to compare teams
12.     def __gt__(self, other_team):
13.             return self.points > other_team.points
14.
15.     # Overload __str__ method to support printing objects
16.     def __str__(self):
17.         return f"{self.name} ({self.points} points)"
18.
19.
20. # This is an example of the usage of Team class
21. swim_team_Yasmina = SchoolTeam("Team A - AlYasmina", 30)
22. swim_team_Raha = SchoolTeam("Team B - Al Raha", 20)
23. soccer_team_Yasmina = SchoolTeam("Team C  - AlYasmina", 15)
24. soccer_team_Raha = SchoolTeam("Team D - Al Raha", 20)
25.
26. # Add two school teams
27. totalPerformance_Al_Yasmina  = swim_team_Yasmina + soccer_team_Yasmina
28. totalPerformance_Al_Raha  = swim_team_Raha + soccer_team_Raha
29.
30. # Compare two school teams
31. if totalPerformance_Al_Yasmina>totalPerformance_Al_Raha:
32.     print(f"Winning School {totalPerformance_Al_Yasmina}")
33. else:
34.     print(f"Winning School {totalPerformance_Al_Raha}")
```

## 5.5 UML CLASS DIAGRAM DESIGN PATTERNS

When developing large systems, complex relationships between classes are realized, and developers usually face problems extending the system's capabilities while maintaining the existing relationships. As a result, experiences in software development led to the design of modeling classes based on patterns that facilitated scalability. Components that require expansion in the future are separated from components that will remain constant. The benefit of these design patterns is that they maintain simplicity and reusability while upgrading system requirements.

We will explore three design patterns in this section – Singleton, factory, and observer – by illustrating their application through examples and detailed explanations.

### 5.5.1 Singleton Design Pattern

The Singleton design pattern ensures that only one object of the class exists in the entire program and is accessible globally. The main reason for using this design technique is to create a single shared resource within the program.

For example, in a client–server application, one may wish to restrict the use of only one server instance. Consider Code 5.9, which restricts the class "Server" to having only one class object. This program defines the class "Server" from Lines 1 to 8.

Line 2 creates a static variable (or class variable) named "_server" that will have a reference to the single object of the class "Server". Lines 4–8 define the "__new__()" method. Usually, the "__new__()" method is used by the "__init__()" method to create an instance of the class. Here, by overriding the "__new__()" method, the class restricts the creation of an object to a single instance, i.e., every other time, the same object instance will be returned by the "__new__()" method. The "__new__()" method receives the reference to the class as "*cls*", the parameter of the method (Line 4). The class reference "*cls*" is used to access the static variable of the class, "*_server*" (Line 5). If the static variable is not yet initialized (Line 5), then the first and only "Server" instance is created and assigned to the static variable "_server" (Line 7). Otherwise, if the "_server" instance already exists, it returns the existing instance of the variable as "self" to the "__new__()" method (Line 8). In other words, it returns the reference to the single instance of the "Server" class that was created during the first and every invocation of the "__new__()" method. At Lines 11 and 12, two instances of the "Server" class are expected to be created. The instance at Line 11 will result in a new "Server" instance. However, in Line 12, the same instance created in Line 11 is returned. In Line 14, when the two instances are compared, the result "True" is returned because both "Server" instances are identical. This design pattern has helped to restrict a single class instance for the entire program. It can be useful for database or network connections, configuration settings, and maintaining log files.

**CODE 5.9 AN IMPLEMENTATION OF THE SINGLETON DESIGN PATTERN**

```
1. class Server: # Define Server class
2.     _server = None # Class variable to store the single instance
3.
4.     def __new__(cls):
5.         if cls._server is None: # Skip this method if _server is not None
6.             # Create a new instance only once in the program using the __new__ method
7.             cls._server = super(Server, cls).__new__(cls)
8.         return cls._server
9.
10. # Create two instances of the Server Class
11. server1 = Server()
12. server2 = Server()
13. # Both server1 and server2 refer to the same instance
14. print("Are they the same instance?", server1 is server2)
```

### 5.5.2 Factory Design Pattern

The factory design pattern is called so because it follows the analogy of a factory. Just like a factory produces different products from a set of raw materials, this pattern creates objects of different types based on provided specifications without explicitly specifying the object creation logic in the code that uses them. The factory design pattern encapsulates object creation within a factory class, isolating the code that uses the objects. The factory provides a central point for initializing different types of objects.

For example, consider an application that facilitates multiple types of users, such as subscribers, premium users, and administrators. For such a system, a factory design pattern can be used to create the different types of users. Such an interface encourages modular design and decouples user creation logic from the main program.

To illustrate this example, consider Code 5.10, which provides a parent "User" class with attributes "username", "password", and "account_info" (Lines 1–5). The "__str__()" method is overridden to return user information to be displayed (Lines 7–11). Three child classes inherit from the parent class "User" and facilitate their unique implementation for "set_info()". "Subscriber" class (Lines 14–16), "PremiumUser" (Lines 19–21), and "Administrator" (Lines 24–26), specify these unique implementations. The factory design pattern is used by creating a class "UserFactory" with a static interface "create_user", which can facilitate user generation based on runtime requirements (Lines 29–42). The "create_user()" method takes three parameters, "user_type" to determine the type of required user, whereas "username" and "password" are required to initialize a user object. The method checks for the type of user at Lines 33, 35, and 37 and generates the respective user, "Subscriber", "PremiumUser", or "Administrator" at Lines 34, 36, and 38. If the required user is not accommodated by the program, then a "ValueError" is raised (Lines 39 and 40). The "set_info()" method is called at Line 41 to initialize the appropriate account information for the user. As the child classes have their own implementation of this method, the respective method of the child class is invoked (method overriding). The user is returned at Line 42, bearing the appropriate type of user. Lines 46–48 generate three types of users by calling the class "UserFactory's" abstract method "create_user()". The last three lines of the code print the details of the three users (Lines 49–51). It is apparent from this code that it facilitates the addition of new user types to the program without modifying the existing code, resulting in a more scalable design.

**CODE 5.10 AN IMPLEMENTATION OF THE FACTORY DESIGN PATTERN**

```
1 class User: # Define a User class to represent a generic user
2.     def __init__(self, username, password): # Initialize user attributes
3.         self.username = username
4.         self.password = password
5.         self.account_info=None # User account info, initially set to None
6.
7.     def __str__(self):
8.         return (
9.         f"Username: {self.username} || "
10.        f"Password={self.password} || "
11.        f"Account Details: {self.account_info}")
12.
13.
14. class Subscriber(User): # Define Subscriber class inherited from User Class
15.     def set_info(self): # Subscriber overrides set_info to customize the account info
16.         self.account_info= "Subscriber: Regular user with viewing access."
17.
18.
19. class PremiumUser(User): # Define Premium User class inherited from User Class
20.     def set_info(self): # Premium User overrides set_info to customize the account info
21.         self.account_info= "PremiumUser: User with subscription for premium content."
22.
23.
```

```
class Administrator(User): # Define Administrator class inherited from User Class
    def set_info(self): # Administrator overrides set_info to customize the account info
        self.account_info= "Administrator: User with administrative privileges."


class UserFactory: # Userfactory class creates user based on their type
    @staticmethod
    def create_user(user_type, username, password):
        user=None
        if user_type == "subscriber":
            user= Subscriber(username, password)
        elif user_type == "premium":
            user= PremiumUser(username, password)
        elif user_type == "administrator":
            user = Administrator(username, password)
        else:
            raise ValueError("Invalid user type")
        user.set_info()# Set additional information for the created user
        return user


# Utilize Userfactory  for Initializing Users
user1 = UserFactory.create_user("subscriber", "Hassan Saeed", "165Dst31")
user2 = UserFactory.create_user("premium", "Mariam Khalil", "#29Bev50")
user3 = UserFactory.create_user("administrator", "Gulbano Ali", "3h!80dem6")
print(user1)
print(user2)
print(user3)
```

### 5.5.3 Observer Design Pattern

The observer design pattern is another commonly used design pattern. With the help of this design pattern, an object can add observers, which are notified whenever there is a change in the object's state. The observer pattern is useful in propagating notifications to relevant objects for appropriate action.

For example, consider a "Blog" that needs to update its followers whenever new posts are created. To illustrate this example, consider Code 5.11. The "Blog" class constitutes one attribute, a list of followers declared in the constructor (Lines 2 and 3). Methods "add_follower()" and "remover_follower()" add and remove follower types of objects from the "followers" list (Lines 5–9), respectively. The "publish_post()" method implements the observer pattern by looping through the list of observers, which in this case are "followers", and notifying each follower of the new post being published (Lines 11–13). The "Follower" class is the user that follows blogs and has two attributes declared in the constructor: "name" and a list of "interests" (Lines 17–19). The "new_post()" method in the "Follower" class is called by the observer whenever a new blog is published (Lines 20–22). This method notifies the user of the newly published post only if the "topic" is of interest to the follower (Lines 22 and 23). From Lines 27 to 31, one blog and two followers are created in the program. In Lines 34 and 35, followers are added to the blog so that whenever a new post is published, the followers are notified. Therefore, in Lines 38 and 39, two new posts are published by calling the "publish_post()" method, resulting in the call to the "new_post()" method of the followers. It is evident from this code that associations can be created between classes such that a change in one object can send notifications to objects of the other class. This design technique is very useful in event-driven programs, user interface design, game design, scheduling apps, and chat notifications.

**CODE 5.11 AN IMPLEMENTATION OF THE OBSERVER DESIGN PATTERN**

```
class Blog: # Class that defines a Blog
    def __init__(self):
        self.followers = [] # Initialize an empty list to store followers

    def add_follower(self, follower): # Method to add a follower to the blog
        self.followers.append(follower)

    def remove_follower(self, follower): # Method to remove a follower to the blog
        self.followers.remove(follower)

    def publish_post(self, title, topic): # Method to publish a post and notify followers
        for f in self.followers:
            f.new_post(title, topic)


class Follower: # Class that defines a Follower
    def __init__(self, name, interests): # Initialize a Follower Instance
        self.name = name
        self.interests = interests

    def new_post(self, title, topic): # Notify a follower about a new post
        if topic in self.interests:
            print(f"User {self.name} notified about new article: {title}")


# Create a Blog
myBlog = Blog()

# Create followers of the Blog
follower1 = Follower("Imran Ali", ["Research", "AI"])
follower2 = Follower("Ibrahim Farooq", ["Education", "Politics"])

# Register followers
myBlog.add_follower(follower1)
myBlog.add_follower(follower2)

# Publish posts
myBlog.publish_post("Bias in AI", "AI")
myBlog.publish_post("Upcoming elections", "Politics")
```

## 5.6 CASE STUDY – A TASK MANAGEMENT SYSTEM

In this case study, we introduce a task management system designed to keep track of various aspects of task management. Here, we highlight the significance of the design process and focus only on the basic requirements.

Consider the following basic requirements: employees are assigned tasks. Managers are employees who manage other employees, and tasks can have subtasks. A task has attributes such as a unique identifier, description, status, and due date. An employee has attributes such as a unique identifier, first name, last name, and position in the company.

From the above description, we have at least four classes, i.e., employee, manager, task, and subtask, with basic attributes outlined in Figure 5.11.

We can start creating associations between classes. For example, a manager manages one or more employees. Therefore, a unidirectional relationship can be created between the manager and employee with cardinality, one-to-many. Similarly, a manager can create and assign tasks. As such, a manager has a relationship with tasks, a unidirectional relationship between the manager and tasks, and cardinality, one-to-many. An employee can be assigned tasks. Therefore, employees and tasks can have a relationship, which can be designed to be unidirectional to avoid circular

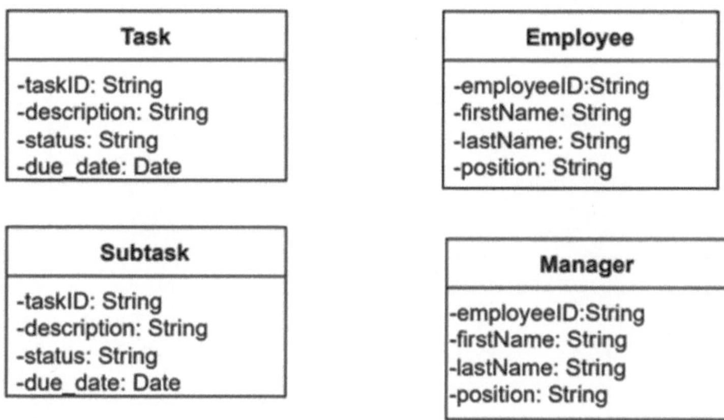


**FIGURE 5.11** Basic class diagram – Iteration 1.

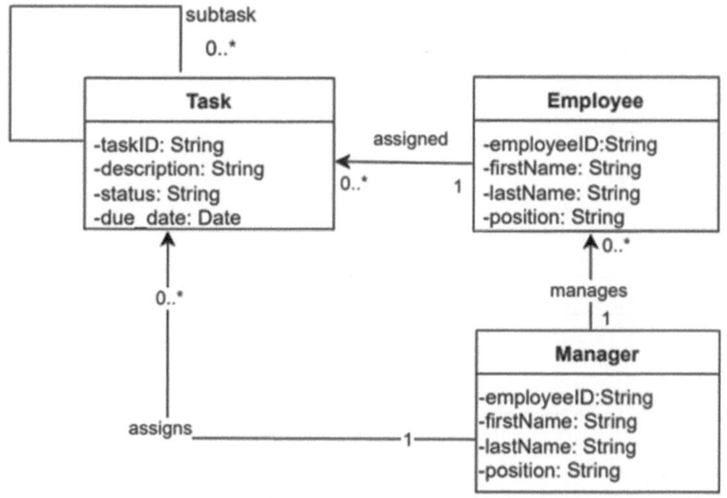


**FIGURE 5.12** Identifying relationships – Iteration 2.

dependency. In this case, if tasks and subtasks have the same attributes and subtasks do not require additional functionality from tasks, then a unary relationship can be created for the task class to show it will contain subtasks. These associations are drawn initially to analyze the system further and understand how many instances are exchanged between classes, as shown in Figure 5.12.

On analyzing Figure 5.12, we observe redundancies, such as managers and employees with common attributes that can be resolved through inheritance. In addition, as a manager manages a collection of employees, we can resolve the simple binary relationship between manager and employee to an aggregate relationship, which will result in a list of employees in the manager class. The two relationships between employee and manager are updated in Figure 5.13. Similarly, the task and subtask relationship can be resolved to show an aggregate relationship, resulting in a collection of subtasks within the task class.

Now, observe additional redundant relationships, such as an employee having a relationship with the task, which is inherited by the manager because it is a type of employee. Therefore, the redundant relationship between task and manager can be removed, as shown in Figure 5.14. In addition, observe any other binary relationship and identify whether it can be represented by an aggregation or composition. For example, the task is created by the employee and can be assigned to an employee

subtask
Task
-taskID: String
-description: String
-status: String
-due_date: Date
assigned
0..*
1
Employee
#employeeID:String
#firstName: String
#lastName: String
#position: Enum
*
manages
0..*
Extends
Manager
assigns
1
1
-employeeID:String
-firstName: String
-lastName: String
-position: String

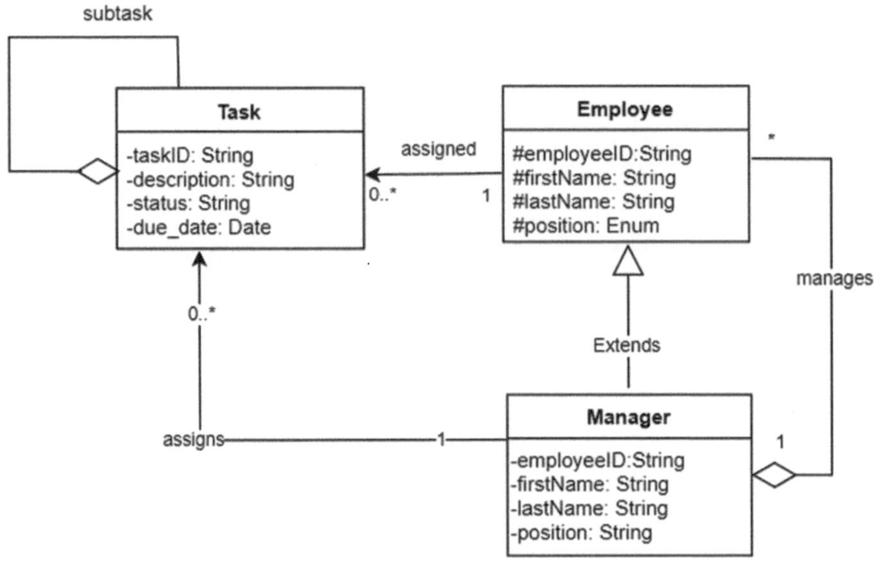

**FIGURE 5.13** Identifying inheritance and aggregation – Iteration 3.

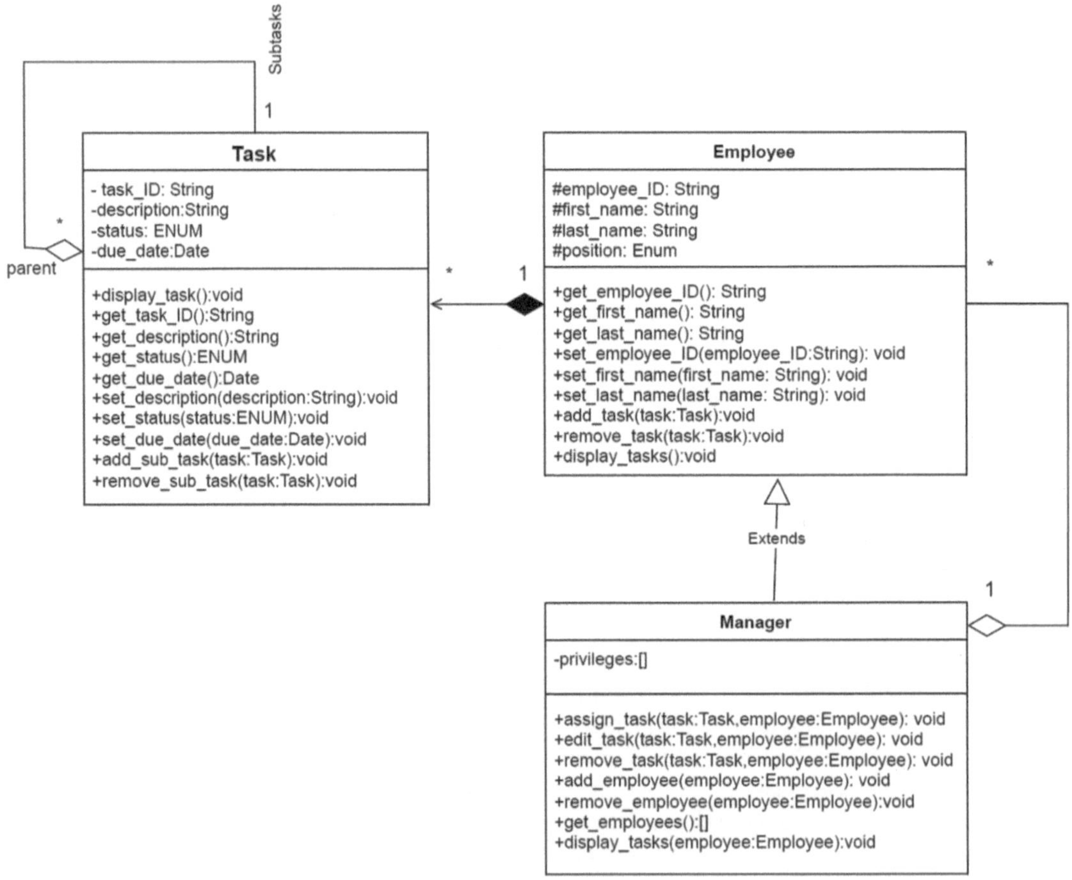

Subtasks
1
Task
- task_ID: String
-description:String
-status: ENUM
-due_date:Date
*
parent
+display_task():void
+get_task_ID():String
+get_description():String
+get_status():ENUM
+get_due_date():Date
+set_description(description:String):void
+set_status(status:ENUM):void
+set_due_date(due_date:Date):void
+add_sub_task(task:Task):void
+remove_sub_task(task:Task):void
*
1
Employee
#employee_ID: String
#first_name: String
#last_name: String
#position: Enum
+get_employee_ID(): String
+get_first_name(): String
+get_last_name(): String
+set_employee_ID(employee_ID:String): void
+set_first_name(first_name: String): void
+set_last_name(last_name: String): void
+add_task(task:Task):void
+remove_task(task:Task):void
+display_tasks():void
*
Extends
1
Manager
-privileges:[]
+assign_task(task:Task,employee:Employee): void
+edit_task(task:Task,employee:Employee): void
+remove_task(task:Task,employee:Employee): void
+add_employee(employee:Employee): void
+remove_employee(employee:Employee):void
+get_employees():[]
+display_tasks(employee:Employee):void

**FIGURE 5.14** The final class diagram of the task management system.

by a manager (a type of employee). Therefore, an employee class can have a composite relationship with tasks, ensuring that employees create, assign, and are assigned tasks. If the composite relationship is selected and if an employee object is deleted, the assigned tasks to the employee will also be deleted. Figure 5.14 is the updated class diagram with a more detailed analysis of the attributes and methods required for the implementation of the classes in Python.

Code 5.12 shows the enumerator types used in this case study. The first two lines import the "Enum" class from the "enum" module and the "datetime" module. Lines 4–8 define the "TaskStatus" class, an enumeration for different task statuses like "SCHEDULED", "ASSIGNED", "ACTIVE", and "COMPLETED" indicating respectively that a task is scheduled but not yet assigned to an employee, a task that is assigned to an employee, a task that is in progress, and a task that is completed. These enumerator types will be used in the "Task" class. Lines 11–15 introduce the "EmployeePosition" class, another enumeration detailing positions like "ACCOUNTANT", "CASHIER", "SALESREP", and "MANAGER", which are the different positions that can be held by an employee. These enumerator types will be used in the "Employee" class. Finally, Lines 17–20 define the "ManagerPrivileges" class, enumerating specific managerial actions such as "ASSIGNTASKS", "EDITTASKS", and "REMOVETASKS", indicating the privileges provided for a manager, respectively, as the ability to assign tasks to employees, edit task details, and remove tasks assigned to employees. These enumerator types will be used in the "Manager" class.

## CODE 5.12 ENUMERATOR TYPE OBJECTS IN THE TASK MANAGEMENT SYSTEM

```
1. from enum import Enum # Import the module enum
2. import datetime
3.
4. class TaskStatus(Enum): # Enum definition for TaskStatus
5.     SCHEDULED = 1
6.     ASSIGNED = 2
7.     ACTIVE = 3
8.     COMPLETED = 4
9.
10.
11. class EmployeePosition(Enum): # Enum definition for EmployeePosition
12.     ACCOUNTANT = 1
13.     CASHIER = 2
14.     SALESREP = 3
15.     MANAGER=4
16.
17. class ManagerPrivileges(Enum): # Enum definition for ManagerPrivileges
18.     ASSIGNTASKS=1
19.     EDITTASKS=2
20.     REMOVETASKS=3
```

Code 5.13 defines a Python class named "Task", designed to model tasks in the task management system application. The "__init__()" method initializes a task with attributes like "task_id", "description", "status", "due_date", and an empty list for "sub_tasks". The class provides various methods for interacting with these attributes.

The "display_task()" method prints the task's details, iterates over, and displays any sub-tasks if they exist. Getter methods like "get_task_ID()", "get_task_description()", "get_status()", "get_due_date()", and "get_sub_tasks()" are provided for accessing the task's attributes. Corresponding setter methods "set_description()", "set_status()", and "set_due_date()" allow modification of the task's "description", "status", and "due_date", respectively.

The methods "add_sub_task()" and "remove_sub_task()" are included to manage the task's sub-tasks, allowing the addition and removal of sub-tasks, respectively. The task attributes are defined with double underscores, indicating they are intended to be private and should be accessed or modified through these provided methods. This class is a comprehensive representation of a task, encompassing its attributes and operations necessary for managing it and its sub-tasks in a task management context.

**CODE 5.13 TASK CLASS OF THE TASK MANAGEMENT SYSTEM**

```
class Task: # Class to represent the tasks in the task management system
    def __init__(self,task_id,description,status,due_date):
        self.__task_id=task_id
        self.__description=description
        self.__status=status
        self.__due_date=due_date
        self.__sub_tasks=[]

    def display_task(self):# Method to display tasks' details
        print(f"Task ID: {self.__task_id}, Task Description: {self.__description},"
              f"Task Status: {self.__status} Task Due Date: {self.__due_date}")
        if len(self.__sub_tasks)>0:
            print("Displaying subtasks")
            for sub_task in self.__sub_tasks:
                sub_task.display_task()

    # Several get methods to retrieve different class attribute
    def get_task_ID(self):
        return self.__task_id

    def get_task_description(self):
        return self.__description

    def get_status(self):
        return self.__status

    def get_due_date(self):
        return self.__due_date

    def get_sub_tasks(self):
        return self.__sub_tasks

    # Several set methods to update different class attributes
    def set_description(self,description):
        self.__description=description

    def set_status(self,status):
        self.__status=status

    def set_due_date(self,due_date):
        self.__due_date=due_date

    def add_sub_task(self,task): # Method to add sub task to list
        self.__sub_tasks.append(task)

    def remove_sub_task(self,task): # Method to remove sub task from list
        self.__sub_tasks.remove(task)
```

Code 5.14 defines a Python class "Employee", which represents an employee in an organization. The "__init__()" method initializes an employee with several attributes: "employee_ID", "first_name", "last_name", "position", and initializes an empty list for "tasks".

The class provides getter methods like "get_employee_ID()", "get_first_name()", "get_last_name()", and "get_tasks()" for accessing the employee's ID, first name, last name, and list of "tasks". Setter methods "set_first_name()" and "set_last_name()" update the employee's first and last names.

The "add_task()" method allows adding a new task to the employee's list of "tasks". The "remove_task()" method removes a task based on its "ID". It iterates through the list of tasks, checking each task's ID against the provided "task_id". If a match is found, the task is removed from the list.

Finally, the "display_tasks" method iterates over the tasks assigned to the employee and calls each task's "display_task" method to print its details. This method also prints a line of underscores after each task's details for better readability.

## CODE 5.14 EMPLOYEE CLASS OF THE TASK MANAGEMENT SYSTEM

```
class Employee: # Class representing an employee in the task management system
    def __init__(self,employee_ID,first_name,last_name,position):
        self._employee_ID=employee_ID
        self._first_name=first_name
        self._last_name=last_name
        self._position=position
        self._tasks=[]

    # Get methods to access employee's ID, first name, last name, and list of task
    def get_employee_ID(self):
        return self._employee_ID

    def get_tasks(self):
        return self._tasks

    def get_first_name(self):
        return self._first_name

    def get_last_name(self):
        return self._last_name

    # Set methods to update first and last names
    def set_first_name(self,first_name):
        self._first_name=first_name

    def set_last_name(self,last_name):
        self._last_name=last_name

    def add_task(self,task): # Method that adds a task
        self._tasks.append(task)

    def remove_task(self,task_id): # Method that removes a task
        for index in range(len(self._tasks)):
            task=self._tasks[index]
            if task.get_task_ID()==task_id:
                self._tasks.pop(index)
                return
    def display_tasks(self): # Method that will print the task with an empty line
        for task in self._tasks:
            task.display_task()
            print("_"*20)
```

Code 5.15 shows the "Manager" class in the code, a subclass of "Employee", indicating that a "Manager" is a specialized type of "Employee". It inherits properties and methods from the "Employee" and extends the functionality with additional attributes and methods. The constructor initializes the "Manager" with standard employee attributes and also includes "privileges" and a list of managed "employees". The class provides methods to manage employees like "add_employee()", "get_employees()", and manage tasks like "assign_task()", "edit_task()", and "remove_task()". Before performing the actions, these task-related methods check if the "Manager" has the required privileges (defined in "ManagerPrivileges"). Finally, "display_tasks()" allows a "Manager" to view the tasks of a specific employee under their supervision.

### CODE 5.15 MANAGER CLASS OF THE TASK MANAGEMENT SYSTEM

```
# Define a Manager class that inherits from the Employee class
class Manager(Employee):
    def __init__(self,employee_ID,first_name,last_name,position,privileges):
        # Call the constructor of the base class (Employee) using super()
        super().__init__(employee_ID,first_name,last_name,position)
        self.__privileges=privileges
        self.__employees=[]

    def add_employee(self,employee): # Method to add an employee
        self.__employees.append(employee)

    def assign_task(self,employee,task): # Method to assign a task to employee
        if ManagerPrivileges.ASSIGNTASKS in self.__privileges:
            employee.add_task(task)

    # Method to edit a task for an employee
    def edit_task(self,employee,task_id,task):
        if ManagerPrivileges.EDITTASKS in self.__privileges:
            for t in employee.get_tasks():
                if t.get_task_ID()==task_id:
                    t=task
    def remove_task(self,employee,task_id): # Method to remove a task from an employee
        if ManagerPrivileges.REMOVETASKS in self.__privileges:
            employee.remove_task(task_id)

    def get_employees(self): # Method to retrieve the list of employees
        return self.__employees

    def display_tasks(self,employee): # Method to display tasks of specific employee
        if employee in self.__employees:
            employee.display_tasks()
```

Code 5.16 illustrates the creation and management of the tasks and employees as instances of the classes "Task", "Employee", and "Manager". It begins by creating several "Task" objects with unique IDs, descriptions, statuses (using the "TaskStatus" enum), and due dates (using "datetime"). Notably, "task_1" is given a sub-task "task_1_1".

Next, employees are instantiated, each with a unique "ID", "name", and "position" (using enumerated "EmployeePosition"). Then, a manager, "manager_1", is created with a unique "ID", "name", "position", and a list of "privileges" (using enumerated "ManagerPrivileges"). This manager adds the previously created employees under their supervision.

The manager assigns various tasks to these employees. For instance, "employee_1" is assigned "task_1" and "task_4", while "employee_2" receives "task_2", "task_3", and "task_10". This assignment is contingent on the manager having the appropriate privileges.

Finally, the code demonstrates the manager's ability to display an employee's tasks and remove a task from an employee. For example, "task_10" is removed from "employee_3"'s list. The "display_tasks()" method showcases the tasks assigned to "employee_1", "employee_2", and "employee_3". This entire code block effectively demonstrates the functionality of task assignment and management within an organizational structure, leveraging the OO principles and the classes and methods defined earlier.

### CODE 5.16 OBJECTS OF THE TASK MANAGEMENT SYSTEM

```
# Create task objects
task_1=Task(
    "T001","check the worksheet",TaskStatus.ACTIVE,datetime.datetime(2023, 11, 30)
)
task_1_1=Task(
    "T001","receive the worksheet from the sales department",
    TaskStatus.ACTIVE, datetime.datetime(2023, 11, 25)
)
task_1.add_sub_task(task_1_1)

task_2=Task(
    "T002","design a marketing plan for the new product",
    TaskStatus.ASSIGNED, datetime.datetime(2023, 12, 5)
)
task_3 = Task(
    "T003", "Review project proposals",
    TaskStatus.ACTIVE, datetime.datetime(2023, 12, 7)
)
task_4 = Task(
    "T004", "Prepare monthly sales report",
    TaskStatus.ASSIGNED, datetime.datetime(2023, 12, 12)
)
task_5 = Task(
    "T005", "Conduct employee training session",
    TaskStatus.ACTIVE, datetime.datetime(2023, 12, 9)
)
task_6 = Task(
    "T006", "Update inventory records",
    TaskStatus.ASSIGNED, datetime.datetime(2023, 12, 11)
)
task_7 = Task(
    "T007", "Plan company picnic",
    TaskStatus.ACTIVE, datetime.datetime(2023, 12, 5)
)
task_8 = Task(
    "T008", "Create marketing materials for the upcoming event",
    TaskStatus.ASSIGNED, datetime.datetime(2023, 12, 14)
)
task_9 = Task(
    "T009", "Review and update customer support guidelines",
    TaskStatus.ACTIVE, datetime.datetime(2023, 12, 8)
)
task_10 = Task(
    "T0010", "Prepare agenda for the board meeting",
    TaskStatus.ASSIGNED, datetime.datetime(2023, 12, 13)
)

# Create Employee and Manager objects
employee_1=Employee("E001","Ahmed","Samir",EmployeePosition.SALESREP)
employee_2=Employee("E002","John","Adams",EmployeePosition.ACCOUNTANT)
employee_3=Employee("E003","Ann","Davis",EmployeePosition.ACCOUNTANT)
employee_4=Employee("E004","Sara","Smith",EmployeePosition.CASHIER)

manager_1=Manager(
    "M001","Paul","Andreson",
EmployeePosition.MANAGER,
[
   ManagerPrivileges.ASSIGNTASKS,
   ManagerPrivileges.EDITTASKS,
   ManagerPrivileges.REMOVETASKS
])

```

```
# Assign employees to manager
manager_1.add_employee(employee_1)
manager_1.add_employee(employee_2)
manager_1.add_employee(employee_3)
manager_1.add_employee(employee_4)

# Assign tasks to employees
manager_1.assign_task(employee_1,task_1)
manager_1.assign_task(employee_1,task_4)
manager_1.assign_task(employee_2,task_2)
manager_1.assign_task(employee_2,task_3)
manager_1.assign_task(employee_2,task_10)
manager_1.assign_task(employee_3,task_5)
manager_1.assign_task(employee_3,task_6)
manager_1.assign_task(employee_3,task_8)
manager_1.assign_task(employee_3,task_9)
manager_1.assign_task(employee_4,task_7)

# Display tasks of specific employees
print("Displaying tasks of employee_1")
manager_1.display_tasks(employee_1)
print("Displaying tasks of employee_2")
manager_1.display_tasks(employee_2)

manager_1.remove_task(employee_3,task_10)
print("Displaying tasks of employee_3")
manager_1.display_tasks(employee_3)
```

## 5.7 CHAPTER SUMMARY

This chapter dives into object-oriented design fundamentals, focusing on effectively planning and structuring OO systems. It introduces UML class diagrams, a visual language representing classes and their relationships. This chapter explores different class relationships, including associations (dependency between objects) and inheritance (where new classes inherit properties from existing ones). Polymorphism, the ability of objects to respond differently to the same message, is also explained, covering method and operator overloading. To showcase design principles, this chapter explores design patterns like Singleton (ensures only one class instance exists), factory (creates objects without specifying the exact class), and observer (propagating notifications to relevant objects for appropriate action).

## 5.8 EXERCISES

### 5.8.1 Test Your Knowledge

1. Describe the components of a UML class diagram.
2. Explain the purpose of access modifiers in a UML class diagram.
3. What are the different relationships between classes, and how are they represented in a UML class diagram?
4. Why is cardinality important when representing associations?
5. Define the purpose of using the cardinality "n..m", where n and m are whole numbers. Provide an example where this cardinality might apply.
6. Explain the concept of cyclic dependency.
7. Explain method-overriding with an example in Python.
8. Describe the process of operator overloading and its role in customizing functionality for user-defined objects.

9. Explain the Singleton Design Pattern and its purpose in software development.
10. Explain the factory design pattern and its purpose in software development.

### 5.8.2 Multiple Choice Questions

1. What is the primary purpose of a UML class diagram in software development?
   a. Execution of code
   b. Testing of applications
   c. Visualization of concepts in the system and their relationships
   d. Database management
2. How are associations between classes represented in a UML class diagram?
   a. Ellipses
   b. Different colors
   c. Lines and connectors
   d. Circles
3. What does the access modifier "+" signify in a UML class diagram?
   a. Private
   b. Public
   c. Protected
   d. Undefined
4. What does a unidirectional association mean in UML?
   a. Both classes have references to each other's objects
   b. One class has a reference to another class
   c. Both classes have no reference to each other
   d. The association is represented by a hollow diamond
5. Which of the following is an example of a whole-part relationship in UML?
   a. Unary association
   b. Bidirectional association
   c. Composition
   d. Binary association
6. In UML cardinalities, what does the notation "0..*" signify?
   a. Exactly one instance
   b. Zero or one instance
   c. Zero to five instances
   d. Zero to many instances
7. What does inheritance refer to in UML class diagrams?
   a. The ability of a child class to inherit only private members from its parent class.
   b. The ability of a child class to inherit public and protected members from the parent class.
   c. The ability of a parent class to inherit attributes and methods from its child class.
   d. The ability of a class to inherit only public members from its superclass.
8. What is the primary advantage of using inheritance in OO programming?
   a. Increased encapsulation
   b. Code duplication
   c. Code reuse and consistency
   d. Limited customization
9. What is polymorphism in OO programming?
   a. Code duplication
   b. Increased encapsulation
   c. Code reuse and consistency
   d. Limited customization

10. Which design pattern is used to ensure only one object of a class exists in the entire program and is accessible globally?
    a. Factory pattern
    b. Singleton pattern
    c. Observer pattern
    d. Prototype pattern

### 5.8.3 Short Answer Questions

1. How is a class name represented in a UML class diagram?
2. Should attributes of a class be private or public by default? Discuss the pros and cons.
3. Explain whole-part associations.
4. Why should bidirectional associations be avoided when writing programs?
5. What does a child class inherit from a parent class?
6. What essential concept in polymorphism involves providing a specific implementation of a method in a subclass for which there is already a definition in the superclass?
7. Explain the composition relationship. How is it denoted in UML class diagrams?
8. Which design pattern establishes a one-to-many dependency between objects?

### 5.8.4 True or False Questions

1. UML class diagrams have no practical impact on software development.
2. An attribute marked with a hash (#) in a UML class diagram is accessible outside the class, allowing other classes to modify its value.
3. The numeric annotations on the lines in a UML class diagram indicate the size of the classes.
4. In a UML class diagram, an arrow pointing from one class to another indicates a bidirectional association.
5. Composition is a weaker form of the whole-part relationship, allowing the parts to exist independently of the whole.
6. Aggregation, represented by a hollow diamond in UML, implies that the parts can exist independently of the whole, making it a weaker form of the whole-part relationship.
7. Cardinalities in UML diagrams specify the number of instances of one class that can be associated with one instance of another class, and they are denoted with numeric and symbolic annotations.
8. Method overriding is a key feature of polymorphism.
9. The observer design pattern defines a one-to-many dependency between objects, ensuring that when one object changes state, all its dependents are notified and updated automatically.
10. The Factory pattern decentralizes object creation in a factory class, hindering modular design and tightly coupling object creation logic with the main program.

### 5.8.5 Fill in the Blanks

1. Fill in the blanks to overload the addition operator in the "ParkingLot" class.

```
class ParkingLot:
    def __init__(self, capacity):
        self.capacity = capacity
        self.available_spaces = capacity

    def __add__(self, other_lot):
```

```
            total_capacity = _______ # Fill in the blank to calculate the combined capacity.
            return ParkingLot(total_capacity)


lot1 = ParkingLot(50)
lot2 = ParkingLot(30)
combined_lot = lot1 + lot2

print(combined_lot.capacity)
```

2. Fill in the blanks to show the whole-part association between Material and Product classes, where a list of Material objects forms a Product.

```
class Material:
    def __init__(self, name, quantity, unit):
        self.name = name
        self.quantity = quantity
        self.unit = unit


class Product:
    def __init__(self, product_id, name, description, materials):
        self.product_id = product_id
        self.name = name
        self.description = description
        self.__________ = ____________
# The rest of the code is omitted for simplicity
```

3. Fill in the blanks in Lines 6, 19, and 21 to illustrate how the method overloading works in the add_numbers method.

```
1. class AddOperation:
2.     def add_numbers(self, *args):
3.         if len(args) == 0:
4.             result = 0
5.             print(f"Result of Addition: {result}")
6.         elif len(args) == ---: # Fill in the blank here
7.             result = sum(args)
8.             print(f"Result of Addition with 2 arguments: {result}")
9.         elif len(args) == 3:
10.            result = sum(args)
11.            print(f"Result of Addition with 3 arguments: {result}")
12.        else:
13.            print("Unsupported number of arguments for addition.")
14.
15.  # Example of using the class
16. addition_operation = AddOperation()
17.
18. # Example of method overloading for adding numbers
19. ________________.add_numbers() # Sum with default values- Fill in the blank here
20. addition_operation.add_numbers(5, 7) # Sum with 2 arguments
21. addition_operation.add_numbers(__,___,__) # Sum with 3 arguments - Fill in the blank here
```

4. Fill in the blanks in Lines 4, 6, and 9 to show the implementation of the Singleton Design Pattern

```
1. class StudentProfile:
2.     _profile_instance = None
3.
4.     def __-----__(cls): # Fill in the blank here
5.         # Check if the instance doesn't exist
6.         if cls._profile_instance is ______: # Fill in the blank here
7.             # Create a new instance using the parent class's __new__ method
8.           cls._profile_instance = super(StudentProfile, cls).__new__(cls)
9.         return cls._----------------
10.
11.    #__init__ omitted for simplicity
12.
13.
14. # Example of using the StudentProfile class
15. student1 = StudentProfile("Alice", "A12345")
16. student2 = StudentProfile("Bob", "B67890")
```

### 5.8.6 Class Diagram Design Problems

1. Design a class Diagram for an Online Shopping system:
   a. Define the classes User, Product, and ShoppingCart:
      - Identify their essential attributes.
      - Identify relevant methods that represent their functionalities.
   b. Relationships between classes:
      - Determine the relationships between the defined classes.
      - Specify the cardinalities for these associations.
   c. Explain how you would use inheritance to represent different types of users within the system.
   d. Use aggregation to model the shopping cart to ensure that changes to the Shopping Cart do not affect the existence of individual Products.
2. Design a class diagram for a Library Management System:
   a. Define the classes Book and Member:
      - Identify their essential attributes.
      - Identify relevant methods that represent their functionalities.
   b. Create a class "ReadingList" with the attributes ID, name, description, and required_books (a list of Book objects).
   c. Utilize composition to:
      - Establish a relationship between the "ReadingList" class and the "Book" class.
      - Explain which whole-part relationship would best suit the relationship between the "Book" class and the "ReadingList" class.
   d. Implement a "Faculty" class that inherits from the "Member" class. Specify any additional attributes or methods that differentiate faculty members from regular members.
   e. Relationships between classes:
      - Determine the relationships between the defined classes.
      - Indicate the cardinalities of these relationships.
3. Design a class diagram for an online course management system.
   a. Build Classes with Attributes and Methods:
      - Course:
        - Attributes: course_id, name, instructor (User object), description, and topics.
        - Methods: add_lecture() and approve_enrollment().

- User:
  - Attributes: user_id, name, email, password, and role.
  - Methods: login() and enroll().
- Enrollment:
  - Attributes: enrollment_id, student (User object), course (Course object), and status.
  - Methods: None.
- Lecture:
  - Attributes: (lecture_id, title, course (Course object), content, and materials.
  - Methods: None.
- Assignment:
  - Attributes: assignment_id, title, course (Course object), due_date, description, and grade.
  - Methods: submit
- Submission:
  - Attributes: submission_id, student (User object), assignment (Assignment object), content, and submitted_date.
  - Methods: None.

b. Describe the relationships between different classes and indicate the different cardinalities.

c. Define and illustrate the whole-part associations within the class diagram. Specifically, focus on the associations involving:
   - The relationship between the Course class and its related components (e.g., Instructor and Enrollment).
   - The relationship between the Course class and Lecture and Assignment classes, representing how a course is composed of multiple lectures and assignments.

### 5.8.7 Coding Problems

1. After designing the online course management system in Problem 3, Section Class Diagram Design Problems, create the corresponding solution in Python:
   a. Define and implement the different classes with constructors to initialize their attributes.
   b. Implement the "__str__()" method for each class to provide a meaningful string representation of the object.
   c. Implement the setter and getter methods for each class.
   d. Implement the methods specified for each class to perform the necessary functionalities.
   e. Create instances of the different classes to illustrate the system's functionality.
2. Enhance the patient management classes introduced in the exercises of the first and second chapters to create a more general base class and apply inheritance and polymorphism to improve the design of the system as follows:
   a. Redefine the "Patient" class:
      - The class attributes: "patient_id", "first_name", "last_name", "age", "contact_details", "patient_status", "diseases", "test_results", "appointment_history", and "visited_doctors".
      - Implement the constructor to initialize these attributes.
      - Implement the "__str__() " method to provide a meaningful string representation of the object.
      - Implement the setter and getter methods for all attributes.
      - Implement the "display_basic_info()" method :

- This method returns the "patient_id", "first_name", "last_name", and "contact_details".
- Add access control logic using the "authorized_user" parameter. Based on the user's role (e.g., "doctor" or "nurse"), return appropriate information and ensure sensitive details (test results and diseases) are only accessible to authorized users.

b. Define the "ChronicPatient" class. This class inherits from the class "Patient" all the attributes and methods.
   - Add specific attributes to the "ChronicPatient" class: "chronic_medicine" and "chronic_disease_list".
   - Implement the "__init__()" to initialize inherited and additional attributes.
   - Include the setter and getter methods for the new attributes.
   - Create the "update_chronic_diseases()" method to manage the list of chronic diseases by adding/removing diseases to/from the list.
   - Define a display_chronic_patient_info() to call the base "Patient" class's method "display_basic_info()" with "authorized_user" for access control to private patient data specific to chronic diseases.

3. Implement the factory design pattern to manage a collection of vehicles while handling the creation of different types of vehicles. The goal is to design a modular and extensible system that accommodates various vehicle types with specific attributes.
   a. Implement a base class named "Vehicle" with attributes "make", "model", and "year". This class should include a "__str__()" method to represent the basic details of a vehicle.
   b. Create three specialized classes that inherit from the "Vehicle" class: "Car", "Motorcycle", and "Truck". Each specialized class should have additional attributes specific to its type, such as "num_doors" for "Car", "num_wheels" for "Motorcycle", and "cargo_capacity" for "Truck".
   c. Develop a "VehicleFactory" class with a static method named "create_vehicle()" that takes parameters for "vehicle_type", "make", "model", "year", and additional type-specific attributes. This factory method should determine the type of vehicle to create based on the "vehicle_type" parameter and return an instance of the appropriate class.
   d. Implement polymorphism by overriding each specialized class's "__str__()" method to display type-specific details and common attributes. For example, a "Car" should display the number of doors, a "Motorcycle" should display the number of wheels, and a "Truck" should display the cargo capacity.

# 6 File Handling, Object Serialization, and Data Persistence

## 6.1 FILE HANDLING

Applications use data to solve problems. Program variables, lists, tuples, arrays, and dictionaries are used by programs as internal storage for temporary data. This data is volatile, i.e., it is in the random-access memory (RAM) during execution; it will be lost after the program finishes execution. This data must be reinitialized or re-entered each time the program runs.

The ability of data to remain available after the program has finished execution is called "Data persistence". Any data that must be persistent and stored permanently must be stored in a file or database outside the program, in a secondary memory storage such as a hard disk. For instance, a university software application would use data related to faculty, students, courses, and grades, which would be stored externally to the programs accessing the data. Whenever the program is executed, the relevant data is retrieved from the files and processed to either display it to the user or be edited and re-stored in the files. Therefore, data stored in a non-volatile storage medium such as a disk file is persistent data storage.

Python provides built-in methods for reading and writing files. Two types of files can be handled in Python: text files (human-readable files) and binary files (machine-readable files).

- **Text files:** This file contains textual data in a human-readable language. Each line is terminated with a special invisible character called the line feed or end of line, represented by "*\n*" in Python.
- **Binary files:** The data stored in this type of file is in a machine-understandable format and language.

## 6.2 TEXT FILES

Any file in the secondary memory must be assigned a filename and access mode. Once the file is created, it needs to be loaded into the primary memory (RAM) to be accessed. Upon loading the file into the primary memory, a file handle unique to the file is provided for reference.

Let's investigate the following Python Syntax:

- file1 = open("myFile1.txt", "a")

Here, the keyword "*open*" is a pre-defined Python method to access a file with two arguments. The first argument is the filename "myFile1.txt", representing the name assigned to the file that needs to reside in memory, and the second argument is the parameter "*a*" that indicates the access mode for the file, which stands for the "Append Only" mode. The file handle returned by the "*open*" method is stored in the variable "file1". The file is created in the same directory location as the Python program.

Code 6.1 is a simple program that reads an email from a text file named "email.txt", as shown in Box 6.1.

DOI: 10.1201/9781032668321-6

**CODE 6.1 OPENING A FILE TO READ CONTENTS**

```
1. # Open the file for read using 'r' mode
2. file1 = open('email.txt', 'r')
3.
4. # Read all the data in the file
5. file_data = file1.read()
6.
7. # Close the file
8. file1.close()
9.
10. # Print all the data read from the file
11. print(file_data)
```

**BOX 6.1 A SAMPLE TEXT FILE NAMED "EMAIL.TXT"**

```
Subject: Greetings

Dear User,
Thank you for visiting our website!

Best regards,
The Wellness Center
```

The "open" method is used in Line 2 of the code, with the first parameter as the name of the file, "email.txt", and the second parameter with the access mode, "r", to open the file for reading. The "open" method checks if the file exists. If it does not exist, the program throws a "FileNotFoundError". Otherwise, it opens the file, and the file handle is saved in the variable named "file1". In Line 5, all the contents of the file are read into the variable "file_data" as one long string as follows:

- file_data = "Subject: Greetings\nDear User,\nThank you for visiting our website!\nBest regards,\nThe Wellness Center"

In Line 8, the file is closed, and the file handle is deleted. Line 11 prints the variable "file_data" containing all the file's contents.

On inspection of the "file_data" variable, some unseen characters or whitespace characters like "\n", "\t", and "\r" may be noticed. When encountering a newline in the text file, the file handler adds a "\n" to the string being read from the file. Similarly, if it encounters a tab space, it adds a "\t" to the string, and a "\r" is added to represent the line feed character, i.e., to move the cursor to the beginning of the next line. However, when the string is printed to the console, the "print()" method identifies these characters and presents the output as multiline text without the whitespace characters, exactly as written in the text file. These characters must be considered when processing the string that is read from a file. A combination of the line feed character "\r" and the newline character "\n" is understood as carriage return and line feed (CRLF).

### 6.2.1 Access Modes

Files can be accessed with different access modes, which are important for understanding how the Python program uses the file. The different access modes are discussed here.

When handling files and directories, the program needs to give relevant, however limited, access to the secondary storage to avoid unintended access and modifications to data. As a result, open the file based on its intended use within the program.

1. **Write Only ("w"):** Opens a file for writing. Creates the file if the file does not exist and positions the handle at the beginning of the file. For an existing file, the file is over-written, erasing the existing contents. The handle is also positioned at the beginning of the file.
2. **Write and Read ("w+"):** Opens a file for reading and writing. If the file does not exist, it creates it and positions the handle at the beginning of the file. If an existing file is overwritten, it erases the existing contents. The handle is also positioned at the beginning of the file.
3. **Read Only ("r"):** Opens a file for reading. The handle is positioned at the beginning of the file. The program raises a "FileNotFoundError" error if the file does not exist. This is also the default mode in which a file is opened if the access mode is not specified.
4. **Read and Write ("r+"):** Opens the file for reading and writing. The handle is positioned at the beginning of the file. The program raises a "FileNotFoundError" error if the file does not exist.
5. **Append Only ("a"):** Opens the file for writing or creates it if it does not exist. In this case, the handle is positioned at the end of the file, and the new data written to the file will be inserted at the end.
6. **Append and Read ("a+"):** Opens the file for reading and writing and creates it if it does not exist. The handle is positioned at the end of the file, and the new data being written to the file will be inserted at the end of the file.

### 6.2.2 Writing to Text Files

Writing to a text file is important for software applications in many instances. For instance, a website's design may require storing user preferences in a cookie file, or the application's errors would have to be logged into a file to ensure proper diagnostics in deployment.

There are two Python functions for writing text to a file.

1. **write():** The "write()" method accepts a string as an argument and writes this string to the text file.
2. **writelines():** The "writelines()" method accepts a string or the list of strings as an argument and writes these strings to the text file.

Consider Code 6.2 as an example of Python code to write text to a file. The example writes information about a database to a configuration file named "config.ini". This demonstrates the use of the "write()" method. In Line 2 of the code, the file is opened with a "w" access mode, which means previous information in the file will be overridden with the new data. From Lines 4 to 6, database access information variables are created. In Line 8, the "write()" method adds the database access information to the configuration file. The "write()" method takes a string as an argument and writes the string to the file. If the newline "\n" character is used appropriately within the string, information is added appropriately to the file. At Line 9, the file configuration file is closed. The result, i.e., the content of the "config.ini" file, is shown in Box 6.2.

**CODE 6.2 OPENING A FILE TO WRITE CONTENT**

```
1. # Open the configuration file
2. config = open('config.ini','w')
3. # Create the Details of the application's database to a file
4. db_name="employees"
5. db_user = "ahmed_alhameli"
6. db_password="&V63fh!r5"
7. # Create a config file that stores the application's database details
8. config.write(f"db_name={db_name}\ndb_user={db_user}\ndb_password={db_password}")
9. config.close()
```

### BOX 6.2 THE CONFIG.INI FILE AND ITS CONTENTS

```
db_name=employees
db_user=ahmed_alhameli
db_password=&V63fh!r5$
```

Now consider the "writelines()" function, which takes a list of values as an argument and adds each list item as a new line in the file. To illustrate the "writelines()" method, consider a scenario of storing data about system users in a file. Box 6.3 illustrates the "users.csv" file, where each line represents a user, and the data is separated by a comma. The first data value is the name, followed by the address, and finally, the phone number. Such comma-separated values (CSV) are stored in a file with the extension ".csv". The advantage of creating a CSV file is the ability to send considerable amounts of structured data in a simple text format over a digital medium. Moreover, spreadsheet programs, such as MS Excel, can read and represent a CSV file as a spreadsheet.

### BOX 6.3 A CSV FILE WITH USER INFORMATION AT EACH LINE

```
Mohamed Ali, Apt 43 SAS Ul Nakhl Village,055-1428435
Steven Dawson, Villa 82 Al Reef Villas,050-57845
Malinda Davidson, Apt 304 Al Raha Gardens,055-3459101
```

Code 6.3 illustrates how to write instances of a class to a CSV file. Lines 2–8 provide a class with three attributes, "full_name", "address", and "phone_number". The "__str__()" method is overridden to provide the instance information in string format. Lines 12–14 create three instances of the "User" class. Line 17 lists the three user instances, each retrieved in string format using the "__str__()" method. Line 19 opens the file with an append mode, "a". As a result, at Line 21, the list of three users will be added to the CSV file in Box 6.3, resulting in six records, as shown in Box 6.4.

### CODE 6.3 WRITING TO A CSV FILE

```
1. # A simple user class for illustration
2. class User:
3.     def __init__(self,full_name,address,phone_number):
4.         self.full_name=full_name
5.         self.address=address
6.         self.phone_number=phone_number
7.     def __str__(self):
8.         return f"{self.full_name},{self.address},{self.phone_number}\n"
9.
10.
11. # Create three Users
12. user1 = User("Ismail Alhamidi", "123 Main St CityA", "555-1234")
13. user2 = User("Foster Smith", "456 Oak St CityB", "555-5678")
14. user3 = User("Stephani Johnson", "789 Pine St CityC", "555-9101")
15.
16. # Create a List of users of the application
17. user_list=[user1.__str__(),user2.__str__(),user3.__str__()]
18. # Append users to an existing csv file
19. csv_file = open("users.csv", 'a')
20. # Write a list of users to the file
21. csv_file.writelines(user_list)
```

**BOX 6.4 UPDATED USERS.CSV FILE AFTER CODE 6.3 APPENDS TO THE FILE**

```
Mohamed Ali,Apt 43 SAS Ul Nakhl Village,055-1428435
Steven Dawson,Villa 82 Al Reef Villas,050-57845
Malinda Davidson,Apt 304 Al Raha Gardens,055-3459101
Ismail Alhamidi,123 Main St CityA,555-1234
Foster Smith,456 Oak St CityB,555-5678
Stephani Johnson,789 Pine St CityC,555-9101
```

### 6.2.3 Reading from Text Files

Reading from text files is the process of accessing and extracting data stored within a file that contains plain text or text in a natural language, with some common file extensions such as ".txt", ".csv", or ".json" files. This process involves opening the file, reading its contents, and then manipulating or using that data within a program.

There are three Python functions for reading text from a file.

1. **read()/read(n):** This will read the whole file at once, and when the number of bytes n is specified, it prints out the first characters that take up as many bytes as specified by n in the parenthesis.
2. **readline()/readline(n):** Reads one line from the file and returns it as a string. When the number of bytes "n" is specified, it reads at most "n" bytes from that line in the file. However, it does not read more than one line, even if "n" exceeds the length of the line.
3. **readlines():** Reads all the lines and returns them as each line is a string element in a list.

Programmers frequently utilize two types of files: batch files (for Windows) and shell files (for Unix-based operating systems). The extension ".bat" designates a batch file, while ".sh" designates a shell file. Several directions for the operating system to follow are included in the file. When a batch file is executed, it opens, and the operating system runs each line at the terminal or command prompt. The "readline()" method can be used in a program to read a batch file.

For example, the batch file "startup.bat" has instructions for opening two applications, as shown in Box 6.5. Code 6.4 provides an example of how the batch file is opened and read. In line 1, Python's operating system "os" module is imported to use the "os.system()" function and send commands to the operating system. Line 4 opens the batch file "startup.bat" with read-only access and assigns the file object to the variable "file". At Line 6, one file line is read and stored in the variable "line". A "while" loop is used in Line 7 that continues if lines are left to read in the file. Line 9 within the loop executes the contents of each line as a system command, and then Line 11 reads another line if a line exists in the file.

**BOX 6.5 THE STARTUP.BAT FILE HAS COMMANDS TO RUN APPLICATIONS**

```
@echo off

REM Open Outlook
start "" "C:\Program Files\Microsoft Office\root\Office16\OUTLOOK.EXE"

REM Open a news website in the default web browser
start "" "https://www.thenationalnews.com/gulf-news/"
```

### CODE 6.4 READING A .BAT FILE

```
# Import the os module helps interact with the operating system
import os
# Open the batch file
file = open('startup.bat', mode='r')
# Read the batch file one line at a time
line=file.readline()
while line:
    # Run each instruction
    exit_code = os.system(line)
    # Read the next line in the batch file
    line=file.readline()
```

Code 6.5 shows how to read data from the CSV file as shown in Box 6.4. The Python code starts by importing the CSV module in Line 1, opening the CSV file in read mode (Line 3), and creating a CSV reader object that allows iterating over the rows of the CSV file (Line 5). This object iterates through each row of the CSV file to print it (Lines 7–8). In Line 9, the file is closed.

### CODE 6.5 READING LINES USING THE CSV MODULE

```
1. import csv
2. # Open the CSV file
3. file = open('users.csv', mode='r')
4. # Read the CSV file
5. csv_file = csv.reader(file)
6. # Display the contents of the CSV file
7. for lines in csv_file:
8.     print(lines)
9. file.close()
```

Code 6.6 revisits the previously defined class "User" in Code 6.3 with the same "__init__()" and "__str__()" methods (Lines 2–8). The CSV file is opened in read mode (Line 12). A list is created where each item in the list is a line read from the CSV file shown in Box 6.3 (Line 15). Subsequently, it iterates through the list of lines using a loop, splitting each line into individual values based on commas and creating "User" instances within the loop. Finally, it prints each "User" instance in string format (Lines 17–22).

### CODE 6.6 READING INSTANCES OF CLASS USER USING THE READLINES() METHOD

```
1. # A basic user class is created for illustration
2. class User:
3.     def __init__(self,fullname,address,phone_number):
4.         self.fullname=fullname
5.         self.address=address
6.         self.phone_number=phone_number
7.     def __str__(self):
8.         return f"{self.fullname},{self.address},{self.phone_number}\n"
9.
10.
11. # Append users to an existing csv file
12. csv_file = open("users.csv", 'r')
```

```

# Create a list with each row of the file as a list item
user_list=csv_file.readlines()
# Loop through the list of rows
for info in user_list:
    # Split the row based on comma
    full_name,address,phone_number=info.split(",")
    # Create the User instance
    user=User(full_name.strip(),address.strip(),phone_number.strip())
    print(user.__str__())
```

### 6.2.4 Handling XML files

Extensible Markup Language (XML) files are a type of text-based file format used to store and arrange data in a structured manner. XML files use tags to define and explain the structure of the data they contain, making them both human-readable and machine-readable. Numerous programming languages support this language-independent and cross-platform text format. Box 6.6 provides an example of an XML file. The file is human-readable and structured with tags like `<person>` or `<team>`. Each markup or tag has a closing tag like `</person>` or `</team>`. Together, the start and end tags form elements in the XML file. The example has the `</team>` tag as the root element containing the entire XML structure. The name, age, and weight elements are hierarchically within the person element, indicating that the details are relative to each person. The `<person>` tag has an attribute "`id`", which uniquely identifies each person. The XML file provides details of two team members.

**BOX 6.6 TEAM DATA IS STORED IN AN XML FILE**

```
1. <?xml version="1.0" ?>
2.
3. <team>
4.     <person id="F8939">
5.         <name>Joseph Mathew</name>
6.         <age>80</age>
7.         <weight>75</weight>
8.     </person>
9.     <person id="F8979">
10.        <name>Sooraj John</name>
11.        <age>43</age>
12.        <weight>95</weight>
13.    </person>
14. </team>
```

Due to their small file size, XML files offer significant benefits for transferring data between programs. They are also a practical solution for transmitting data where the structured nature of the language is required for easy processing and searchability. This makes XML files ideal for storing and retrieving data in various applications.

A "well-formed" XML file adheres to the following rules:

- XML file must have a root element.
- XML tags must have a closing tag.
- XML elements are case-sensitive.
- XML elements must be properly nested.
- XML attribute values must be in quotations (such as Line 4 in Box 6.6 for the value of `id`).

A "well-formed" XML file differs from a "valid" XML file. A "valid" XML file must be well-formed, and in addition, it must conform to a document type definition (DTD), which defines the schema for the elements in an XML file. More recently, the World Wide Web Consortium (W3C) recommends XML schema definition (XSD), a more advanced schema definition that formally outlines how to describe the elements in an XML file. It confirms that every element in an XML file follows the necessary description and order.

### BOX 6.7 AN XML SCHEMA DEFINITION (XSD)

```
1. <?xml version="1.0" encoding="utf-8"?>
2. <xs:schema attributeFormDefault="unqualified" elementFormDefault="qualified" xmlns:xs= "http://www.w3.org/2001/XMLSchema">
3.   <xs:element name="team">
4.     <xs:complexType>
5.       <xs:sequence>
6.         <xs:element maxOccurs="unbounded" name="person">
7.           <xs:complexType>
8.             <xs:sequence>
9.               <xs:element name="name" type="xs:string" />
10.              <xs:element name="age" type="xs:unsignedByte" />
11.              <xs:element name="weight" type="xs:unsignedByte" />
12.            </xs:sequence>
13.            <xs:attribute name="id" type="xs:string" use="required" />
14.          </xs:complexType>
15.        </xs:element>
16.      </xs:sequence>
17.    </xs:complexType>
18.  </xs:element>
19. </xs:schema>
```

In simpler terms, an XSD document is a template that outlines the requirements for formatting data in an XML document. The XSD in Box 6.7 is a template for the XML document provided earlier in Box 6.6. Line 1 in the XSD file is a processing instruction for the system to identify the version of XML used and character encoding for the text file. Line 2 sets the default requirements for the attributes and elements to avoid namespace conflicts. For `attributeFormDefault` is set to "unqualified" and `elementFormDefault` is set to "qualified", which means that elements can use variables and data types from the XML Schema namespace is accessible with the URL (http://www.w3.org/2001/XMLSchema). However, a prefix will be required to use variables from this namespace, which is defined with `xmlns:xs`. An element is created at Line 3, with the prefix `xs` (XML Schema) followed by the tag `element` and then the `name="team"`. It signifies that the root element will be named "team" in the XML documents that intend to follow this XSD. In Line 4, the team element is declared as a `complexType`, which means that the team elements can contain other elements. The other option is `simpleType`, which means that an element will not contain sub-elements. In this case, the team element can have a sub-element created at Line 6 with a proper indentation for a sub-element and is given a name and person. The `xs:sequence` tag at Lines 5 and 8 means that the sub-elements that follow this tag must be based on the sequence provided after this tag. Line 8 has sub-elements name, age, and weight. As a result, the "person" element must have a "name", "age", and "weight" in the correct sequence in the corresponding XML document. In Lines 9, 10, and 11, the sub-elements, "name", "age", and "weight" are declared with their required datatypes, `xs:string, xs:unsignedByte, and xs:unsignedByte`, respectively. These datatypes are taken from the XML schema document included in Line 2. The XSD has data that can be assigned to XML elements or attributes. The string data type can contain any sequence of characters, including letters, numbers, symbols, and spaces. The "unsignedByte", data type is a numeric datatype representing a positive integer between 0 and 255.

The method of reading the information from an XML file and further analyzing its logical structure is known as XML Parsing. The XML Document Object Model (DOM) defines the properties and methods for accessing and editing XML. Four different Python modules to parse XML documents are discussed here.

1. Minimal Document Object Model (MiniDOM)
2. BeautifulSoup
3. Element tree
4. Simple API for XML (SAX)

#### 6.2.4.1 Process XML with MiniDOM

The `MiniDOM` is a package part of the Python standard library and facilitates the parsing of XML documents. To allow MiniDOM to parse an XML document correctly, it must follow the DOM structure, a standard that considers all elements of a document as objects or nodes. Therefore, terms such as elements, objects, and nodes can be used interchangeably when discussing documents that follow the DOM structure. For example, in the case of HTML pages, the head, body, table, and div are document elements. As a result, to include an element to the document, such as a head, the code includes an opening tag, `<head>`, and closes it with a tag, including a backslash, `</head>`. Similarly, an XML document that follows the MiniDOM structure can contain nodes with opening and closing tags.

Let's consider information about visitors to an amusement park, where visitors scan their tickets before taking a ride. This information can be stored in an XML document (shown in Box 6.8) and shared across computers in the amusement park. The XML document follows a tree structure with the root node representing the amusement park. This root node contains child nodes of the type "visitor", shown with an indentation. Each visitor contains further details with indented child nodes, including "name", "age", and "rides". The "rides" tag contains a list of child nodes representing each ride the visitor takes.

**BOX 6.8 XML DOCUMENT BASED ON MINIDOM FORMAT**

```
<?xml version="1.0" ?>
<amusement_park>
    <visitor>
        <name>Abraham Reynard</name>
        <age>15</age>
        <rides>
            <ride>Haunted House</ride>
            <ride>Roller Coaster</ride>
        </rides>
    </visitor>
    <visitor>
        <name>Rawdha Ismail</name>
        <age>7</age>
        <rides>
            <ride>Teacups</ride>
            <ride> Carousel</ride>
        </rides>
    </visitor>
 </amusement_park>
```

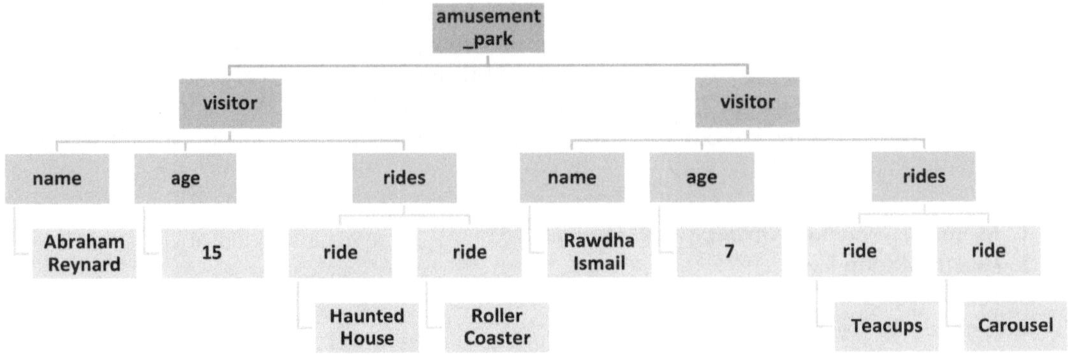


**FIGURE 6.1** DOM Tree Structure.

In Figure 6.1, the XML data from Box 6.8 is depicted in the form of a tree structure.

The Python module "xml.dom.minidom" can be used to parse an XML document containing a hierarchy of nodes represented as class objects. To understand how to read the amusement park XML file, consider Code 6.7. Line 1 imports the "xml.dom.minidom" module, and Lines 3–10 create a "Visitor" class with attributes "name", "age", and "rides". At Line 14, the module's "parse()" method reads the XML file given as an argument, creates the tree structure illustrated in Figure 6.1, and saves the structure in a variable called "dom". In Line 17, the root node is stored in a variable called "amusement_park", which contains all the other nodes in the XML document. Using the "getElementByTagName()" method in Line 20, we can access all nodes with the tag name "visitor" in the root node. A list of all nodes with the specified tag is returned and stored in the list variable "visitor_nodes". In Line 23, an empty list called "visitors_list" is created to add visitor objects. The "for" loop in Line 26 iterates through the "visitor_nodes". Within the visitor node, we can use the "getElementByTagName()" method to retrieve the name node at Line 28, the age node at Line 31, and the rides node at Line 35. Observe that at the end of Lines 28 and 31, the index "[0]" is used to retrieve the first element with the tag "name", and "age", respectively, because the "getElementByTagName()" method returns a list of nodes that have the same tag. In Line 30, we access the first "childNode" within the "name_node" and access its data to retrieve the visitor's name. Similarly, in Line 32, we access the "age" of the visitor. In Line 37, we access all tags named "rides", iterate through each ride, and add it to the rides list from Lines 38 to 40. After retrieving all the data of a visitor, we create the "visitor" object in Line 43 and append it to a "visitors_list" in Line 44. Lines 46 and 47 loop through each "visitor" object and print each visitor's information to the console.

**CODE 6.7 PYTHON CODE TO PARSE AN XML FILE USING MINIDOM**

```
1. import xml.dom.minidom
2.
3. class Visitor: # Create a Visitor class
4.     def __init__(self, name, age, rides):
5.         self.name = name
6.         self.age = age
7.         self.rides = rides
8.
9.     def __str__(self):
10.        return f"Visitor Info: name={self.name}, age={self.age}, rides taken={self.rides}"
11.
12.
13. # Parse the xml document and store the tree structure in a variable
14. dom = xml.dom.minidom.parse("amusement_park.xml")
15.
```

```
# Get the root element of the XML document
amusement_park = dom.documentElement

# Find all nodes with the visitor tags in the XML document
visitor_nodes = amusement_park.getElementsByTagName("visitor")

# Create a list to store Visitor objects
visitors_list = []

# Iterate over each visitor node
for each_visitor_node in visitor_nodes:
    # Find a node called "name" within each_visitor_node & access the first occurrence
    name_node=each_visitor_node.getElementsByTagName("name")[0]
    # Find the first childNode to get the text entered within the name node
    name =name_node.childNodes[0].data
    age_node = each_visitor_node.getElementsByTagName("age")[0]
    age=int(age_node.childNodes[0].data)

    # Find the list of rides taken by the visitor
    rides_node = each_visitor_node.getElementsByTagName("rides")[0]
    # Find all the nodes starting with the tag "rides", a list will be returned
    ride_nodes = rides_node.getElementsByTagName("ride")
    rides = []
    for ride in ride_nodes:
        rides.append(ride.childNodes[0].data )

    # Create a Visitor object and add it to the list
    visitor = Visitor(name, age, rides)
    visitors_list.append(visitor)

for visitor in visitors_list:
    print(visitor)
```

The XML documents that follow the MiniDom structure are easy to read and parse. However, they become unmanageable and slow when large amounts of data are shared and parsed.

### 6.2.4.2 Process XML with BeautifulSoup

BeautifulSoup is a Python module that can parse XML documents. Its strength lies in extracting information from the web and is robust in handling broken HTML and XML documents. However, BeautifulSoup relies on external parsers to parse the web content, and one of the most used parsers is "lxml".

For example, XML data containing information about plants in the US is extracted from a plant catalog (https://www.w3schools.com/xml/plant_catalog.xml), and the first two nodes of the XML document are provided in Box 6.9. Each plant in the document has sub-nodes to represent plant details. The COMMON tag represents the plant's common name, the BOTANICAL tag provides its scientific name, the ZONE tag represents the areas in the US where it can thrive, the LIGHT tag represents the amount of light it can receive, the PRICE tag is the plant's average price, and the AVAILABILITY tag is the code that represents the date and location of availability.

**BOX 6.9 XML FILE WITH PLANT DATA**

```
<CATALOG>
   <script>window._wordtune_extension_installed = true;</script>
   <PLANT>
      <COMMON>Bloodroot</COMMON>
      <BOTANICAL>Sanguinaria canadensis</BOTANICAL>
```

```
            <ZONE>4</ZONE>
            <LIGHT>Mostly Shady</LIGHT>
            <PRICE>$2.44</PRICE>
            <AVAILABILITY>031599</AVAILABILITY>
        </PLANT>
        <PLANT>
            <COMMON>Columbine</COMMON>
            <BOTANICAL>Aquilegia canadensis</BOTANICAL>
            <ZONE>3</ZONE>
            <LIGHT>Mostly Shady</LIGHT>
            <PRICE>$9.37</PRICE>
            <AVAILABILITY>030699</AVAILABILITY>
        </PLANT>
...
</CATALOG>
```

The first step toward working with BeautifulSoup is to install the necessary packages with the following Python code:

```
pip install beautifulsoup4
pip install lxml
```

Code 6.8 is an example of extracting XML data from a website, parsing the data, and creating objects in Python. At Line 1 in the code, we imported the requests module to be able to send a request to a website for the XML file. In Line 2, we import the "BeautifulSoup" package. To keep the code simple, from code Lines 4 to 18, the "Plant" class is defined with the "__init__()" method and the "__str__()" method to define and display plant information. The code in Line 22 stores the website URL in a variable, and in Line 25, a "get()" request is sent to the website to retrieve information from the link provided. A variable "response" stores the answer to this request, and in Line 26, the response text is stored in the variable named "xml:content". The "BeautifulSoup" object, called "soup", is recreated in Line 29 and is initialized to parse the "xml:content" using the XML parser. Line 32 defines an empty list of "plants" to add plant objects. The "soup.find_all("Plant")" in Line 33 retrieves all the nodes with the name "Plant" as a list. Therefore, in Line 33, we iterate through this list of plant nodes. For each plant object, "plant_data", we access the attributes using the ".find(node_name).text" method to locate the attribute and the text stored in it is based on the node name. For example, to access the common name of the plant, the following method is used:

```
Plant_data.find("Common").text
```

After retrieving all the child nodes within the plant node, we create the "Plant" object in Line 41 and initialize the object with the attributes retrieved from Lines 34 to 39. In Line 42, we add the "plant" to the "plants" list. Code Lines 45–47 print all the "plants" extracted from the plant's catalog that was extracted from the Web. Box 6.10 shows the first few lines of the output for Code 6.8.

### CODE 6.8 CREATING PYTHON OBJECTS FROM AN XML FILE ON THE WEB

```
import requests
from bs4 import BeautifulSoup

class Plant: # Create Plant Class
    def __init__(self, name, botanical, zone, light, price, availability):
        self.name = name
        self.botanical = botanical
        self.zone = zone
        self.light = light
        self.price = price
        self.availability = availability

    def __str__(self):
        return (
            f"Plant(Name: {self.name}, Botanical: {self.botanical},"
            f"Zone: {self.zone}, Light: {self.light},"
            f"Price: {self.price}, Availability: {self.availability})"
        )


# URL of the XML file
xml:url = "https://www.w3schools.com/xml/plant_catalog.xml"

# Make a request to the URL and getting the content
response = requests.get(xml:url)
xml:content = response.text

# Parse the XML content with BeautifulSoup and lxml parser
soup = BeautifulSoup(xml:content, features="xml")

# Extract information using Beautiful Soup and create Plant objects
plants = []
for plant_data in soup.find_all('PLANT'):
    name = plant_data.find('COMMON').text
    botanical = plant_data.find('BOTANICAL').text
    zone = plant_data.find('ZONE').text
    light = plant_data.find('LIGHT').text
    price = plant_data.find('PRICE').text
    availability = plant_data.find('AVAILABILITY').text

    plant = Plant(name, botanical, zone, light, price, availability)
    plants.append(plant)

# Print Plant objects
print("Plant Information:")
for plant in plants:
    print(plant)
```

### BOX 6.10 FIRST FEW LINES OF THE OUTPUT FOR CODE 6.8

```
Plant Information:
Plant(Name: Bloodroot, Botanical: Sanguinaria canadensis, Zone: 4, Light: Mostly Shady, Price: $2.44,
Availability: 031599)
Plant(Name: Columbine, Botanical: Aquilegia canadensis, Zone: 3, Light: Mostly Shady, Price: $9.37,
Availability: 030699)
```

Although BeautifulSoup is simple and effective for web scraping, it is not specifically designed to manipulate XML documents.

#### 6.2.4.3 Process XML with Element Tree

Element Tree is another module included within Python's standard library that parses the complete XML document as a tree structure. This structure is very useful for complex data that contains multilevel hierarchies. For example, consider an example of a family tree with a parent having a hierarchy of children, who have their children, and those grandchildren have further children, as illustrated in the XML document in Box 6.11. All tags are named "person", but they are at a different level in the hierarchy.

**BOX 6.11 FAMILY TREE IN AN XML FILE**

```
<family_tree>
    <person name="Mustafa" gender="male">
        <person name="Hassan" gender="male">
            <person name="Ali" gender="male">
                <person name="Layla" gender="female">
                </person>
            </person>
        </person>
        <person name="Zubaida" gender="female">
            <person name="Ibrahim" gender="male">
            </person>
        </person>
    </person>
 </family_tree>
```

Based on our previous examples of parsing XML documents, we can use the ".find('person')" method to locate elements based on the name tag. However, as we have hierarchical data, we will use recursion to find children, grandchildren, and great-grandchildren and create a tree of Python objects.

Code 6.9 imports the "ElementTree" module as "ET" in Line 1. Lines 3–7 define the class "Person" with attributes, "name", "gender", and "children" to store the person's children in a list. In Line 11, the XML document "familytree.xml" is given as a parameter to the "parse()" method of the "ElementTree" and the resulting tree structure is stored in the variable named "tree". The root node of the "tree" is stored in the variable "root" at Line 12. A recursive method "create_person_from_element()" is created at Line 15 to iterate through the tree structure and create a tree of "person" objects in Python. It takes the "ElementTree" node and creates a "person" object at Line 16. In Line 17, it loops through each child node "child_element". Line 18 calls for each child node to be converted into a "person" object. In Line 19, the "person" object returned by the recursive call is added to the person's "children" list. Finally, at Line 20, the "person" node is returned to where the method call was made. This results in a tree structure of a person's objects, depicted in Figure 6.2 (a). The method "print_family_tree()" (Lines 24–29) is another recursive method that traverses through the family tree, starting at the root person with the name "Mustafa". The second parameter of the method is the level in the family tree. Line 25 prints spaces based on the level in the family tree and then the person's information. Line 28 iterates through the person's children and recursively calls itself to print each child as the person. Line 32 makes the first call to the method "create_person_from_element()" with the root node of the "ElementTree". The methods return the person to the root of the family tree. Line 35 makes the first call to the "print()" method with the first parameter as the "root" person and the second parameter with Level 0.

## CODE 6.9 PARSING AN XML DOCUMENT WITH ELEMENT TREE

```
import xml.etree.ElementTree as ET

class Person: # Create Person Object
    def __init__(self, name, gender):
        self.name = name
        self.gender = gender
        self.children = []


# Parse the XML file
tree = ET.parse('familytree.xml')
root = tree.getroot()

# A Method to create a Person from ElementTree node
def create_person_from_element(person_element):
    person = Person(person_element.attrib['name'], person_element.attrib['gender'])
    for child_element in person_element:
        child_person = create_person_from_element(child_element)
        person.children.append(child_person)
    return person


# A Method to print the family tree recursively
def print_family_tree(person, level):
    print(".." * level + f"{person.name} ({person.gender})")

    # Print recursively the information for each child
    for child in person.children:
        print_family_tree(child, level + 1)

# Create a Person object representing the entire family tree
person_tree_root = create_person_from_element(root[0])

# Print the family tree starting from the root Person object
print_family_tree(person_tree_root, 0)
```

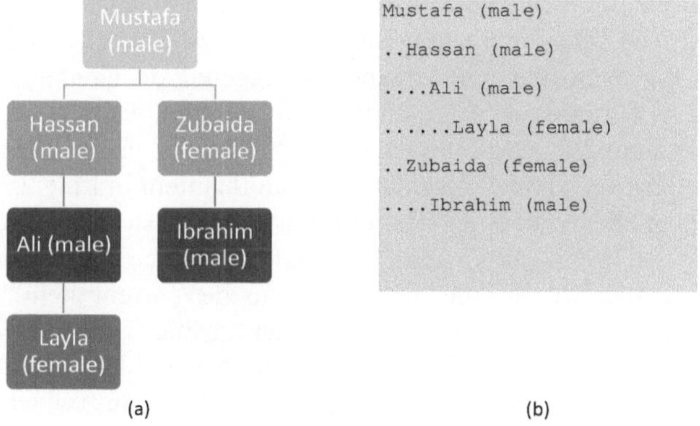


**FIGURE 6.2** (a) Conceptual representation of the family tree. (b) Output of the program.

The ElementTree is a robust library because its tree structure makes it easy to add and remove elements from the XML document without breaking the existing formatting. However, it works well with small-to-medium-sized documents but becomes slow with large amounts of data.

#### 6.2.4.4 Process XML with SAX

SAX is also known as the simple API for parsing XML files. It is very useful for processing large XML documents because it can trigger events while parsing the document and does not need to load the entire content in memory. Moreover, it is very efficient for streaming data. To use SAX, content handlers are defined to handle the events triggered while parsing the document. To illustrate the use of SAX, consider the XML document shown in Box 6.12, which contains a list of menu items with sub-element categories, names, and prices. Code 6.10 parses the data using the SAX module.

**BOX 6.12 THE XML FILE (MENU.XML) CONTAINS A LIST OF MENU ITEMS**

```
<menu>
    <item>
        <category>Appetizers</category>
        <name>Hummus with Pita</name>
        <price>$5.99</price>
    </item>
    <item>
         <category>Main Courses</category>
        <name>Shawarma Platter</name>
        <price>$12.99</price>
    </item>
    <item>
        <category>Desserts</category>
        <name>Baklava</name>
        <price>$6.99</price>
    </item>
</menu>
```

After importing the "SAX" module at Line 1 of Code 6.10, the class "Item" is defined from Lines 5 to 11. In Line 15, the "MenuHandler" class is defined as inheriting from the "ContentHandler" class as a customized version to handle the "menu.xml" file. The "__init__()" method of the "MenuHandler" initializes a list of "menu_items" that will store the menu items read from the XML file (Line 17). The "current_item" variable will contain the current tag being parsed, and the "buffer" variable stores the text read between opening and closing tags (Lines 18 and 19). When the parser encounters a start tag, the "startElement()" method is triggered, declared at Line 22. It receives the "name" of the tag and its attributes as parameters "attrs". Within this method, we include code that will initialize storing an item. We check if the tag is an "item" tag, create a new object of the "Item" class, and clear the buffer to read new information about this item in Line 25. The "endElement()" method is triggered at the closing of a tag. Therefore, we need to store data from the "buffer" to the appropriate variable in the "Item" class within this method. In Line 29, we check if it is the closing tag for "category". If yes, then we save the "buffer" data to the "current_item"'s "category" variable. We make similar checks in Lines 31 and 33 for the other attributes, "name" and "price", and store them accordingly in Lines 32 and 34 to the "current_item"'s variables "name" and "price", respectively. Finally, we check in Line 35 if it is the closing tag for the "item", which means that the item's data has been read and stored, and now it is ready to be added to the item list "menu_items" (Line 36). The "characters()" method at Line 39 is triggered when the parser encounters text between the tags. In this case, the "content" is stripped of the newline and space characters and added to the "buffer". In Line 44, we create an instance of the "MenuHandler" class. The XML filename "menu.xml" and "handler" (the instance of the "MenuHandler" class) are given as parameters when creating the SAX parser at Line 46. Lines 49 and 50 iterate through the "item" list and print all the items.

### CODE 6.10 PARSING AN XML DOCUMENT USING SAX

```
1. # Import xml.sax module for XML parsing
2. import xml.sax
3.
4. # Define a class representing an item in the menu
5. class Item:
6.     def __init__(self, category, name, price):
7.         self.category = category
8.         self.name = name
9.         self.price = price
10.    def __str__(self):
11.            return f"Category: {self.category} || Name: {self.name} || Price: {self.price}"
12.
13.
14. # Define a content handler for parsing XML
15. class MenuHandler(xml.sax.ContentHandler):
16.     def __init__(self):
17.         self.menu_items = []
18.         self.current_item = None
19.         self.buffer = ""
20.
21.     # Sets up MenuHandler for a new XML element, creating Item and resetting buffer
22.     def startElement(self, name, attrs):
23.         if name == "item":
24.             self.current_item = Item("", "", "")
25.         self.buffer = ""
26.
27.     # Handles the closure of XML elements, updating Item attributes.
28.     def endElement(self, name):
29.         if name == "category":
30.             self.current_item.category = self.buffer
31.         elif name == "name":
32.             self.current_item.name = self.buffer
33.         elif name == "price":
34.             self.current_item.price = self.buffer
35.         elif name == "item":
36.             self.menu_items.append(self.current_item)
37.
38.     # Accumulate the character data in the buffer
39.     def characters(self, content):
40.         self.buffer += content.strip()
41.
42.
43. # Create an instance of MenuHandler
44. handler = MenuHandler()
45. # Parse the XML file "menu.xml" using the handler
46. xml.sax.parse("menu.xml",handler)
47.
48. # Print the menu details
49. for item in handler.menu_items:
50.      print(item)
```

All four libraries for reading XML files have simple commands for reading tags or iterating through the DOM structure. Choosing the right library depends on the project requirements and the required methods. SAX or ElementTree may be more suitable while working with large XML files. However, BeautifulSoup and MiniDOM would be better for extracting data from HTML files.

## 6.3 BINARY FILES

While text files are beneficial for storing textual content and are lightweight, they are not useful when complex data structures must be stored and accessed. For instance, consider a nested dictionary with details of all employees in a company, like ID, name, age, and band, as shown in Box 6.13. To write the details to a text file, the values of all the attributes need to be written as strings because only strings can be written to text files.

**BOX 6.13 SAMPLE DICTIONARY**

```
employees = {
  'Emp1': {
        'Name': "Austin", 'Age':21, 'Band':4,
    },
  'Emp2': {
        'Name': 'Khulood', 'Age':22, 'Band':4
    },
  'Emp3': {
        'Name': 'Emma', 'Age':22, 'Band':3
    }
}
```

Now, the data in the text file cannot be read back as a dictionary, and trying to access the key/value pairs of the dictionary would throw an error. Reconstructing the dictionary from the data in the text file would require a very lengthy process. In such cases, storing the data as objects in binary files is better.

### 6.3.1 Object Serialization

Serialization is the process of converting an object's state into a form that can be persisted or transferred. Lists, tuples, dictionaries, and even objects of user-defined classes can be converted into a sequence of bytes or a stream of bytes that the computer can process and understand. The process is called serialization. The same data structures can be later accessed when the stream of bytes is converted back to their respective data structures. This process is called deserialization.

Object serialization maintains type fidelity, meaning an exact copy of the object is obtained upon deserialization, preserving the object's entire state. Typically, serialization helps maintain an object's state between several states of an application. It is possible to serialize objects as objects to a file, a memory location, or a network location. Examine Figure 6.3, where an object is serialized into a byte stream and stored in a file. After the object is deserialized, the replica can be recovered with all its states preserved.

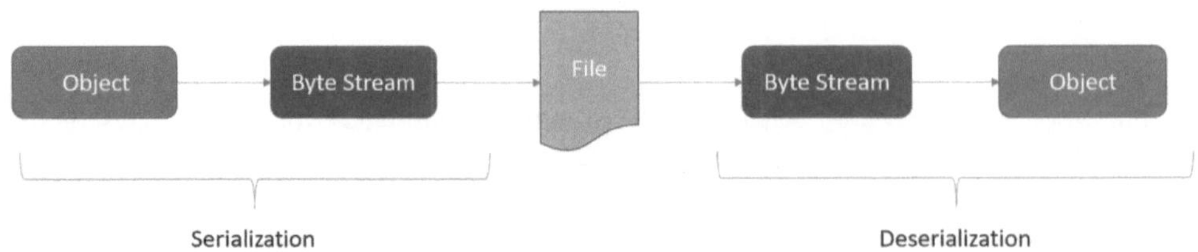


**FIGURE 6.3** Maintaining object structure using Serialization and Deserialization.

### 6.3.2 Serialization with Python Pickle

In Python, the process of serializing an object is called pickling. "Pickle" is a module provided within the Python standard library with various methods for serializing or pickling objects. Pickled objects cannot be unpickled using any other programming language since this format is native to Python. When it comes to serialization formats, Pickle has advantages as well as disadvantages.

#### 6.3.2.1 Advantages of Using Pickle

- Pickle can serialize almost every commonly used built-in Python data type. It also retains the exact state of the object.
- Pickle can store recursive structures as it writes an object only once.
- Pickle is efficient when deserializing objects because it can easily save different variables and load them back in different sessions, recovering the data exactly as it was.

#### 6.3.2.2 Disadvantages of Using Pickle

- When deserializing an object, Pickle cannot distinguish between a malicious callable and a non-malicious one, and, therefore, Pickle can be unsafe.
- Pickle is a Python-specific module and cannot be deserialized using a different language.
- Pickle is generally slower and produces larger serialized values than other formats.

### 6.3.3 Pickling – Saving Objects to Files

To save objects in a file, first, the Pickle module is imported, and then the following steps are followed:

1. Open a file for binary writing (wb).
2. Use the "dump()" method to Pickle the object and write it to the opened file.
3. After all objects are pickled and saved into the file, close the file.

Consider Code 6.11, a simple example of how to serialize a dictionary. Line 1, the "pickle" library is imported. Lines 4–7 define the "phonebook" dictionary, which contains names as keys and phone numbers as values. At Line 10, a file handler opens the "phonedata.dat" file with a binary writing mode "wb". The extension ".dat" is commonly used and lets users know that this file contains data. Another common extension is ".pickle". The extension is irrelevant to how data is stored because the data contained within the file is in binary format and not human-readable. In Line 13, the dictionary is serialized using the "pickle.dump()" method, which takes the data structure "phonebook" as the first argument and the file handler of the "output_file" as the second. At Line 16, the file is closed.

**CODE 6.11 USING PICKLE TO SERIALIZE A DICTIONARY**

```
1. import pickle
2.
3. # Dictionary for phone book
4. phonebook = {
5.             'Javed': '507937682',
6.             'James': '559038281',
7.             'Joanne': '509073672'}
8.
```

```
# Open file for binary writing
output_file = open('phonedata.dat', 'wb')

# Pickle the object to the file
pickle.dump(phonebook, output_file)

# Close the file
output_file.close()
```

Consider Code 6.12 to differentiate between how to serialize objects to file or serialize objects to variables in memory for further use in the program. In Line 2, we imported the "pickle" library, and from Lines 5 to 12, we defined the class "Item" to be used for serialization. In Lines 16 and 17, two instances, "item1" and "item2" of the "Item" class are created. In Line 20, a file "data.dat" is opened in binary writing mode to store serialized "Item" objects. In Line 21, the "dump()" method is used from the Pickle library to serialize the "item1" object into binary data and then write this data to the file "file_handler". Notice that in Line 24, a different method from the library Pickle is used, the "dumps()" method, which serializes the object in memory and stores it in the variable called "serialized_data". In Lines 27 and 28, to confirm the data type of the serialized information, we check if "serialized_data" is of type "bytes"; if it is, a message is printed to indicate the binary format. In Line 30, the file is closed.

## CODE 6.12 SERIALIZE AN ITEM USING THE DUMP AND DUMPS METHOD

```
1. # Import the pickle module
2. import pickle
3.
4. # Create a class named "Item"
5. class Item:
6.     def __init__(self, category, name, price):
7.         self.category = category
8.         self.name = name
9.         self.price = price
10.
11.     def __str__(self):
12.         return f"Category: {self.category} || Name: {self.name} || Price: {self.price}"
13.
14.
15. # Create 2 instances of the Item Class
16. item1 = Item(category="Electronics", name="Laptop", price=3260.99)
17. item2 = Item(category="Electronics", name="SmartPhone", price=2300)
18.
19. # Use pickle.dump to serialize an object as binary data
20. file_handler= open('data.dat', 'wb')
21. pickle.dump(item1, file_handler)
22.
23. # Use pickle.dumps to serialize an object to a binary string
24. serialized_data = pickle.dumps(item2)
25.
26. # Check the type of the object
27. if isinstance(serialized_data, bytes):
28.     print("Serialized data is of type bytes")
29. # Close the file
30. file_handler.close()
```

### 6.3.4 Unpickling – Reading Objects from Files

To read the pickled object from a file, the following steps are to be followed:

1. Import the Pickle library.
2. Open the file in binary reading mode (**rb**).
3. Use the `load()` method to read the file and unpickle/deserialize the object.
4. After unpickling/deserialization all the objects from the file, close the file.

In Code 6.11, a dictionary called "phonebook" was serialized to the file "phonedata.dat". In Code 6.13, the serialized dictionary will be read from the file and printed to the console, and this process is called deserialization. In Line 1, we import the "pickle" library. In Line 4, we open the file "phonedata.dat" file that contains serialized data in binary reading mode "rb". The "load()" method in Line 7 deserializes the object stored in the file (the serialized dictionary from Code 6.11) by converting it back into a dictionary object and assigning it to the variable "pb". In Line 10, the dictionary ("pb") of the phone book is printed to the console, and the file is closed in Line 13.

**CODE 6.13 CODE TO DESERIALIZE A DICTIONARY OBJECT**

```
1. import pickle
2.
3. # Open file for binary reading
4. input_file = open('phonedata.dat', 'rb')
5.
6. # Unpickle the object from the file
7. pb = pickle.load(input_file)
8.
9. # Print the dictionary
10. print(pb)
11.
12. # Close the file
13. input_file.close()
```

In Code 6.12, two items are serialized. The "item 1" to a file named "data.dat" and "item2" to a variable called "serialized_data". Code 6.14 is an updated version of the code to include deserialization, which is in Lines 29–37. To explain deserialization, consider Line 33, where the file "data.dat" is opened for reading. In Line 34, the "load()" method is used to read data from a file, which deserializes the item object and prints it in Lines 35 and 36. Notice that in Line 39, deserialization here is performed using the "loads()" method that receives the "serialized_data" variable as a parameter and returns a Python object for display at Line 41. The "load()" method deserializes data from a file, whereas the "loads()" method deserializes data in memory.

**CODE 6.14 DESERIALIZING ITEM OBJECTS**

```
1. # Import the pickle module
2. import pickle
3.
4. # Create a class named "Item"
5. class Item:
6.     def __init__(self, category, name, price):
7.         self.category = category
8.         self.name = name
9.         self.price = price
10.
```

```
    def __str__(self):
        return f"Category: {self.category} || Name: {self.name} || Price: {self.price}"


# Create 2 instances of the Item Class
item1 = Item(category="Electronics", name="Laptop", price=3260.99)
item2 = Item(category="Electronics", name="SmartPhone", price=2300)

# Use pickle.dump to serialize an object as binary data
file_handler= open('data.dat', 'wb')
pickle.dump(item1, file_handler)

# Use pickle.dumps to serialize an object to a binary string
serialized_data = pickle.dumps(item2)

# Check the type of the object
if isinstance(serialized_data, bytes):
    print("Serialized data is of type bytes")
# Close the file
file_handler.close()

# Deserialize using pickle.load to read from a file
file_handler= open('data.dat', 'rb')
loaded_item1 = pickle.load(file_handler)
print("\nLoaded Item from file:")
print(loaded_item1)

# Deserialize using pickle.loads to read from a string
loaded_item2 = pickle.loads(serialized_data)
print("\nLoaded Item from string:")
print(loaded_item2)
```

The output in Box 6.14 for Code 6.14 shows that the loaded dictionary data matches the original dictionary defined in Code 6.12.

**BOX 6.14 OUTPUT FOR CODE 6.14**

```
Serialized data is of type bytes

Loaded Item from file:
Category: Electronics || Name: Laptop || Price: 3260.99

Loaded Item from string:
Category: Electronics || Name: SmartPhone || Price: 2300
```

## 6.4 Data Persistence with JSON Module

JavaScript Object Notation (JSON) is a lightweight data-interchange format like XML. It is a language-independent and cross-platform text format supported by many programming languages. Data exchange between web servers and clients usually uses JSON.

JSON is suitable for data interchange between systems and languages and human-readable configuration files. Python's JSON module methods and interfaces are comparable to those of Pickle. JSON can be used to serialize basic Python data types, such as dictionaries, tuples, lists, and arrays. However, user-defined class objects must be converted into JSON-compatible data types. The data in Box 6.15 provides a sample JSON file with two items. JSON stores data as key-value pairs, similar to dictionaries in Python.

### BOX 6.15 SAMPLE FILE THAT CONTAINS TWO ITEM DETAILS IN JSON FORMAT

```
[
  {
    category: "Electronics",
    name: "Laptop",
    price: 3260.99
  },
  {
    category: "Electronics",
    name: "SmartPhone",
    price: 2300
  }
]
```

Code 6.15 demonstrates how to serialize and deserialize class objects into JSON. In this example, two item objects are serialized: one to a JSON file and the other to a variable in memory. Line 19 opens a file to serialize data in write mode. In Line 21, the JSON library's "dump()" method is used with the first parameter as an "item1" object converted into a dictionary and then written to the file given "data.json" as the second parameter. In Line 22, the file is closed. In Line 25, the use of the "dumps()" method is shown where it takes an "item2" object in dictionary format and serializes it as a "json_variable".

From Line 27 onward, the code includes functionality for deserialization. The "data.json" file is opened again in read mode. The JSON data is read using the "load()" method at Line 28, and the read item is stored in the "file_item" variable in Line 29. In Line 32, the use of the "loads()" method is shown to deserialize the "json_variable" and save it as a "memory_item". Both items are used as dictionaries and not yet defined as objects of the "Item" class. Lines 35 and 36 convert these dictionaries into "Item" objects using "**" to unpack the dictionary into keys and values, which are given as arguments to create "Item" objects and are then printed in Lines 39 and 43.

### CODE 6.15 SERIALIZING AND DESERIALIZING CLASS OBJECTS INTO JSON

```
1. # Import the json5 library as json
2. import json5 as json
3.
4. # Define a class named Item with attributes category, name and price
5. class Item:
6.     def __init__(self, category, name, price):
7.         self.category = category
8.         self.name = name
9.         self.price = price
10.
11.     def __str__(self):
12.         return f"Category: {self.category} || Name: {self.name} || Price: {self.price}"
13.
14.
15. # Create instances of the Item class
16. item1 = Item(category="Electronics", name="Laptop", price=3260.99)
17. item2 = Item(category="Electronics", name="SmartPhone", price=2300)
18.
19. file_handler= open('data.json', 'w')
20. # Serialize to JSON and write to a file using json.dump
21. json.dump(item1.__dict__, file_handler)
22. file_handler.close()
23.
```

```
# Serialize to JSON variable using json.dumps
json_variable = json.dumps(item2.__dict__)

# Deserialize JSON  from file using json.load
file_handler= open('data.json', 'r')
file_item = json.load(file_handler)

# Deserialize JSON variable using json.loads
memory_item = json.loads(json_variable)

# Create new instances from the deserialized data
deserialized_item1 = Item(**file_item)
deserialized_item2 = Item(**memory_item)

# Print the deserialized items
print("\nItem1 deserialized from file:")
print(deserialized_item1)

print("\nItem2 deserialized from a variable:")
print(deserialized_item2)
```

### 6.4.1 The JSONDecoder Class

The code in Code 6.15 explains how to convert JSON data to dictionary objects and then to user-defined Python objects. However, this becomes problematic with complex user-defined classes that contain more complex data types, such as lists and dictionaries. Therefore, it is recommended that a decoder class be created that can read JSON data and convert it into the relevant class objects.

Consider Code 6.16 to illustrate how to create your custom decoder class. Lines 4–11 have the "Item" class. In Line 15, we created the "ItemDecoder" class, which has a method called "decode()". The "decode()" method receives one JSON item at a time as a parameter (Line 19). The "if" statement at Line 20 checks if the JSON item contains the keys: "category", "name", and "price", and if True, then it retrieves their values and creates an "item" object at line 21. The JSON item is added to the "item_list" at Line 22. The main program begins at Line 27 by opening the JSON file with the read mode in binary format and stores access to the data as "json_data". In Line 30, the custom "item_decoder" object is created. In Line 33, the "json.load()" method is called with the "json_data" as the first argument. The second argument uses the keyword "object_hook" and assigns it to the "ItemDecoder"'s "decode()" method. This registers the "decode()" method to be used for deserializing the "json_data" in the "load()" method. Therefore, the "load()" method iterates through each "JSON" object and calls the "decode()" method on each "JSON" object. The "item" objects created by the "load()" method are stored in the "item_decoder"'s "item_list". Lines 36–37, iterate through this "item_list" and print each item.

**CODE 6.16 USING A CUSTOM DECODER CLASS TO DESERIALIZE JSON DATA**

```
1. import json5 as json
2.
3. # Define a class named Item with attributes category, name, and price
4. class Item:
5.     def __init__(self, category, name, price):
6.         self.category = category
7.         self.name = name
8.         self.price = price
9.
```

```
    def __str__(self):
        return f"Category: {self.category} || Name: {self.name} || Price: {self.price}"


# Create a custom decoder class for Item objects
class ItemDecoder:
    def __init__(self):
        self.item_list=[]

    def decode(self, json__item):
        if "category" in json__item and "name" in json__item and "price" in json__item:
            item=Item(json__item["category"], json__item["name"], json__item["price"])
            return self.item_list.append(item)
        return json__item


# Open the JSON file in binary read mode
json_data = open("data.json", "rb")

# Create an instance of the custom decoder class
item_decoder = ItemDecoder()

# Decode JSON data using json.load with the custom decoder
json.load(json_data, object_hook=item_decoder.decode)

# Print the Item objects
for item in item_decoder.item_list:
    print(item)
```

## 6.5 CASE STUDY

The case study in Chapter 1 used the object-oriented approach to create bank transactions and maintain an account balance. The same example is presented in Code 6.17, now including data serialization using the "pickle" and "JSON" modules to serialize and deserialize the "TransactionManager" objects. Lines 1–5 import the necessary libraries, including "pickle" and "json". From Lines 8–11, define the "Enum" class "TransactionType". The class "Currency" is defined from Lines 14 to 19. The class "Transaction" is defined from Lines 22 to 83. A method "to_dict()" has been added to the "Transaction" class (61–69) to return each transaction as a dictionary, with attribute names as keys and attribute values as the corresponding values. This is required to serialize the "Transaction" object as a JSON string. Although each object, such as "Transaction", inherits a built-in "__dict__ ()" method that returns the object as a dictionary. However, this method will not work for the "Transaction" class, as it includes custom attributes such as "TransactionType", "Currency", and "datetime".

The class "TransactionManager" is created from Lines 86 to 185. An additional "to_dict()" method has been added to the "TransactionManager" class (157–164) to convert the object to a dictionary for JSON serialization. At Line 158, a new list is created to store dictionary-type transactions, "transactions_list". An empty dictionary, "manager_dict" is initialized in Line 159, representing the Transaction Manager attributes as a dictionary. At Line 160, a "for" loop iterates through each transaction in the "__transactions" list to call the "to_dict()" method and retrieve the transaction as a dictionary-type object. These dictionary-type transactions are appended to a new list (162), called "transactions_list". The list is added to a dictionary against the key "transactions" and returned by the method (163 and 164).

Another method, "decode()", is added to the "TransactionManager" class (167–185), which is used to decode each JSON object during the deserialization of the transaction manager object read from a JSON file. This method takes one "json_item" as a parameter. Lines 168–175 ensure that the JSON item contains all transaction attributes by using attribute names as keys to retrieve the values.

At Lines 177–182, the Transaction object is created from retrieved attribute values using the attribute names as keys. In Line 184, the transaction is added to the Transaction Manager's list, and in Line 185, the JSON item is returned. The "decode()" method is called every time a new JSON object is retrieved by the file handler.

Line 189 creates a "TransactionManager" object named "trans_manager". From Lines 192 to 196, the "trans_manager" creates two "Transaction" objects by taking user input. The transactions are displayed, and the total balance is printed.

Data serialization starts from Line 199; a file, "transaction_data.pk", is opened in binary write mode to serialize the "trans_manager", which includes a list of the two transaction objects created earlier. In Line 200, the "dump()" method is used from the Pickle library to serialize the "trans_manager" object into binary data, then write this data to the file "file_handler" and then close the file in Line 201. Deserialization of the same object is performed in Lines 204–206 using the "load()" method from the Pickle library. Lines 209–211 print the deserialized Transaction Manager, named "loaded_trans_manager". The output of this "print()" statement will include the two transaction objects created earlier.

From Lines 214 to 216, the "loaded_trans_manager" object is serialized using the JSON library. In Line 215, the "dump()" method of JSON is used to store the "trans_manager" in a file named "transactions.json". The "dump()" method requires a dictionary-type object for JSON serialization. As a result, the "to_dict()" method is called on the "loaded_trans_manager", which retrieves the object as a dictionary. The file is closed at Line 216.

We deserialized the transaction manager from Lines 219 to 225. At Line 219, we create a new transaction manager named "loaded_trans_manager_json". At Line 221, we open the same file, "transactions.json", in read mode. In Line 223, the serialized JSON data is loaded using the "load()" method. The "object_hook" parameter value is given the "decode()" method of the transaction manager to ensure that the customer attributes of the transaction are loaded correctly and added to the "loaded_trans_manager_json".

From Lines 227 to 229, the transactions of the "loaded_trans_manager_json" are printed, followed by the total account balance.

### CODE 6.17 OBJECT SERIALIZATION FOR TRANSACTIONS.

```
1. import datetime
2. from enum import Enum
3. import pickle
4. import json5 as json
5.
6.
7. # The definition of class TransactionType
8. class TransactionType(Enum):
9.     """An enumerator type class that defines the types of transactions"""
10.     INCOME = 1  # An income type defines a transaction of a gained amount of money
11.     EXPENSE = 2  # An expense type defines a transaction of a spent amount of money
12.
13.
14. class Currency(Enum):
15.     """An enumerator type class that defines the types of currencies"""
16.     USD = 1  # US Dollars
17.     EUR = 2  # Euro
18.     GBP = 3  # Great Britain Pound
19.     AED = 4  # Arab Emirati Dirham
20.
21.
22. class Transaction:
23.     """A class that represents a financial transaction"""
24.     # A static variable that keeps track of the number of transactions
```

```
    transaction_count = 0

    # Initialize the Transaction class
    def __init__(self, transaction_type, amount, currency, description, date_time):
        # Increment the number of transactions with the creation of a new transaction
        Transaction.transaction_count += 1

        # Assign a unique transaction ID based on the number of transactions
        self.__trans_id = "T" + str(Transaction.transaction_count)
        self.__trans_type = transaction_type
        self.__trans_amount = amount
        self.__trans_description = description
        self.__trans_currency = currency
        self.__trans_date_time = date_time

    def get_trans_id(self):
        return self.__trans_id

    def get_transaction_type(self):
        return self.__trans_type

    def get_trans_amount(self):
        return self.__trans_amount

    def get_trans_date_time(self):
        return self.__trans_date_time

    def get_trans_currency(self):
        return self.__trans_currency

    def get_trans_description(self):
        return self.__trans_description

    def set_trans_description(self, description):
        self.__trans_description = description

    def to_dict(self):
            return {
                "__trans_id": self.__trans_id,
                "__trans_type": self.__trans_type.name,  # Convert Enum to string
                "__trans_amount": self.__trans_amount,
                "__trans_description": self.__trans_description,
                "__trans_currency": self.__trans_currency.name,  # Convert Enum to string
                "__trans_date_time": self.__trans_date_time.strftime("%Y-%m-%d %H:%M:%S")
            }


    # Define a string representation of the transaction
    def __str__(self):
        return (
            'Transaction ID: {id}, Amount: {amount} {currency}, '
            'Type: ({type}), Description: {description}, '
            'Date & Time: {date_time}'.format(
                id=self.__trans_id,
                amount=self.__trans_amount,
                currency=Currency(self.__trans_currency).name,
                type=TransactionType(self.__trans_type).name,
                description=self.__trans_description,
                date_time=self.__trans_date_time))


class TransactionManager:
    """
    This class keeps track of transactions and allows users
    to create transactions, and calculate their total amount
```

```
    """

    def __init__(self):
        # Transactions managed by TransactionManager as a list
        self.__transactions = []

    # A method that allows the creation of a transaction
    # User Input is collected for each transaction
    def create_transaction(self):
        # Transaction type
        trans_type_input = input("Transaction type (Expense/Income)? ")
        assert trans_type_input in ["Expense", "Income"], "Invalid Input"
        trans_type = TransactionType.EXPENSE  # Defaulting the type to Expense
        if trans_type_input == "Income":
            trans_type = TransactionType.INCOME
        trans_amount = float(input("Transaction amount? "))
        if trans_type == TransactionType.EXPENSE:
            trans_amount *= -1

        # Input the transaction description
        trans_description = input("Transaction description: ")

        # Input the transaction currency
        trans_currency_input = input("Currency of Transaction (USD/EUR/GBP/AED)? ")
        assert trans_currency_input in ["USD", "EUR", "GBP", "AED"], "Invalid Input"
        if trans_currency_input == "USD":
            trans_currency = Currency.USD
        elif trans_currency_input == "EUR":
            trans_currency = Currency.EUR
        elif trans_currency_input == "GBP":
            trans_currency = Currency.GBP
        else:
            trans_currency = Currency.AED

        # Input transaction date & time
        trans_date_time_input = input("Transaction date - 'now' or YYYY-MM-DD hh:mm:ss: ")
        if trans_date_time_input == "now":
            trans_date_time = datetime.datetime.now()
        else:
            trans_date_time = datetime.datetime.strptime(
                trans_date_time_input,
                "%Y-%m-%d %H:%M:%S")

        # Create an object of class Transaction
        transaction = Transaction(transaction_type=trans_type,
                                  amount=trans_amount,
                                  currency=trans_currency,
                                  description=trans_description,
                                  date_time=trans_date_time)

        # Add the newly created transaction to the list of transactions
        self.__transactions.append(transaction)
        return transaction

    def get_total_amount(self):
        total = 0
        # For each transaction, read the amount and then add it to the total
        for transaction in self.__transactions:
            total += transaction.get_trans_amount()
        return total

    # Print the transactions
    def print_transactions(self):
        for transaction in self.__transactions:
            print(transaction)

```

```
    # Convert transactions into a dictionary for json serialization
    def to_dict(self):
        transactions_list=[]
        manager_dict={}
        for transaction in self.__transactions:
           # Convert each transaction into a string format
           transactions_list.append(transaction.to_dict())
        manager_dict["transactions"] =transactions_list
        return manager_dict

    # Decode JSON data and convert to transaction list
    def decode(self, json__item):
        if (
            "__trans_id" in json__item and
            "__trans_type" in json__item and
            "__trans_amount" in json__item and
            "__trans_description" in json__item and
            "__trans_currency" in json__item and
            "__trans_date_time" in json__item
         ):

            transaction = Transaction(
                TransactionType[json__item["__trans_type"]],
                json__item["__trans_amount"],
                Currency[json__item["__trans_currency"]],
                json__item["__trans_description"],
                datetime.datetime.strptime(json__item["__trans_date_time"], "%Y-%m-%d %H:%M:%S")
            )
            return self.__transactions.append(transaction)
        return json__item


# Create a Transaction Manager object
trans_manager = TransactionManager()

# Create two transactions defined by the user
for counter in range(2):
    trans_manager.create_transaction()
    trans_manager.print_transactions()
    print("Balance Amount: {amount}".format(
        amount=trans_manager.get_total_amount()))

# Serialize data using Pickle for Transaction object
file_handler = open('transaction_data.pkl', 'wb')
pickle.dump(trans_manager, file_handler)
file_handler.close()

# Unpickle data using pickle to get back the Transaction object
file_handler = open('transaction_data.pkl', 'rb')
loaded_trans_manager = pickle.load(file_handler)
file_handler.close()

# Print deserialized data
print("\nDeserialized Data from Pickle:")
loaded_trans_manager.print_transactions()
print("Balance Amount: {amount}".format(amount=loaded_trans_manager.get_total_amount()))

# Serialize data using JSON
file_handler= open('transactions.json', 'w')
json.dump(loaded_trans_manager.to_dict(), file_handler)
file_handler.close()

# Create a new TransactionManager object to load data from json
loaded_trans_manager_json = TransactionManager()
# Deserialize JSON data from the file
file_handler_json = open("transactions.json", "rb")
```

```
# Decode JSON data using json.load with the custom decoder
json.load(file_handler_json, object_hook=loaded_trans_manager_json.decode)
file_handler_json.close()

# Print deserialized data from JSON
print("\nDeserialized Data from JSON:")
loaded_trans_manager_json.print_transactions()
print("Balance Amount:{amount}".format(amount=loaded_trans_manager_json.get_total_amount()))
```

## 6.6 CHAPTER SUMMARY

This chapter lays the foundation for working with external data storage in software applications. It begins with file handling, a core skill for managing text and binary files. Readers learn about various access modes for files, techniques for writing and reading text data, and even handling structured data formats like XML. This chapter then dives deeper into storing complex data structures persistently. Object serialization is introduced, explaining how objects can be converted into a format suitable for file storage. Python's Pickle module is explored in detail, demonstrating how to serialize (Pickle) objects to files and deserialize (unpickle) them back into memory when needed. This chapter also explores data persistence using JSON, a popular human-readable format for data exchange across systems.

## 6.7 EXERCISES

### 6.7.1 Test Your Knowledge

1. Compare how Python handles text files and binary files.
2. Why is it necessary to store data externally, such as in files or databases, for programs requiring data persistence?
3. What is the significance of access modes in file handling? Explain the different access modes provided by Python.
4. Describe the impact of the "\n", "\t", and "\r" characters when reading a file in Python. How are they handled in the output?
5. Explain the difference between the "write()" and "writelines()" functions in Python file handling.
6. What are the benefits of using XML files for storing data?
7. Compare and contrast the different XML parsing approaches in Python.
8. Explain the purpose of object serialization in Python.
9. Describe the steps involved in pickling and unpickling an object to a file using Python's Pickle module.
10. With an example, explain your understanding of JSONDecoder.

### 6.7.2 Multiple Choice Questions

1. What is the access mode "r+" used for file handling in Python?
   a. Read only
   b. Write only
   c. Append only
   d. Read and Write

2. Which of the following access modes creates the file if it does not exist?
   a. "r"
   b. "w"
   c. "a+"
   d. 'w+"
3. What does the Python program do if the file specified in the "open()" method does not exist when using access mode "r"?
   a. Raises a "FileNotFoundError"
   b. Creates a new file
   c. Opens the file for reading
   d. Overwrites the existing file
4. Which Python module is used for reading CSV files?
   a. OS
   b. CSV
   c. SYS
   d. JSON
5. What does the "csv.reader()" method do when reading CSV files?
   a. Reads the entire CSV file at once
   b. Reads a specific line in the CSV file
   c. Reads each line in the CSV file as a separate element in a list
   d. Reads n lines from the CSV file
6. Which XML module in Python is part of the standard library and facilitates parsing of XML documents following the DOM structure?
   a. BeautifulSoup
   b. MiniDOM
   c. ElementTree
   d. SAX
7. What does the XSD primarily provide for an XML document?
   a. A set of XML elements to be used
   b. Instructions for system processing
   c. Template outlining requirements for data formatting
   d. Namespaces for XML elements
8. What is the disadvantage of using Python's Pickle module for serialization?
   a. It cannot serialize basic data types.
   b. It is slower compared to other formats.
   c. It is not Python-specific.
   d. It cannot distinguish between malicious and non-malicious callable.
9. Why might one prefer JSON over Pickle for object serialization in Python?
   a. JSON is faster.
   b. JSON is Python-specific.
   c. JSON is a universal standard.
   d. JSON cannot serialize complex data structures.

### 6.7.3 Short Answer Questions

1. Explain data persistence.
2. For which two types of files does Python provide built-in methods to handle?
3. Which file access mode positions the handle at the end of the file?
4. With an example, explain how Python reads and writes a whole text file.

5. Which libraries are included within the Python standard library that parses XML documents?
6. How do we ensure XML documents are well-formed and valid?
7. Which Python library can process XML files but requires external parsers?
8. What is the name of the process of converting the state of an object into a form that can be persisted?
9. Write the open() method for writing to a text file and a binary file. Explain how the methods work.
10. What is the name of the method in the JSON module that can help with the deserialization of files?

### 6.7.4 True or False Questions

1. Data stored in RAM is volatile and gets erased after a program finishes its execution.
2. The "w" access mode in file handling overwrites the existing content of the file with new data.
3. Element Tree becomes slow with large amounts of data and is not suitable for processing large XML documents.
4. Reading from text files involves opening the file, reading its contents, and manipulating or using that data within a program.
5. XML files must have a root element, and XML tags are not case-sensitive.
6. SAX is suitable for processing large XML documents because it loads the entire content into memory.
7. Pickle is a Python-specific module, and its serialized objects can be easily deserialized using other programming languages.
8. JSON is a language-independent, cross-platform text format commonly used to exchange data between web servers and clients.
9. Binary files are suitable for storing and accessing complex data structures, such as dictionaries with nested attributes.
10. The "json.loads()" method in Python is used to deserialize JSON data stored in a file.

### 6.7.5 Fill in the Blanks

1. Fill in the blanks to open a file in read and write mode "r+" with the name "records.txt".

```
file_record = "records.txt"
access_mode = 'r+'
file_handler = open(________, ______)
```

2. Fill in the blanks to write a list of strings to a file named "output.txt" using "writelines()" method.

```
output_file = open(_____, _____)
string_list = ["Line 1", "Line 2", "Line 3"]
output_file._____(_____)
output_file.close()
```

3. Complete the code to open a file named "log.txt" in append mode "a" and write a log entry using the "write()" method.

```
log_file = open(_____, _____)
log_entry = "Error: Connection Timeout"
log_file._____(_____)
log_file.close()
```

4. Complete the code to open an XML file named "users.xml" using the "ElementTree" module:

```
import xml.etree.ElementTree as ET
file_name = 'users.xml'
tree = ET._______(____________)
```

5. Fill in the blanks to deserialize an object from a file named "data.pkl" using Python's Pickle module:

```
import pickle
file_handler = open('data.pkl', '___')
deserialized_object = _________._____(file_handler)
file_handler.close()
```

6. Fill in the blanks to deserialize JSON data from a string

```
import json
json_string = '{"key": "value"}'
deserialized_data = json._____(json_string)
print(deserialized_data)
```

### 6.7.6 Coding Problems

1. Write a Python program that reads a text file and performs the following tasks:
   a. Print the number of characters in the file.
   b. Calculate and display the number of words in the file.
   c. Implement error handling to gracefully handle cases where the file does not exist, or there are issues with file reading.
2. Develop a Python program to support the university registrar officer in efficiently managing and storing student details in a CSV file. The program should include the following features:
   a. Allow the registrar to enter student details (name, phone number, student ID, date of birth, and major) into the CSV file.
   b. Accept continuous entry of student details until the registrar chooses to exit.
   c. Implement a search feature that enables the registrar to find a student using their student ID.

   d. Display the contents of the CSV file, including an option
      - View all student details.
      - Filter the display based on specific criteria (e.g., display all students from a particular major).
3. Consider the XML file below, which represents a collection of books. Each book has attributes such as title, author, and publication year.

```
<library>
    <book>
        <title>Culinary Magic Of The Emirates</title>
        <author>Alexandra Vohn Hahn</author>
        <year>2019</year>
    </book>
    <book>
        <title>The Indian Cookery Course</title>
        <author>Monisha Bharadwaj</author>
        <year>2016</year>
    </book>
    <book>
        <title>The Mediterranean Dish</title>
        <author> Suzy Karadsheh </author>
        <year>2022</year>
    </book>
</library>
```

   a. Create a Python code using the SAX module to parse the given XML file and extract information about each book.
   b. Create another Python code to parse this XML file using "ElementTree".
4. Consider the dictionary given below.

```
myscores = {
    'Courses': ['IDS100', 'IDS200', 'ICS300', 'ICS220'],
    'Assignments': {
        'IDS100': [95, 88, 92],
        'IDS200': [80, 75, 88],
        'ICS300': [85, 90, 82],
        'ICS220': [96, 94, 88]
        }
    'Scores': [87, 76, 86, 98],
}
```

   a. Write a Python program to Pickle the dictionary into a file named "myscores.dat".
   b. Add to your program code to unpickle the dictionary from the file "myscores.dat" created in the previous question and print the total of all the scores.
5. You are to develop a Python program to parse the XML file below, which contains tourism data for various destinations in the UAE.

```
<tourism>
    <destination>
        <name>Burj Khalifa</name>
        <location>Dubai</location>
        <description>The iconic Burj Khalifa, the tallest building in the world, offers
        breathtaking views of Dubai's skyline.</description>
    </destination>
```

```
    <destination>
        <name>Sheikh Zayed Grand Mosque</name>
        <location>Abu Dhabi</location>
        <description>The Sheikh Zayed Grand Mosque is a stunning architectural marvel and a key
        landmark in Abu Dhabi.</description>
    </destination>
    <destination>
        <name>Al Noor Island</name>
        <location>Sharjah</location>
        <description>Experience the beauty of Al Noor Island, a unique destination in Sharjah
        featuring art, nature, and stunning landscapes.</description>
    </destination>
</tourism>
```

a. Create a Python class named "Destination" to represent a tourist destination. The class attributes would be: "name", "location", and "description".
b. Use the "ElementTree" module to parse the provided XML file and extract information about each destination.
c. Create objects of the "Destination" class for each destination in the XML file and store the instantiated objects in a suitable data structure.
d. Write a method to display the information for each destination with its name, location, and description. Then, call this method to print the details of all destinations.

6. Consider the problem from previous chapters where you were asked to develop a program using object-oriented programming to serve as an Emirati food ordering system. The program asked to present the Emirati food menu along with corresponding prices. Subsequently, it allowed the user to choose items from the menu, generated an invoice detailing the selected items along with their prices and descriptions, and provided the total amount payable. The program contained different classes such as "Menu", "FoodItem", "Order", and "Invoice" classes with corresponding attributes.
   a. Order history storage and retrieval:
      - After the user places an order and the invoice is generated, save the order details (order instances with selected items, prices, and total amount) to a binary file using the "pickle" module. This ensures persistent storage of order history.
      - Implement a method to retrieve and display previous orders from the Pickle file, enabling users to view their order history.
   b. Menu Management with JSON storage:
      - Extend the program to include the option to save and load the menu information (items, prices, and descriptions) to and from a JSON file.
      - Implement a method to update the menu, allowing the addition or modification of food items and their details. Save the updated menu to the JSON file for future use.
   c. For tasks a and b, ensure you utilize "try" and "except" blocks to handle any file-related errors. This includes handling scenarios where files might not exist or cannot be read or written to.
7. Consider the below XML file containing information about a company and its employees:

```
<company>
    <employee>
        <id>101</id>
        <name>John Doe</name>
        <position>Software Engineer</position>
        <department>Engineering</department>
    </employee>
```

```
    <employee>
        <id>102</id>
        <name>Alice Smith</name>
        <position>Marketing Specialist</position>
        <department>Marketing</department>
    </employee>
</company>
```

a. Create a Python class named "Employee" to represent an employee in a company. The class attributes can include "ID", "name", "position" and "salaries".
b. Create a method to display the parsed information, then call it to display the employee's details.
c. Use the "ElementTree" module to parse the provided XML file, create an "Employee" class object for each entry, and assign the corresponding information from the parsed XML to its attributes.
d. Extend the program to serialize these "Employee" objects using the "pickle" module to save them to a binary file.
e. Ensure proper exception handling for file operations and parsing errors.

# 7 Graphical User Interface with Tkinter

## 7.1 GRAPHICAL USER INTERFACE (GUI)

In the early days of computer programs, interaction was limited to text-based input. This meant that only a select few computer specialists with knowledge of writing computer commands could interact with computer systems through a terminal. However, with the advent of GUIs, the landscape of software interaction changed dramatically. Users could now interact with software by simply clicking on visual elements, such as icons, menus, buttons, and windows. This shift in interaction style opened up the use of computers to a broader category of non-technical users from various fields. It allowed them to harness the capability of computing systems in creating, modifying, and sharing data. This virtual interaction provided an intuitive, user-friendly means of navigating and manipulating a computer, making it feasible for general-purpose applications, such as word processing, graphic design, and web browsing.

Python offers Tkinter, a built-in library commonly used for creating standalone GUIs. Tkinter provides different types of widgets, which are building blocks of user interface design. These widgets are the visual, interactive components one sees in everyday applications – text entry boxes, drop-down combo boxes, selection boxes, buttons, and more. Each widget has specific functionality; for example, a button is clicked to trigger an action, and a text box is used to provide information. Widgets are not just visual elements but powerful objects that can interact with the program's core functionalities. Creating a GUI involves arranging these widgets on a main window. This main window acts like a container, holding all the other widgets. Interestingly, unlike other widgets, this main window doesn't need a parent; it sits at the top of the hierarchy, like the foundation upon which the entire user interface is built. Combining widgets into a user-friendly interface and integrating it with business logic and long-term file storage results in an end-to-end interactive and engaging Python program for users.

## 7.2 THE CONTAINERS

Containers are elements used in GUIs to organize and structure other visual components. They hold and arrange numerous GUI elements, such as buttons, text fields, menus, images, and other containers, to produce a clear and orderly layout. Containers have certain important aspects to ensure an aesthetically pleasing and intuitive interface. One such aspect is to position and resize child components placed within the container's area, which ensures a consistent and responsive layout for various screen sizes. Another aspect is to manage the visibility and behavior of their child components. They can function as a single entity to display or hide child elements or enable and disable them. Containers can group related items or create a hierarchy of child components. Some containers do not have a visual appearance and only exist to organize and control their child components.

Windows, frames, panels, and dialogs are the most frequently used GUI containers. Windows are top-level containers with adjustable borders and title bars. They often contain additional UI components. Frames are like windows but commonly group related components with borders and captions inside the main window. In contrast, panels are rectangular containers within windows that can hold and group numerous items. Dialogs are windows placed on top of the main window and are frequently used for alerts, notifications, confirmation messages, or error messages. Dialogs can be modal, meaning they appear on top of the main window and require an interaction with the user,

DOI: 10.1201/9781032668321-7

hindering the user from returning to the main window, whereas non-modal dialog boxes can be minimized or closed without interaction. With the help of containers, structured, organized, and user-friendly GUIs can be developed that are simple to navigate and understand.

### 7.2.1 Window

The Tkinter window is the container of all GUI elements (widgets); it forms the foundation for the entire GUI application. The window can be customized with different dimensions and other properties. An example of a basic window is provided in Code 7.1.

**CODE 7.1 SIMPLE TKINTER GUI WINDOW**

```
1. import tkinter as tk
2.
3. class MyGUI:
4.     """Class to create simple GUI window"""
5.
6.     # Constructor
7.     def __init__(self):
8.         # Create the main window
9.         self.main_window = tk.Tk()
10.
11.        # Main loop to display window
12.        self.main_window.mainloop()
13.
14.
15. # Create an instance of the MyGUI class.
16. my_gui = MyGUI()
```

In Code 7.1, the Python "tkinter" GUI library is imported at Line 1. A new class to create the simple GUI window named "MyGUI" is defined in Line 3. Lines 7–12 represent the class constructor that sets up the initial state of the "MyGUI" class's instance. The "main_window" refers to the window created by initializing the Tkinter class's instance at Line 9. The main event loop for this window is initiated by the method "mainloop()", which displays the window and waits for events to happen, such as the user clicking buttons. This main loop keeps the GUI window open and responsive to user interactions. Line 16 creates an instance of the "MyGUI" class, which automatically triggers the execution of the constructor "__init__()" method. This, in turn, creates the main window and starts its "mainloop()" method, displaying the GUI. The "main window" created by this code is displayed in Figure 7.1.

**FIGURE 7.1** A simple GUI window.

### 7.2.2 Frame

The Python class Frame represents a rectangular area on the screen that can be added as a child component to other frames or top-level windows. The purpose of the Frame is to contain other widgets. The "highlightthickness" property of a frame defaults to 0, which means that the frame has no border. If "highlightthickness" is assigned the value "2", the frame outline will be visible.

Code 7.2 demonstrates the creation of a GUI with two frames using the Tkinter library in Python. The Tkinter library is imported in Line 1. Line 2 introduces a new class called "MyFrames", which is designed to showcase the use of Frames in Windows.

The constructor defined in Line 7 initializes the "MyFrames" class instance by creating the main window using the Tkinter instance in Line 9. The main window is then given the title "My window" in Line 10. Lines 13–18 create two frames within the main window: "top_frame" and "bottom_frame", each with a border highlighted in blue and red, respectively. The frames are arranged and positioned within the main window with specified padding in Lines 20 and 21. Lines 24–27 create labels ("top_label" and "bottom_label") using the Tkinter Label class. These labels are added to the respective frames, giving the frame as the first parameter for the Label instance. Line 29 starts the Tkinter main event loop to display the GUI. Finally, in Line 33, an object of the class "MyFrames" is created to display the GUI.

**CODE 7.2 TKINTER GUI WITH TWO FRAMES**

```
1. import tkinter as tk
2.
3.
4. class MyFrames:
5.     """Class to demonstrate the use of the Frames in Windows"""
6.
7.     def __init__(self):
8.         # Create the main Window
9.         self.main_window = tk.Tk()
10.        self.main_window.title("My Window")
11.
12.        # Create two frames for the window
13.        self.top_frame = tk.Frame(
14.            self.main_window, highlightthickness=2, highlightbackground="blue")
15.
16.        self.bottom_frame = tk.Frame(
17.        self.main_window, highlightthickness=2, highlightbackground="red")
18.
19.        # Pack the two frames into the window
20.        self.top_frame.pack(padx=10, pady=10)
21.        self.bottom_frame.pack(padx=10, pady=10)
22.
23.        # Create a label in each frame and pack it in
24.        top_label = tk.Label(self.top_frame, text="Frame 1", width=70, height=20)
25.        top_label.pack()
26.        bottom_label = tk.Label(self.bottom_frame, text="Frame 2", width=70, height=20)
27.        bottom_label.pack()
28.
29.        self.main_window.mainloop()
30.
31.
32. # Create an object to show the GUI
33. show_window = MyFrames()
```

Figure 7.2 shows the created frames within "My Window" with "Frame 1" at the top with a blue border and "Frame 2" at the bottom with a red border.

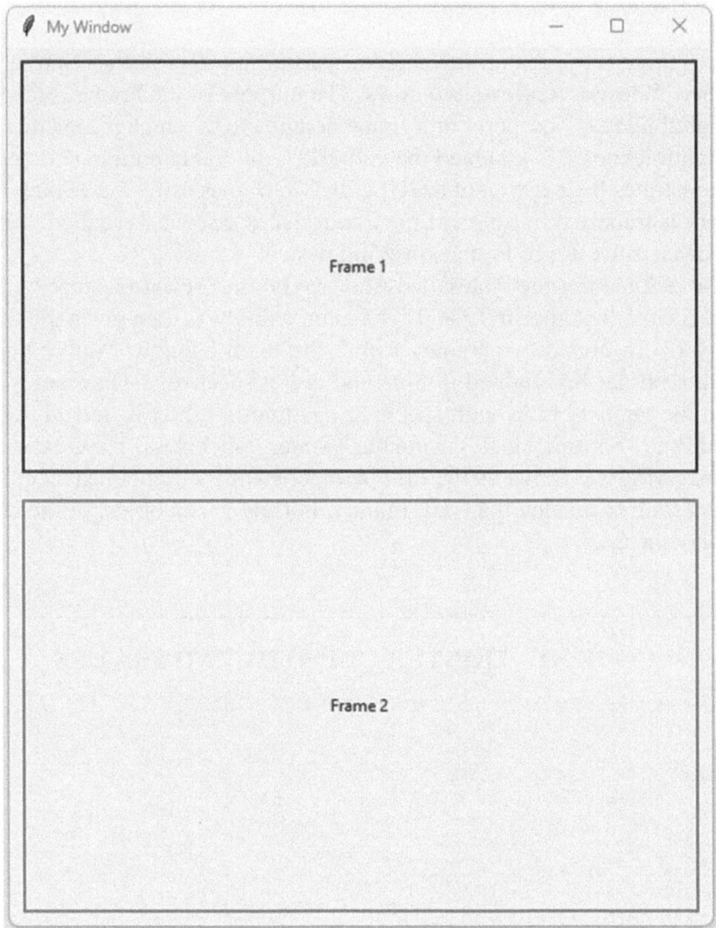


**FIGURE 7.2** Two frames in a window.

## 7.3 THE WIDGETS

Tkinter facilitates a user-friendly interface with the help of a variety of widgets for visual interaction. These elements are the building blocks of interface design and perform specific roles. For example, labels display text, an entry field for entering single-line information, a list box to display a list to select from, and a button to trigger actions/methods. Combining these widgets allows one to design user-friendly interfaces that cater to different user needs and make Python programs interactive.

### 7.3.1 Label

A label is a widget that displays text or images that can be viewed without user interaction. Labels are typically used to guide the user in identifying controls or other parts of the user interface. Code 7.3 shows an example of a simple Label in Tkinter. It is an enhancement of Code 7.1, where the class "MyGUI" was created to facilitate the creation of a simple window. In Code 7.3, the window is given the title of "My window" in Line 10. Line 13 is added, which creates a label widget using the Tkinter "Label" class. The label is added to the "main_window", given as the first parameter, and the second parameter text is set to "My first Label". This text is displayed on this label. Additionally, Line 16 automatically organizes the label widget within the main window using the "pack()" method.

**CODE 7.3 SIMPLE LABEL WIDGET IN TKINTER**

```
import tkinter as tk

class MyGUI:
    """Class to create a simple GUI window with a label window"""

    # Constructor
    def __init__(self):
        # Create the main window
        self.main_window = tk.Tk()
        self.main_window.title("My Window")

        # Create a label widget
        self.label=tk.Label(self.main_window, text="My first Label")

        # Pack the label widget into the main window
        self.label.pack()

        # Main loop to display window
        self.main_window.mainloop()


# Create an instance of the MyLabel class.
my_gui = MyGUI()
```

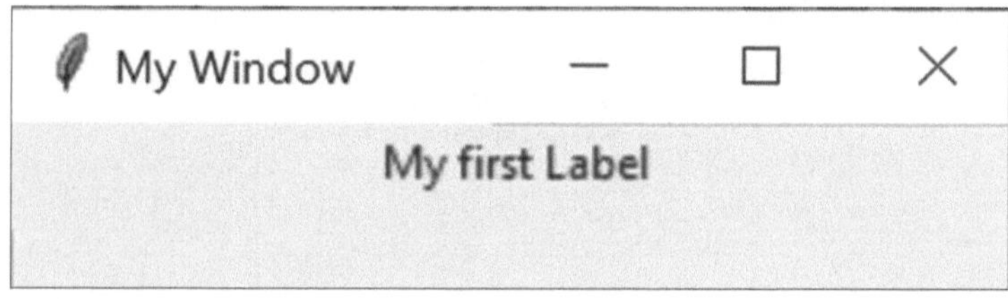


**FIGURE 7.3** A label widget in a window.

Figure 7.3 shows the GUI output of Code 7.3. The "My Window" window contains a label widget with the text "My first Label". The label widget is packed into the main window.

### 7.3.2 Entry

An entry widget gives users a single-line text field to type in a string value. Code 7.4 demonstrates the use of the "Entry" widget, building upon the previous example in Code 7.3, where the simple label was created within the main Window. The label text is updated to "*What is your name?*" in Line 13. Additionally, Line 17 introduces an "Entry" widget with a border thickness of "3" pixels. This entry widget is automatically positioned within the main window using "pack()" on Line 20.

**CODE 7.4 SIMPLE LABEL AND ENTRY WIDGETS IN TKINTER**

```
1. import tkinter as tk
2.
3. class MyGUI:
4.     """Class to create a simple GUI window with a label and Entry in a window. """
5.
```

```
    # Constructor
    def __init__(self):
        # Create the main window
        self.main_window = tk.Tk()
        self.main_window.title("My Window")

        # Create a label widget and pack it into the main window
        self.label=tk.Label(self.main_window, text="What is your name?")
        self.label.pack()

        # Create an entry widget with a border thickness of 3
        self.entry = tk.Entry(self.main_window, bd=3)

        # Pack the entry into the main window
        self.entry.pack()

        # Main loop to display window
        self.main_window.mainloop()


# Create an instance of the MyLabel class.
my_gui = MyGUI()
```

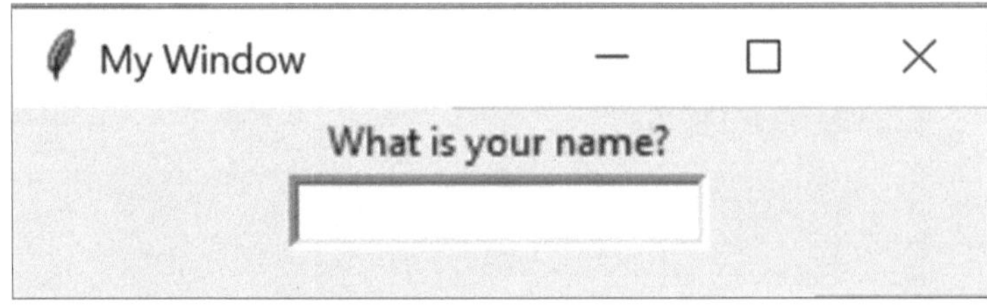


**FIGURE 7.4** A label and an entry widget in a window.

Figure 7.4 shows the GUI output of Code 7.4. The window titled "My Window" contains a label widget and an entry widget. The label packed into the main window has the text "*What is your name?*" The user is expected to interact with the GUI to answer the question in the entry widget.

### 7.3.3 Text

The text widget allows the user to edit multiline text. Its display can be formatted by changing its font style or color. Code 7.5 illustrates the creation of a feedback form using Tkinter, showcasing a text widget alongside labels and an entry widget. The code defines a new class named "FeedbackForm" on Line 3 to encapsulate the form's functionality. Lines 8 and 9 create a Tkinter main window titled "*Feedback Form*". At the top of the window, a centered label with the text "Feedback Form" is created that represents the form's title (Lines 12 and 13). Another label and an entry widget are created to prompt the user to enter their name (Lines 16–21). They are positioned to the left, with the property "anchor" set to "w" to mean west (Lines 21 and 25). A third label, with the text "Enter your Feedback": in Lines 24 and 25, is added to guide users to enter feedback. The text widget is introduced in Lines 28 and 29, which allows users to enter their feedback in several lines. Specifications for the text's width, height (Line 28), and padding (Line 29) are provided. Line 32 starts the main loop to keep the window open. An instance of the "FeedbackForm" is created in Line 36.

### CODE 7.5 TEXT WIDGET IN TKINTER IN A FEEDBACK FORM

```
import tkinter as tk

class FeedbackForm:
    """Class to create a feedback form GUI"""

    def __init__(self):
        # Create the main window
        self.main_window = tk.Tk()
        self.main_window.title("Feedback Form")

        # Create a label for the form title (centered)
        title_label = tk.Label(self.main_window, text="Feedback Form")
        title_label.pack(pady=(10, 0))  # Centered with 10 pixels padding at the top

        # Create a label for the name entry
        name_label = tk.Label(self.main_window, text="Enter your name:")
        name_label.pack(anchor='w', padx=10)  # Label's positioned left with padding

        # Create an entry widget for the user's name
        self.name_entry = tk.Entry(self.main_window, width=30)
        self.name_entry.pack(anchor='w', padx=10)  # Entry's positioned left with padding

        # Create a label for the feedback entry
        feedback_label = tk.Label(self.main_window, text="Enter your feedback:")
        feedback_label.pack(anchor='w', padx=10)  # Label's positioned left with padding

        # Create a text widget for the user's feedback
        self.feedback_text = tk.Text(self.main_window, width=50, height=10)
        self.feedback_text.pack(anchor='w', padx=10)  # Text's positioned left with padding

        # Main loop to display window
        self.main_window.mainloop()


# Create an instance of the FeedbackForm class
feedback_form = FeedbackForm()
```

Figure 7.5 shows the GUI output of Code 7.5 displaying the Feedback Form window. It uses a label to display the heading "Feedback form". There is an entry widget with the label "*Enter your name*" and a text widget with the label "*Enter your feedback*".

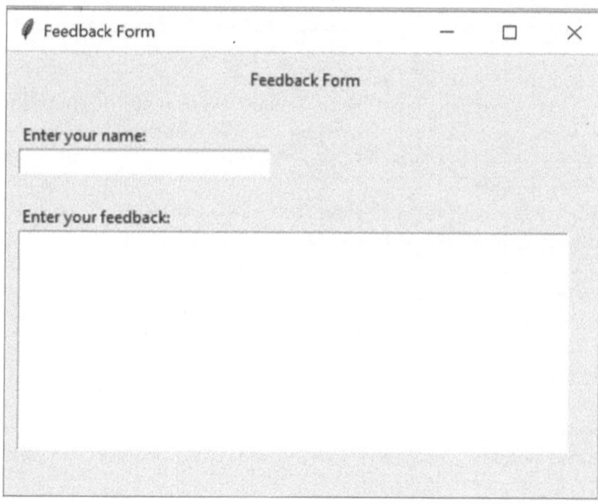


**FIGURE 7.5** Using text widget in Tkinter for feedback form GUI.

### 7.3.4 Button

Unlike a frame or label, a button allows user interaction, i.e., the user presses the button to trigger an action. Like labels, a button can display text or images. Code 7.6 represents a modification of Code 7.5, with an additional "Button" widget to facilitate the form submission. The button with the text "Submit" is introduced in Lines 32 and 33, and clicking it triggers the execution of the command/method, "submit_feedback()". This method, defined in Lines 38–45, retrieves the user's input from the entry widget and stores it in a variable called "name". Similarly, the feedback entered by the user in the text widget is captured and stored in a variable named "feedback"(Line 41). The parameter "1.0" in Line 41 indicates the position where the text selection starts, beginning from Line 1 entered in the text widget and the first character (at index 0). The parameter "end-1c" indicates where the text selection ends, stopping just before the last character of the text widget's contents, excluding the newline character typically present at the end of the text in the text widget. These values are then printed in the code output.

**CODE 7.6 A FEEDBACK FORM WITH A TKINTER'S BUTTON WIDGET**

```
1. import tkinter as tk
2.
3. class FeedbackForm:
4.     """Class to create a feedback form GUI"""
5.
6.     def __init__(self):
7.         # Create the main window
8.         self.main_window = tk.Tk()
9.         self.main_window.title("Feedback Form")
10.
11.        # Create a label for the form title (centered)
12.        title_label = tk.Label(self.main_window, text="Feedback Form")
13.        title_label.pack(pady=(10, 0))  # Centered with 10 pixels padding at the top
14.
15.        # Create a label for the name entry
16.        name_label = tk.Label(self.main_window, text="Enter your name:")
17.        name_label.pack(anchor='w', padx=10)  # Label's positioned left with padding
18.
19.        # Create an entry widget for the user's name
20.        self.name_entry = tk.Entry(self.main_window, width=30)
21.        self.name_entry.pack(anchor='w', padx=10)  # Entry's positioned left with padding
22.
23.        # Create a label for the feedback entry
24.        feedback_label = tk.Label(self.main_window, text="Enter your feedback:")
25.        feedback_label.pack(anchor='w', padx=10)  # Label's positioned left with padding
26.
27.        # Create a text widget for the user's feedback
28.        self.feedback_text = tk.Text(self.main_window, width=50, height=10)
29.        self.feedback_text.pack(anchor='w', padx=10)  # Text's positioned left with padding
30.
31.        # Create a button widget
32.        submit_button = tk.Button(self.main_window, text="Submit", command=self.submit_feedback)
33.        submit_button.pack(pady=10)  # Add some padding below the button
34.
35.        # Main loop to display window
36.        self.main_window.mainloop()
37.
38.    def submit_feedback(self):
39.        # Method to handle button click event
40.        name = self.name_entry.get()
41.        feedback = self.feedback_text.get("1.0", "end-1c")  # Retrieve text widget's text
42.        print("Name:", name)
43.        print("Feedback:", feedback)
```

```
        # Main loop to display window
        self.main_window.mainloop()


# Create an instance of the FeedbackForm class
feedback_form = FeedbackForm()
```

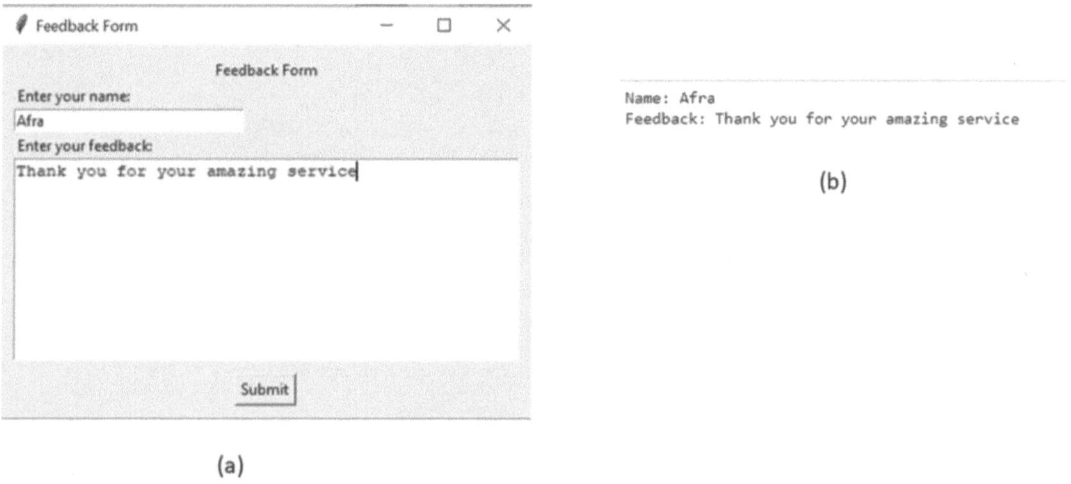

(a)

(b)

**FIGURE 7.6** Feedback form with button widget and the output. (a) Feedback form with a Button widget, (b) The form's expected output when the Submit button is pressed.

Figure 7.6 represents the GUI output of Code 7.6 and displays the Feedback Form window with the button widget. In Figure 7.6 (a), the form with a sample user input in the entry widget and text widget is shown, and how they are saved and printed after the user clicks the button widget is shown in Figure 7.6 (b).

### 7.3.5 Checkbutton

A checkbutton widget requires user selection and holds a binary value, i.e., checked or unchecked. When pressed, a checkbutton flips the toggle and invokes a callback method. The "Checkbutton" widgets are frequently used to allow users to turn an option on or off. They are ideally used in situations where users must select from a list of independent options that are not mutually exclusive.

Code 7.7 demonstrates the use of the checkbutton widget in a simple form where users can select a checkbox to indicate "yes" or clear it to indicate "no". The code creates a window to contain all the widgets (Lines 8 and 9). A label prompts the user with the text "*Which of the following statements do you agree with?*" (Lines 12 and 13). The "BooleanVar()" method (in Lines 17 and 24) creates special Tkinter variables associated with "Checkbutton" widgets, allowing the state of the buttons to be tracked (True for checked, False for unchecked). Two checkbuttons are created to get the user a yes/no selection. The first checkbutton widget, "option0_button" prompts the user to select an option in response to the question, "*Is it easy to create a user interface with Tkinter?*" (Lines 18–21). The option text is added to the text attribute of the checkbutton widget. The second checkbutton widget allows selecting, "*Tkinter provides a variety of widgets for UI design*" (Lines 24–28). This option text is also added to the text attribute of the widget.

The code also introduces the method "submit_response()" (Lines 37–47) to handle the output when the "Submit" button widget is clicked (Lines 31 and 32), saving the user's options provided in both checkbuttons.

## CODE 7.7 A SIMPLE FORM WITH A TKINTER'S CHECKBUTTON WIDGET

```
import tkinter as tk

class CheckButtonForm:
    """Class to create a simple GUI with check buttons"""

    def __init__(self):
        # Create the main window
        self.main_window = tk.Tk()
        self.main_window.title("Check Button Form")

        # Create a label to ask a  question
        title_label = tk.Label(self.main_window,
                               text="Which of the following statements do you agree with.")
        title_label.pack(anchor='w', padx=20, pady=10)  # Position Left with 10 pixels padding

        # Create a check button for getting a Yes or No answer to an option
        self.option0_var = tk.BooleanVar()
        option0_button = tk.Checkbutton(self.main_window,
                                text="It is easy to create a UI with Tkinter?",
                                variable=self.option0_var, compound="right", padx=15)
        option0_button.pack(anchor='w', padx=20)  # Check button positioned left with padding

        # Create a check button for getting a Yes or No answer to an option
        self.option1_var = tk.BooleanVar()
        option1_button = tk.Checkbutton(self.main_window,
                                text="Tkinter provides a variety of widgets for UI design.",
                                variable=self.option1_var,compound="left", padx=15)
        option1_button.pack(anchor='w', padx=20)  # Check button positioned left with padding

        # Create a button to submit response
        submit_button = tk.Button(self.main_window,
                                text="Submit", command=self.submit_response)
        submit_button.pack(pady=10)  # Add some padding below the button

        # Main loop to display window
        self.main_window.mainloop()

    def submit_response(self):
        # Method to handle button click event
        if self.option0_var.get():
            print(f"For Option 1 User Selected: Yes")
        else:
            print(f"For Option 1 User Selected: No")

        if self.option1_var.get():
            print(f"For Option 2 User Selected: Yes")
        else:
            print(f"For Option 2 User Selected: No")


# Create an instance of the CheckButtonForm class
check_button_form = CheckButtonForm()
```

Figure 7.7 represents the output of Code 7.7. Figure 7.7 (a) displays the simple form window, with two checkbutton widgets with their corresponding options. Once checked, the user clicks the "Submit" button to save the answers and print the selections on the console. The user can select either of the two checkbuttons or both. The console output is displayed in Figure 7.7(b), where the first option was selected.

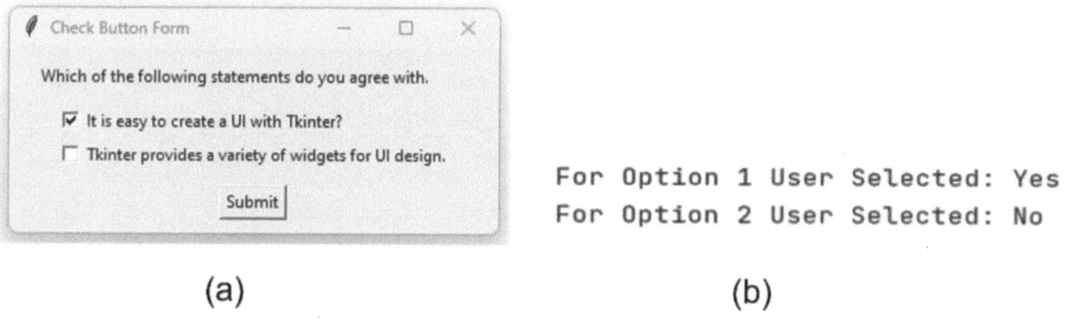


**FIGURE 7.7** A simple form with checkbutton widgets and output. (a) A simple form with Checkbutton widgets, (b) The form's output when selecting one choice and pressing the Submit button.

### 7.3.6 RADIOBUTTON

A radiobutton widget requires user interaction to select one option from multiple choices. Radiobuttons are always used together in a set, where multiple radiobutton widgets are tied to enable the user to make a single choice or preference.

Code 7.8 illustrates the usage of the "Radiobutton" widget in a simple GUI application. The code creates a window for all widgets (Lines 8 and 9). A label prompts the user, "*Which Tkinter widget is easier to use?*" (Lines 12–14). The user is prompted to select among three options: "Button", "Label", and "Entry". Each of these widget options is represented by a "Radiobutton" with corresponding "text" and "value" attributes (Lines 20–25). These radiobuttons are aligned to the left with padding and linked to a common variable "self.widget_choice", an object of type "StringVar", a special variable in Tkinter that can hold a string (Line 17). It stores the value of the selected radiobutton. This variable is then used by the "submit_button()" method (Lines 34–37) to retrieve the selected radio button and print it to the console.

**CODE 7.8 A SIMPLE GUI WITH A TKINTER'S RADIOBUTTON WIDGET**

```
1. import tkinter as tk
2.
3. class EasyWidget:
4.     """Class to create a GUI for selectin easiest widget"""
5.
6.     def __init__(self):
7.         # Create the main window
8.         self.main_window = tk.Tk()
9.         self.main_window.title("Widget Preference")
10.
11.         # Create a label for the question
12.         question_label = tk.Label(self.main_window, text="Which Tkinter widget "
13.                                   "is easier to use?")
14.         question_label.pack(pady=10)
15.
16.         # Variable to store the selected widget
17.         self.widget_choice = tk.StringVar()
18.
19.         # Create radio buttons for widget choices
20.         tk.Radiobutton(self.main_window, text="Button", variable=self.widget_choice,
21.                        value="Button").pack(anchor='w', padx=10)
22.         tk.Radiobutton(self.main_window, text="Label", variable=self.widget_choice,
23.                        value="Label").pack(anchor='w', padx=10)
24.         tk.Radiobutton(self.main_window, text="Entry", variable=self.widget_choice,
25.                        value="Entry").pack(anchor='w', padx=10)
26.
```

```
        # Create a button to submit the selected radio button
        submit_button = tk.Button(self.main_window,
                       text="Submit",command=self.submit_choice)
        submit_button.pack(pady=10)

        # Main loop to display window
        self.main_window.mainloop()

    def submit_choice(self):
        # Method to handle button click event
        chosen_widget = self.widget_choice.get()
        print("Your selected widget:", chosen_widget)


# Create an instance of the EasyWidget class
widget_easy = EasyWidget()
```

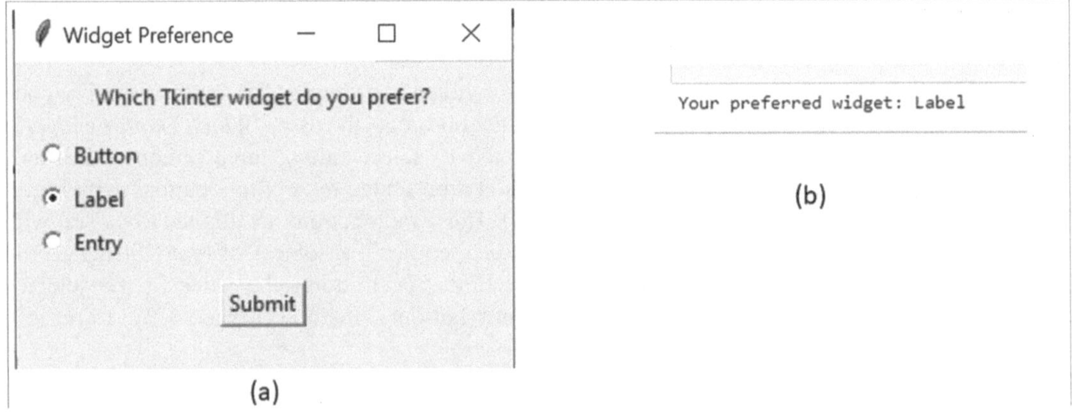


**FIGURE 7.8** A simple window with Radiobutton Widget Output. (a) A simple window with Radiobutton Widget, (b) The expected output after selecting one Radiobutton and pressing the Submit button.

The output of Code 7.8 is illustrated in Figure 7.8. Part (a) showcases a basic GUI window output featuring three radiobuttons as a list of available options. On selecting an option and then clicking the submit button, the selected choice is printed, as illustrated in Figure 7.8 (b).

### 7.3.7 Combobox

A Combobox widget combines an entry with a list of choices, allowing users to select predefined values from a drop-down list. A predefined set of choices minimizes the chance of a user mistyping or entering invalid data and ensures consistency. Moreover, selection is a faster interaction than typing when there are limited choices.

Code 7.9 represents a modification of Code 7.8 where the Radiobutton is replaced with a "Combobox" widget to create a simple GUI to prompt the user, "*Which Tkinter widget is easier to use?*". The modifications are in Lines 19–21. Note the use of the module "ttk" in Line 19, which was imported in Line 2. The module "ttk" is a module within the Tkinter library that provides access to more advanced widgets such as "Combobox" and "Treeview".

### CODE 7.9 A SIMPLE GUI WITH A TKINTER'S COMBOBOX WIDGET

```
import tkinter as tk
from tkinter import ttk


class EasyWidget:
    """Class to create a GUI for widget preference"""

    def __init__(self):
        # Create the main window
        self.main_window = tk.Tk()
        self.main_window.title("Widget Preference")

        # Create a label for the question
        question_label = tk.Label(self.main_window, text="Which Tkinter widget"
                                  " is easier to use?")
        question_label.pack(pady=10,padx=10)

        # Create a ComboBox for widget choices
        self.widget_choices = ttk.Combobox(self.main_window,
                                           values=["Button", "Label", "Entry"])
        self.widget_choices.pack(anchor='center', padx=10)

        # Create a button to submit the choice
        submit_button = tk.Button(self.main_window, text="Submit",
                      command=self.submit_choice)
        submit_button.pack(pady=10)

        # Main loop to display window
        self.main_window.mainloop()

    def submit_choice(self):
        # Method to handle button click event
        chosen_widget = self.widget_choices.get()
        print("Your preferred widget:", chosen_widget)


# Create an instance of the WidgetPreference class
widget_easy = EasyWidget()
```

Figure 7.9 (a) showcases a basic GUI window output featuring a "Combobox" widget with a drop-down menu of options. Figure 7.9 (b) displays the drop-down choices, which include button, label, and entry. Figure 7.9 (c) displays the output printed to the console when the user selects an option and clicks the submit button.

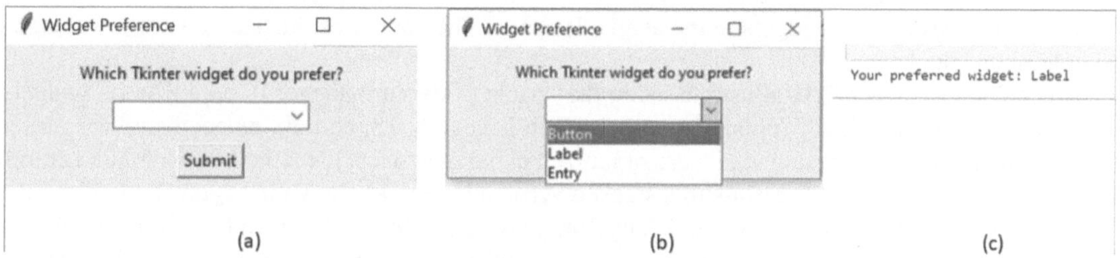


**FIGURE 7.9** A simple window with Combobox widget output. (a) A simple window with a combobox widget, (b) The drop-down choices of a combobox, (c) The output resultant after selecting one option and pressing the Submit button.

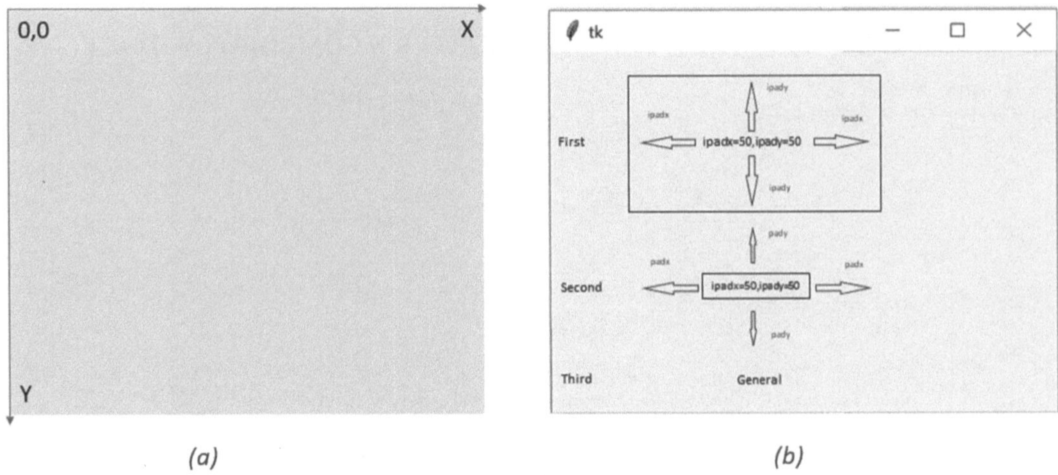


**FIGURE 7.10** GUI Window Layout. (a) GUI Window coordinate, (b) GUI Window layout showing the padding difference of padx/pady and ipadx/ipady

## 7.4 LAYOUT MANAGERS

To design an application's GUI, the organization of the widgets in a window (container) must be decided. Non-visible objects, called layout managers, are used to organize the widgets. Three built-in layout managers are available in Tkinter: the pack, the grid, and the place managers. The pack geometry manager organizes widgets in rows. The grid geometry manager places widgets in a two-dimensional grid. The place manager positions widgets using absolute positioning.

### 7.4.1 The GUI Window

As shown in Figure 7.10, the window coordinates start from the top left corner with (x, y) as (0,0), which is the origin. The x-coordinate increments from left to right, and the y-coordinate increments from top to bottom. The attributes *padx* and *pady* provide padding space outside the widget, and the attributes *ipadx* and *ipady* provide padding space inside the widget.

### 7.4.2 The Pack Layout

The pack layout manager organizes widgets using a linear approach, adding widgets on each row. The "padx" and "pady" parameters with the "pack()" method describe the number of pixels surrounding the widget to create horizontal (padx) or vertical (pady) padding between widgets. The pack layout manager has the simplest approach to creating quick prototypes, as it automatically organizes and resizes the widgets in the available space. The code uses the pack layout to manage a simple phone book.

Lines 28–35 in Code 7.10 illustrate using the "pack()" layout manager to organize the widgets. The label widgets ("name", "phone", and "city") in Lines 16, 18, and 20, respectively, are placed into the main window. These widgets are organized using the "pack()" method with default settings in Lines 28, 30, and 32. This results in a stacked vertical layout, as shown in Figure 7.11. The entry widgets ("nameBox", "phoneBox", and "cityBox") created in lines 17, 19, and 21, respectively, are placed into the window using "pack()" with padx and pady parameters at Lines 29, 31, and 32. These parameters add "12" pixels of horizontal padding and "3" pixels of vertical padding around each widget to adjust placement and create space between the widgets. Similarly, in Lines 34 and 35, "printbutton" and "clearbutton" are packed into the window using pack(), with additional "padx" and "pady" parameters to control the spacing between them. The background property of the button is set using the parameter "bg"[1] in windows systems.

### CODE 7.10 A SIMPLE GUI PHONE BOOK ORGANIZED USING PACK()

```
import tkinter as tk

class PackedGUI:
    """Class to create simple GUI window"""

    # Constructor
    def __init__(self):
        # Create the main window
        self.main_window = tk.Tk()
        self.main_window.title("Phone Book")
        # Setting the size of the screen (300 width and 300 height)
        self.main_window.geometry("300x300")

        # Create widgets for the window
        # Create Label and Entry for data
        self.namelabel = tk.Label(text="Name:")
        self.nameBox = tk.Entry()
        self.phonelabel = tk.Label(text="Phone: ")
        self.phoneBox = tk.Entry()
        self.citylabel = tk.Label(text="City: ")
        self.cityBox = tk.Entry()
        # Create buttons
        self.printbutton = tk.Button(text="Print!",
                                     fg="black", bg="white", command=self.printbox)
        self.clearbutton = tk.Button(text="Clear!", fg="white",
                                     bg="black", command=self.clearbox)
        # Insert the GUI element into the window using the pack layout manager
        self.namelabel.pack()
        self.nameBox.pack(padx=12, pady=3)
        self.phonelabel.pack()
        self.phoneBox.pack(padx=12, pady=3)
        self.citylabel.pack()
        self.cityBox.pack(padx=12, pady=3)
        self.printbutton.pack(padx=12, pady=3)
        self.clearbutton.pack(padx=12, pady=3)
        # main loop to display window
        self.main_window.mainloop()

    # Print the output
    def printbox(self):
        line = self.nameBox.get() + "," + self.phoneBox.get() + "," + self.cityBox.get()
        print(line)

    # Create clear button
    def clearbox(self):
        # Clear the entry boxes
        self.nameBox.delete(0, tk.END)
        self.phoneBox.delete(0, tk.END)
        self.cityBox.delete(0, tk.END)

gui=PackedGUI()
```

Figure 7.11 showcases a basic Phone Book GUI output featuring a "pack()" layout manager with padding to adjust the widget placement and the space between them. The label widgets are shown to be stacked vertically, and the entry widgets are placed based on the specified padding within the window. Figure 7.11 (a) shows the use of the data entered into the entry widgets, which will be printed (c) when the user clicks on the "Print!" button and cleared (b) when the user presses the "Clear!" button.

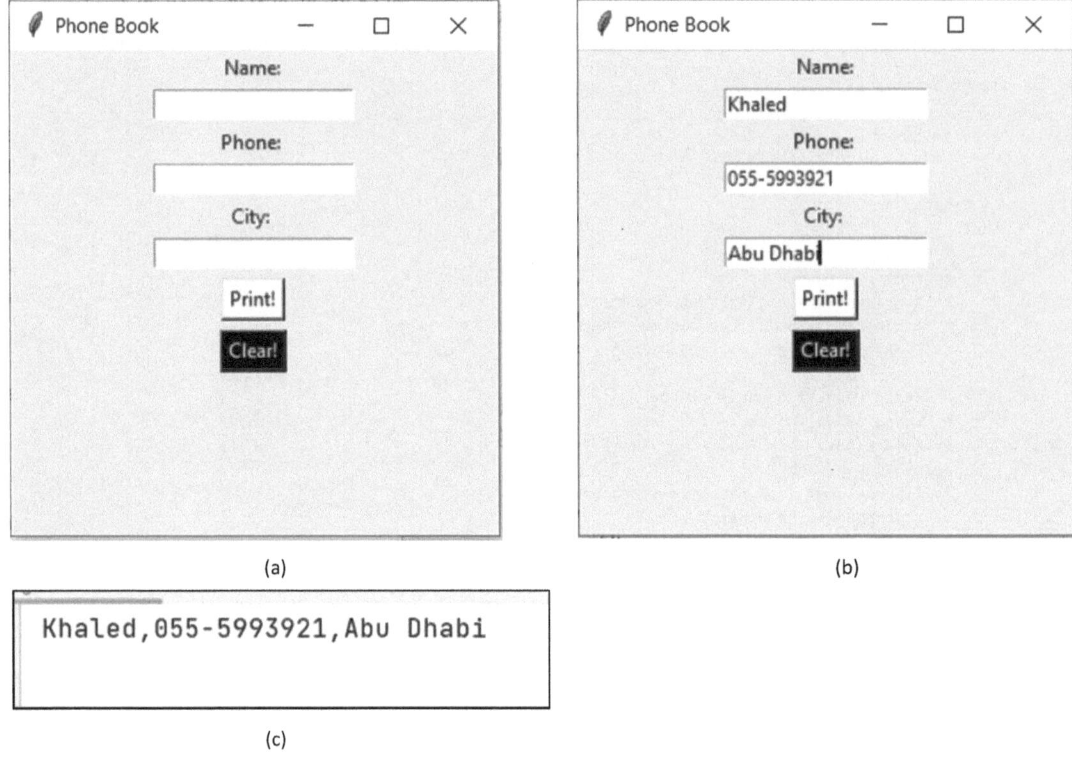


**FIGURE 7.11** A simple GUI Phone Book Output using pack(). (a) A simple GUI Phone Book Form using pack(), (b) The form filled with sample data, (c), The output when clicking the Print button.

### 7.4.3 The Grid Layout

The grid geometry manager places widgets in a two-dimensional grid. The grid's number of rows and columns is decided, and widgets are placed in specific cells. To illustrate the application of the grid layout, consider the previous phone book example modified using the grid layout.

Lines 31–38 in Code 7.11 introduce the use of the "grid()" Layout manager to arrange the widgets in a grid-like fashion within the main window. Each widget is positioned using the "grid()" method with specified row and column values. The label widgets ("name", "phone", and "city") in Lines 37, 39, and 41 are placed in rows "0", "1", and "2", respectively, but placed vertically in the same column 0. The entry widgets ("nameBox", "phoneBox", and "cityBox") in Lines 31, 33, and 35 are positioned in the corresponding rows of the labels ("0", "1", and "2") but in column 1. This column separation ensures that the widgets are aligned in columns to design an aesthetically pleasing layout. Similarly, the "printbutton" in Line 37 and "clearbutton" in Line 38 are placed in row 3, within the same column 1. However, the "sticky" parameter ensures they do not overlap in the column. The "printbutton" is placed east ("e"), and the "clearbutton" is placed west ("w") of the column. The "grid()" method facilitates more precise control over the positioning of widgets compared to the "pack()" method, enabling customization of their placement within the window.

**CODE 7.11 A SIMPLE GUI PHONE BOOK ORGANIZED USING GRID()**

```
1.import tkinter as tk
2.
3.class GridGUI:
```

```
    """Class to create simple GUI window"""

    # Constructor
    def __init__(self):
        # Create the main window
        self.main_window = tk.Tk()
        self.main_window.title("Phone Book")
        # Set the size of the screen (250 width and 200 height)
        self.main_window.geometry("250x200")

        # Create widgets for the window
        # Create Label and Entry for data
        self.namelabel = tk.Label(text="Name:")
        self.nameBox = tk.Entry()
        self.phonelabel = tk.Label(text="Phone: ")
        self.phoneBox = tk.Entry()
        self.citylabel = tk.Label(text="City: ")
        self.cityBox = tk.Entry()

        # Create a print button:
        self.printbutton = tk.Button(text="Print!", fg="black",
                                     bg="white", command=self.printbox)
        # Create clear button
        self.clearbutton = tk.Button(text="Clear!", fg="white",
                                     bg="black", command=self.clearbox)

        # Insert the GUI element into the window using the grid layout manager
        self.namelabel.grid(row=0, column=0, padx=10, pady=10)
        self.nameBox.grid(row=0, column=1, padx=10, pady=10)
        self.phonelabel.grid(row=1, column=0,padx=10, pady=10)
        self.phoneBox.grid(row=1, column=1,padx=10, pady=10)
        self.citylabel.grid(row=2, column=0,padx=10, pady=10)
        self.cityBox.grid(row=2, column=1,padx=10, pady=10)
        self.printbutton.grid(row=3, column=1,padx=10, pady=10, sticky="e")
        self.clearbutton.grid(row=3, column=1,padx=10, pady=10, sticky="w")

        # main loop to display window
        self.main_window.mainloop()

    def printbox(self):
        line = self.nameBox.get() + "," + self.phoneBox.get() + "," + self.cityBox.get()
        print(line)

    def clearbox(self):
        # Clear the entry boxes
        self.nameBox.delete(0, tk.END)
        self.phoneBox.delete(0, tk.END)
        self.cityBox.delete(0, tk.END)

# Create an instance of the MyGUI class.
gui=GridGUI()
```

Figure 7.12 presents the output of a basic Phone Book GUI organized using the "grid()" layout manager. Unlike Figure 7.11, which uses the "pack()" layout manager, this layout provides a structured grid-like arrangement for the widgets. Labels for "Name", "Phone", and "City" are placed in distinct rows alongside their respective entry fields. Upon entering data into the entry widgets (b), users can utilize the "Print!" button to display the entered information displayed in (c). The "Clear!" button enables users to reset the entry fields.

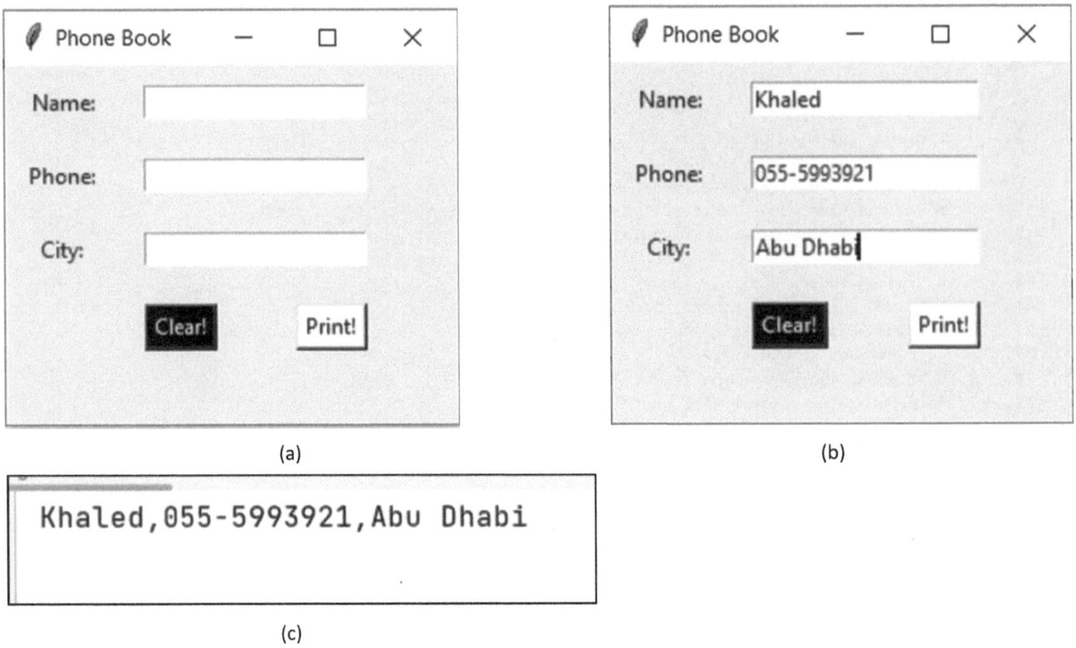

(a)

(b)

(c)

**FIGURE 7.12** A simple GUI Phone Book Output using grid(). (a) A simple GUI Phone Book Form using grid(), (b) The form filled with sample data, (c), The output when clicking the Print button.

### 7.4.4 The Place Layout

The place layout organizes widgets using absolute positioning, i.e., exact coordinates (x and y pixels) on the screen. This gives designers precise control over the widget's position on the screen. To illustrate the place layout's application, consider the previous phone book example updated to apply the place layout.

Lines 28–36 in Code 7.12 introduce the "place()" layout manager to position the widgets within the main window. Each widget is explicitly positioned using the "place()" method with specified x and y coordinates. The label widgets ("name", "phone", and "city") in Lines 29, 31, and 33 are placed at "x=0", whereas the y coordinates values are "0", "40", and "80", respectively, for the vertical arrangement of 40 pixels apart. The entry widgets ("nameBox", "phoneBox", and "cityBox") in Lines 30, 32, and 34 are positioned at "x=50" to align horizontally with their corresponding labels, and at y coordinates matching their respective labels ("0", "40", and "80", respectively). These coordinates ensure alignment between the labels and entry boxes, providing a clear and organized layout. Similarly, the "clearbutton" in Line 35 and "printbutton" in Line 36 are placed at x coordinates, "x=80" and "x=130", however, vertically at the same level at "y=120", respectively, allowing for precise control over their placement. The "place()" method offers flexibility in positioning widgets but may require manual adjustments to ensure proper alignment and spacing.

**CODE 7.12 A SIMPLE GUI PHONE BOOK ORGANIZED USING PLACE()**

```
1.import tkinter as tk
2.
3.class PlaceGUI:
4.   """Class to create simple GUI window"""
5.
6.   # Constructor
7.   def __init__(self):
```

```
        # Create the main window
        self.main_window = tk.Tk()
        self.main_window.title("Phone Book")

        # Create widgets for the window
        self.namelabel = tk.Label(text="Name:")
        self.nameBox = tk.Entry()
        self.phonelabel = tk.Label(text="Phone: ")
        self.phoneBox = tk.Entry()
        self.citylabel = tk.Label(text="City: ")
        self.cityBox = tk.Entry()

        # Create the print button: The background property for windows is "bg"
        # However, the background systems in Mac is "highlightbackground"
        self.printbutton = tk.Button(text="Print!", fg="black",
                                     bg="white", command=self.printbox)
        # Create the clear button
        self.clearbutton = tk.Button(text="Clear!", fg="white",
                                     bg="black", command=self.clearbox)

        # Insert the GUI element into the window using the place layout manager
        self.namelabel.place(x=0,  y=0)
        self.nameBox.place(x=50,  y=0)
        self.phonelabel.place(x=0, y=40)
        self.phoneBox.place(x=50, y=40)
        self.citylabel.place(x=0, y=80)
        self.cityBox.place(x=50, y=80)
        self.clearbutton.place(x=80, y=120)
        self.printbutton.place(x=130, y=120)

        # main loop to display window
        self.main_window.mainloop()

    def printbox(self):
        line = self.nameBox.get() + "," + self.phoneBox.get() + "," + self.cityBox.get()
        print(line)


    def clearbox(self):
        # Clear the entry boxes
        self.nameBox.delete(0, tk.END)
        self.phoneBox.delete(0, tk.END)
        self.cityBox.delete(0, tk.END)


# Create an instance of the MyGUI class.
gui=PlaceGUI()
```

Figure 7.13 presents the output of a basic Phone Book GUI organized using the “place()” layout manager. This layout places the different label widgets, entry widgets, and buttons based on specific x and y as indicated in the code. Figure 7.13 (a), (b), and (c) shows the entering data output, cleared data output, and displayed output, respectively.

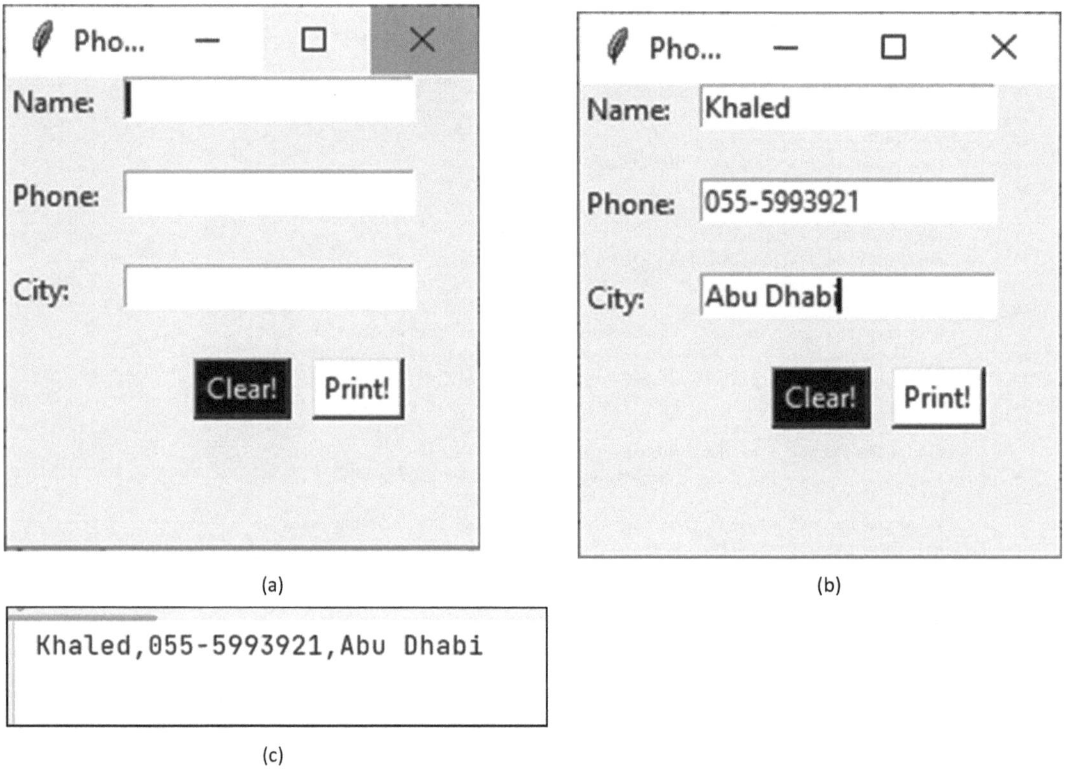

FIGURE 7.13 A simple GUI Phone Book Output using place(). (a) A simple GUI Phone Book Form using place(), (b) The form filled with sample data, (c), The output when clicking the Print button.

## 7.5 GUI APPLICATION – THE CALCULATOR

An interface for a calculator application is a good example of understanding how widgets are organized in a window. Figure 7.14 displays the GUI that facilitates basic mathematical calculations.

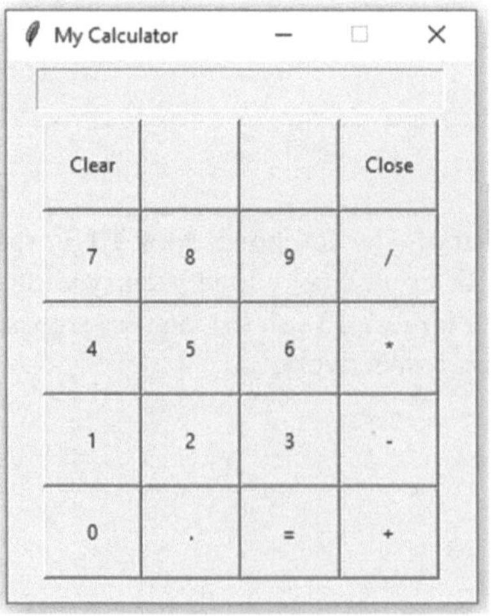

FIGURE 7.14 GUI for a calculator.

Code 7.13 uses both the "grid()" and the "pack()" layout to create a calculator. In the code, the main window contains two frames. The top frame contains only the entry widget where the numbers should be visible, and the bottom frame contains the calculator buttons. The bottom frame uses the "grid()" layout. The two frames are placed in the window using the "pack()" layout.

### CODE 7.13 A SIMPLE GUI CALCULATOR

```
import tkinter as tk

class MyCalc:
    """Class to represent calculator layout"""

    # Constructor
    def __init__(self):
        # Create the window
        self.main_window = tk.Tk()
        self.main_window.title("My Calculator")

        # Create two frames inside the window
        self.top_frame = tk.Frame(self.main_window)
        self.bottom_frame = tk.Frame(self.main_window)

        # Entry in the top frame
        self.entry = tk.Entry(self.top_frame, width=25, justify="right")
        self.entry.grid(row=0, columnspan=4, sticky="W")

        # Buttons in the bottom frame and in grid() layout
        self.cls = tk.Button(self.bottom_frame, text="Clear", width=4)
        self.cls.grid(row=1, column=0)
        self.bck = tk.Button(self.bottom_frame, text="", width=4)
        self.bck.grid(row=1, column=1)
        self.lbl = tk.Button(self.bottom_frame, text="", width=4)
        self.lbl.grid(row=1, column=2)
        self.clo = tk.Button(self.bottom_frame, text="Close")
        self.clo.grid(row=1, column=3)
        self.sev = tk.Button(self.bottom_frame, text="7", width=4)
        self.sev.grid(row=2, column=0)
        self.eig = tk.Button(self.bottom_frame, text="8", width=4)
        self.eig.grid(row=2, column=1)
        self.nin = tk.Button(self.bottom_frame, text="9", width=4)
        self.nin.grid(row=2, column=2)
        self.div = tk.Button(self.bottom_frame, text="/", width=4)
        self.div.grid(row=2, column=3)

        self.fou = tk.Button(self.bottom_frame, text="4", width=4)
        self.fou.grid(row=3, column=0)
        self.fiv = tk.Button(self.bottom_frame, text="5", width=4)
        self.fiv.grid(row=3, column=1)
        self.six = tk.Button(self.bottom_frame, text="6", width=4)
        self.six.grid(row=3, column=2)
        self.mul = tk.Button(self.bottom_frame, text="*", width=4)
        self.mul.grid(row=3, column=3)

        self.one = tk.Button(self.bottom_frame, text="1", width=4)
        self.one.grid(row=4, column=0)
        self.two = tk.Button(self.bottom_frame, text="2", width=4)
        self.two.grid(row=4, column=1)
        self.thr = tk.Button(self.bottom_frame, text="3", width=4)
        self.thr.grid(row=4, column=2)
        self.mns = tk.Button(self.bottom_frame, text="-", width=4)
        self.mns.grid(row=4, column=3)

        self.zer = tk.Button(self.bottom_frame, text="0", width=4)
```

```
57.         self.zer.grid(row=5, column=0)
58.         self.dot = tk.Button(self.bottom_frame, text=".", width=4)
59.         self.dot.grid(row=5, column=1)
60.         self.equ = tk.Button(self.bottom_frame, text="=", width=4)
61.         self.equ.grid(row=5, column=2)
62.         self.pls = tk.Button(self.bottom_frame, text="+", width=4)
63.         self.pls.grid(row=5, column=3)
64.
65.         # Pack both the frames
66.         self.top_frame.pack()
67.         self.bottom_frame.pack()
68.
69.         # Display window and keep focus
70.         self.main_window.mainloop()
71.
72.
73. # Create an object of the calculator
74. mycal = MyCalc()
```

Code 7.13 initializes the simple calculator interface using the Tkinter library in Python. It defines a class named "MyCalc" on Lines 3–70 to represent the calculator layout. Within the class constructor "__init__()" on Lines 7–70, the calculator window is created on Line 9, titled "My Calculator" (Line 10). Two frames are defined on Lines 13–14 to organize the widgets. The top frame holds an "entry" widget on Line 17 to display input/output, while the bottom frame contains buttons representing numbers, arithmetic operators, and a close button organized as shown in Figure 7.14. These widgets are arranged using the "grid()" method on Lines 18–63 to position them in a grid layout. After setting up the frames and widgets within the frames, the frames are arranged using the "pack()" method on Lines 66–67 to make them visible in the window. Finally, the "mainloop()" method is called on Line 70 to display the window and keep it active until the user closes it.

## 7.6 CASE STUDY – A GUI FOR A BANKING SYSTEM

This case study presents a banking system built using Python's Tkinter library. Users can add and view bank accounts and transfer funds between them. The application prioritizes user experience by implementing error-handling mechanisms, ensuring informative messages are displayed in case of invalid inputs or unexpected situations. In Figure 7.15, the application's landing page facilitates adding bank accounts, transferring funds, and viewing existing accounts.

Figure 7.16 shows the system architecture of the banking system. System architecture provides a high-level view of different parts of the system and how they interact. The diagram comprises three components: BankAccount, BankSystem, and BankUI. Segregating functionality into three components, model, view, and controller, is a common design pattern for developing software applications. For simplicity of understanding, the three components in this case study are individual classes. The BankAccount represents the model that should constitute the business logic related to the banking system incorporated as object-oriented design, the complete set of classes, their relationships, and functionality. The BankSystem class represents the controller, a class that serves as an intermediary between the model and the view. Finally, the BankView class represents the view responsible for presenting the model for interaction.

The BankSystem manages an aggregation of BankAccounts. In the case of a complete system, it would have access to more than one class of the model. The BankSystem also facilitates the persistence of the bank account data by saving and loading BankAccounts from a file. The BankUI class utilizes Tkinter's capabilities to create the application's interface, as depicted in Figure 7.16. This interface allows users to interact with the banking system through entry boxes, buttons, and tables.

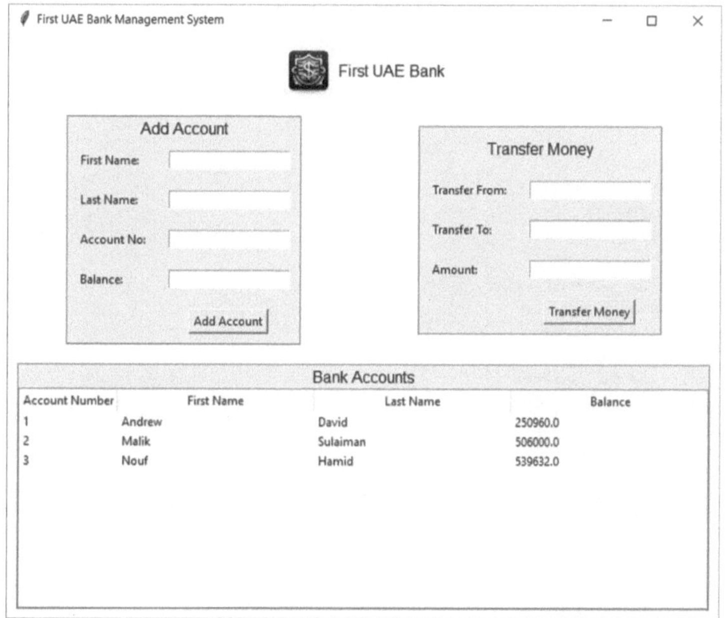


**FIGURE 7.15** The landing screen of the banking system.

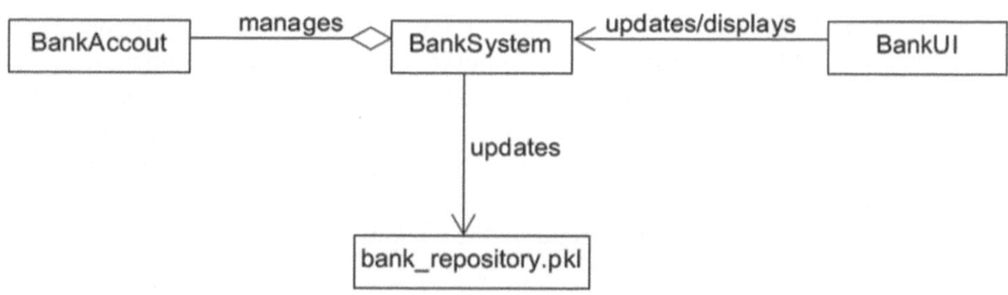


**FIGURE 7.16** System architecture of the Banking Management System.

Code 7.14 defines a "BankAccount" class to represent individual bank accounts (Lines 1–8). When creating a new account, the constructor ensures the initial balance is at least 0 (Line 7). Methods like "getFirstName()", "getLastName()", "getAccountNumber()", and "getBalance()" allow access to the account holder's information and current balance (Lines 10–17). The "withdraw()" method ensures that the withdrawal amount does not exceed the balance before updating the balance (Lines 18–21). However, if the withdrawal amount is greater, then an exception is raised (Lines 19–20). Similarly, the "deposit()" method ensures the deposit amount is positive before updating the balance (Lines 22–25). This "BankAccount" class definition is then saved as "bank_account.py", creating a module that can be imported and used in another Python script.

**CODE 7.14 THE BANKACCOUNT CLASS**

```
1. class BankAccount:
2.     """this class represents a bank account"""
3.     def __init__(self,firstName="",lastName="",accountNumber="",balance=0.0):
4.         self.__firstName=firstName
```

```
        self.__lastName=lastName
        self.__accountNumber=accountNumber
        assert balance>=0 # ensure the balance is at least 0 AED
        self.__balance=balance

    def getFirstName(self):
        return self.__firstName
    def getLastName(self):
        return self.__lastName
    def getAccountNumber(self):
        return self.__accountNumber
    def getBalance(self):
        return self.__balance
    def withdraw(self,amount):
        if amount>self.__balance:
            raise Exception("The amount is greater than the balance")
        self.__balance-=amount
    def deposit(self,amount):
        if amount<=0:
            raise Exception("Amount should be greater than zero")
        self.__balance+=amount
```

Code 7.15 defines a "BankSystem" class in Python, which manages operations for bank accounts. This class interacts with bank account data stored in files and provides methods for manipulating these accounts. First, Line 1 imports the "BankAccount" class from the "bank_account" module that was created by saving Code 7.14 into a file named "bank_account.py". Line 2 imports "pickle" for object serialization, and Line 3 imports the "os" module for file management operations.

The "Bank" class is initialized with a "name" and a "filename". The "name" represents the bank's name, while the "filename" stores and retrieves bank account data from a file. The accounts are loaded into a private dictionary "__accounts" during initialization using the "readAccountsFromFile()" method (Line 9).

Several methods are provided for manipulating the bank data:

- "setBankName()" and "getBankName()" allow setting and getting the bank's name (Lines 11–15).
- The "getAccounts()" method returns the dictionary of accounts (Lines 17 and 18).
- The "addAccount" method creates a new "BankAccount" and adds it to the "__accounts" dictionary if the provided account number does not already exist; otherwise, it raises an exception (Lines 20–25).
- The "findAccount()" method searches for an account by number and returns it if found; otherwise, it raises an exception (Lines 27–30).
- The "deleteAccount()" method removes an account by its number if it exists; otherwise, it raises an exception (Lines 32–36).
- The "transfer()" method moves a specified amount of money from one account to another if both accounts exist and the withdrawal is successful; errors in this process raise exceptions (Lines 38–46).
- The "readAccountsFromFile()" method checks if the specified file exists. If not, it creates a new file and initializes an empty dictionary. If the file exists, it reads the accounts using Pickle (Lines 48–65). This method ensures that the account data is loaded into the bank object on instantiation (called within the "__init__()" method).
- The "writeCustomersToFile()" method saves the current state of "__accounts" to the file using Pickle. This method ensures that all account changes have persisted (Lines 67–73).

To test the bank system class, an instance of "BankSystem" is created with specific attributes (Line 77). Several accounts are added, and a transaction is made between them (Lines 78–80). The account data is then saved to the file, and "The End" is printed if all operations succeed (Line 82). If any exceptions occur during these processes, they are caught and printed, indicating an error in the operation (Lines 84 and 85).

## CODE 7.15 THE BANKSYSTEM CLASS

```
 1.from bank_account import BankAccount
 2.import pickle
 3.import os
 4.
 5.class BankSystem:
 6.    def __init__(self,name, filename):
 7.        self.__bank_name=name
 8.        self.__filename=filename
 9.        self.__accounts=self.read_accounts_from_file()
10.
11.    def setName(self, name):
12.        self.__bank_name=name
13.
14.    def getName(self):
15.        return self.__bank_name
16.
17.    def getAccounts(self):
18.        return self.__accounts
19.
20.    def addAccount(self,firstName,lastName,accountNumber,balance):
21.            if not accountNumber in self.__accounts:
22.                account= BankAccount(firstName=firstName,
                lastName=lastName,
                accountNumber=accountNumber,balance=balance)
23.                self.__accounts[accountNumber]=account
24.            else:
25.                raise Exception(f"Sorry. This Account Number ='{accountNumber}' already exists.")
26.
27.    def findAccount(self,accountNumber):
28.            if accountNumber in self.__accounts:
29.                return self.__accounts[accountNumber]
30.            raise Exception(f"Account Number ='{accountNumber}' not found")
31.
32.    def deleteAccount(self,accountNumber):
33.            if accountNumber in self.__accounts:
34.                del self.__accounts[accountNumber]
35.                return "Deleted Successfully"
36.            raise Exception(f"Account Number ='{accountNumber}' not found")
37.
38.    def transfer(self,fromAccountNumber,toAccountNumber,amount):
39.        try:
40.            fromAccount=self.findAccount(fromAccountNumber)
41.            toAccount=self.findAccount(toAccountNumber)
42.            fromAccount.withdraw(amount)
43.            toAccount.deposit(amount)
44.
45.        except Exception as e:
46.            raise e
47.
48.    def read_accounts_from_file(self):
49.        try:
50.            # Check if the file exists
51.            if not os.path.exists(self.__filename):
52.                # If the file doesn't exist, create it and write an empty dictionary
53.                with open(self.__filename, 'wb') as file:
```

```
                    self.__accounts={}
                    file.close()

            else:
                # Now open the file in read mode ('rb')
                with open(self.__filename, 'rb') as file:
                    # Load the data from the file
                    self.__accounts = pickle.load(file)
                    file.close()
            return self.__accounts
        except:
            raise Exception("Unable to read from the Bank Accounts File")

    def write_customers_to_file(self):
            try:
                with open(self.__filename, 'wb') as f:
                    pickle.dump(self.__accounts,f)
                    f.close()
            except:
                raise Exception("Unable to write to the Bank Accounts File")


try:
    main_bank=BankSystem("First UAE Bank","bank_accounts_details.pkl")
    main_bank.addAccount("Rauda","Said","7",10000)
    main_bank.addAccount("Ahmed","Ali","2",500)
    main_bank.transfer("1","2",400)
    main_bank.write_customers_to_file ()
    print("The End")

except Exception as e:
    print(e)
```

Code 7.16 shows the Python implementation of the “BankUI” class, which serves as the main interface for the banking system. Initially, “Tkinter” is imported as “tk” in Line 1, and the “ttk” module is imported from “Tkinter” to utilize the themed widget “Treeview” (Line 2), which is essential for displaying tabular data in a hierarchical structure, making it suitable for representing bank account information in a clear and organized manner. In Line 3, the “messagebox” module is imported from “Tkinter” to provide interactive message boxes for user notifications. In Line 4, from “PIL” Python Imaging Library, the “Image” and “Imagetk” are imported to display the bank logo as shown in Figure 7.15. In Line 5, the “BankSystem” class is imported from the “bank_system” module, which was created by saving Code 7.15 into a file named “bank_system.py”.

The “BankUI” class is the primary class that creates the main window of the banking system. It initializes the GUI, sets up the window attributes such as title and background color, and creates various frames and widgets within the window for user interaction (Lines 8–187). The constructor (“__init__()”) receives an instance of the “BankSystem” class, which manages bank account operations.

Within the constructor, several frames are set up: one for displaying the bank logo and name, another for adding new accounts, a third for transferring funds between accounts, and a fourth to display the account details (Lines 22–49). Components such as images, labels, entry boxes, and buttons are positioned within these frames to allow the user to input data like first name, last name, account number, balance for new accounts, account numbers, and amount for fund transfers (Lines 51–187). Specifically, labels and entry widgets from Lines 69 to 184 collect the necessary data for creating new accounts and performing transfers. These labels indicate what information is required (e.g., “First Name:”, “Last Name:”, “Account No:”, “Balance:”, “Transfer From:”, “Transfer To:”, and “Amount:”), while the entry widgets provide the fields where users can input the corresponding data.

Methods like "transfer_funds()" and "add_account()" are linked to buttons to handle corresponding operations triggered by user actions (Lines 189–215), incorporating error handling to display messages in case of exceptions using the "messagebox" widget. For example, the "add_account()" method is linked to the button defined in Lines 115–119, which, when pressed, triggers the process of adding a new account with the entered details.

The "on_closing()" method is linked to the window's close event (at Line 182) to perform necessary cleanup or final operations, such as saving customer data to a file, that are performed before the application closes (Lines 217–224).

Upon initiation of the BankUI, the "display_accounts()" method is invoked (Line 179). This method generates a table with appropriate headings to display all bank accounts in the system (Lines 226–254). After an account is created or an amount is transferred from one account to another, the "update_account_view()" method refreshes the bank accounts table with updated information (Lines 256–268).

Finally, an instance of the "BankSystem" class is created with a specific bank name and a file for storing account details. This instance is used to initialize a "BankUI" object, thus starting the GUI application with access to the banking system (Lines 272–273).

### CODE 7.16 THE BANKUI CLASS

```
1.import tkinter as tk
2.from tkinter import ttk
3.from tkinter import messagebox
4.from PIL import Image, ImageTk
5.from bank_system import BankSystem
6.
7.
8.class BankUI:
9      """A user interface Class to manage Bank Accounts"""
10.
11.    def __init__(self, bank_system):
12.        try:
13.            # The bank object that manages the accounts
14.            self.bank_system = bank_system
15.
16.            self.window = tk.Tk()
17.            self.window.title(
18.                self.bank_system.getName() + " Management System"
19.            )
20.            self.window.configure(bg="#FFFFFF")
21.
22.            # Frame to display bank logo and Name
23.            self.frame_bank_info = tk.Frame(
24.                self.window, width=700, bg="#FFFFFF"
25.            )
26.            self.frame_bank_info.grid(
27.                row=0, column=0, columnspan=2, padx=10, pady=10
28.            )
29.            # Frame to create a new account
30.            self.frame_add_account = tk.Frame(
31.                self.window, highlightthickness=1, highlightbackground='grey'
32.            )
33.            self.frame_add_account.grid(
34.                row=1, column=0, padx=10, pady=10
35.            )
36.            # Frame to transfer money from one account to another
37.            self.frame_transfer_money = tk.Frame(
38.                self.window, highlightthickness=1, highlightbackground='grey'
39.            )
40.            self.frame_transfer_money.grid(
```

```
            row=1, column=1, padx=10, pady=10
        )
        # Frame to display Account details
        self.frame_bank_details = tk.Frame(
            self.window, highlightthickness=1, highlightbackground='grey'
        )
        self.frame_bank_details.grid(
            row=2, columnspan=2, padx=10, pady=10
        )

        # Add Logo and Company Title to the Window
        original_image = Image.open("logo.jpeg")
        resized_image = original_image.resize(
            (50, 50), Image.ANTIALIAS
        ) # re-sizing the image
        self.logo_image = ImageTk.PhotoImage(resized_image)
        self.logo_label = tk.Label(
            self.frame_bank_info, image=self.logo_image, bg="#FFFFFF"
        )
        self.logo_label.grid(row=0, column=0)
        self.company_title = tk.Label(
            self.frame_bank_info,
            text=self.bank_system.getName(),
            font=("Helvetica", 12),
            bg="#FFFFFF"
        )
        self.company_title.grid(row=0, column=1)

        # Labels and Entry Widgets in frame_add_account
        # Heading for the frame
        self.frame1_label = tk.Label(
            self.frame_add_account,
            text="Add Account",
            fg="black",
            font=("Arial", 12)
        )
        self.frame1_label.grid(row=0, column=0, columnspan=2)

        self.fName_label = tk.Label(
            self.frame_add_account, text="First Name:"
        )
        self.fName_label.grid(
            row=1, column=0, sticky='w', padx=10, pady=10
        )
        self.fName_box = tk.Entry(self.frame_add_account)
        self.fName_box.grid(row=1, column=1, padx=10, pady=10)

        self.lName_label = tk.Label(
            self.frame_add_account, text="Last Name:"
        )
        self.lName_label.grid(
            row=2, column=0, sticky='w', padx=10, pady=10
        )
        self.lName_box = tk.Entry(self.frame_add_account)
        self.lName_box.grid(row=2, column=1, padx=10, pady=10)

        self.account_no_label = tk.Label(
            self.frame_add_account, text="Account No:"
        )
        self.account_no_label.grid(
            row=3, column=0, sticky='w', padx=10, pady=10
        )
        self.account_no_box = tk.Entry(self.frame_add_account)
        self.account_no_box.grid(row=3, column=1, padx=10, pady=10)

```

```
        self.balance_label = tk.Label(
            self.frame_add_account, text="Balance:"
        )
        self.balance_label.grid(
            row=4, column=0, sticky='w', padx=10, pady=10
        )
        self.balance_box = tk.Entry(self.frame_add_account)
        self.balance_box.grid(row=4, column=1, padx=10, pady=10)

        self.add_account_Btn = tk.Button(
            self.frame_add_account,
            text="Add Account",
            command=self.add_account
        )
        self.add_account_Btn.grid(
            row=5, column=1, padx=10, pady=10
        )

        # Labels and Entry Widgets in frame_transfer_money
        # Add Heading for the frame
        self.frame2_label = tk.Label(
            self.frame_transfer_money,
            text="Transfer Money",
            fg="black",
            font=("Arial", 12)
        )
        self.frame2_label.grid(
            row=0, column=2, padx=10, pady=10, columnspan=2
        )

        self.transfer_from_label = tk.Label(
            self.frame_transfer_money, text="Transfer From:"
        )
        self.transfer_from_label.grid(
            row=1, column=2, sticky='w', padx=10, pady=10
        )
        self.transfer_from_box = tk.Entry(self.frame_transfer_money)
        self.transfer_from_box.grid(
            row=1, column=3, padx=10, pady=10
        )

        self.transfer_to_label = tk.Label(
            self.frame_transfer_money, text="Transfer To:"
        )
        self.transfer_to_label.grid(
            row=2, column=2, sticky='w', padx=10, pady=10
        )
        self.transfer_to_box = tk.Entry(self.frame_transfer_money)
        self.transfer_to_box.grid(
            row=2, column=3, padx=10, pady=10
        )

        self.amount_label = tk.Label(
            self.frame_transfer_money, text="Amount:"
        )
        self.amount_label.grid(
            row=3, column=2, sticky='w', padx=10, pady=10
        )
        self.amount_box = tk.Entry(self.frame_transfer_money)
        self.amount_box.grid(
            row=3, column=3, padx=10, pady=10
        )

        self.transfer_Btn = tk.Button(
            self.frame_transfer_money,
```

```
                text="Transfer Money",
                command=self.transfer_funds
            )
            self.transfer_Btn.grid(
                row=4, column=3, padx=10, pady=10
            )

            # Display current Bank details in the frame_bank_details
            self.display_accounts()

            # Link the on_closing method to the window close event
            self.window.protocol("WM_DELETE_WINDOW", self.on_closing)
            # Display the window
            self.window.mainloop()

        except Exception as e:
            messagebox.showinfo("Error", e.__str__())

    def transfer_funds(self):
        try:
            fromAccount = self.transfer_from_box.get()
            toAccount = self.transfer_to_box.get()
            amount = float(self.amount_box.get())
            # Transfer money from one account to another
            self.bank_system.transfer(fromAccount, toAccount, amount)
            # The bank acount details view
            self.all_Accounts_View = self.update_account_view()

        except Exception as e:
            messagebox.showinfo("Error", e.__str__())

    def add_account(self):
        try:
            fName = self.fName_box.get()
            lName = self.lName_box.get()
            accountNo = self.account_no_box.get()
            balance = float(self.balance_box.get())
            # Create the account within the Bank
            self.bank_system.addAccount(fName, lName, accountNo, balance)
            print("The account has been added successfully")
            # The bank acount details view
            self.all_Accounts_View = self.update_account_view()

        except Exception as e:
            messagebox.showinfo("Error", e.__str__())

    def on_closing(self):
        try:
            self.bank_system.write_customers_to_file()
            self.window.destroy()

        except Exception as e:
            messagebox.showinfo("Error", e.__str__())
            self.window.destroy()

    def display_accounts(self):
        try:
            self.frame3_label = tk.Label(
                self.frame_bank_details, text="Bank Accounts", fg="black", font=("Arial", 12)
            )
            self.frame3_label.pack()

```

```
            # Create the table to display current bank accounts in the system
            self.table = ttk.Treeview(
                self.frame_bank_details,
                columns=('Account Number', 'First Name', 'Last Name', 'Balance'),
                show='headings'
            )
            self.table.heading('Account Number', text='Account Number')
            self.table.heading('First Name', text='First Name')
            self.table.heading('Last Name', text='Last Name')
            self.table.heading('Balance', text='Balance')
            # Set the width of the columns
            self.table.column("#1", width=100)
            self.table.pack()

            all_accounts = self.bank_system.getAccounts()
            for id, account in all_accounts.items():
                self.table.insert(", 'end', values=(
                    id, account.getFirstName(), account.getLastName(), account.getBalance()
                ))

        except Exception as e:
            messagebox.showinfo("Error", e.__str__())

    def update_account_view(self):
        try:
            # Clear the table and display updated values
            self.table.delete(*self.table.get_children())
            # Displaying current bank accounts in the system
            all_accounts = self.bank_system.getAccounts()
            for id, account in all_accounts.items():
                self.table.insert(", 'end', values=(
                    id, account.getFirstName(), account.getLastName(), account.getBalance()
                ))

        except Exception as e:
            messagebox.showinfo("Error", e.__str__())


try:
    main_bank = BankSystem("First UAE Bank", "… bank_repository.pkl")
    bank_system = BankUI(main_bank)

except Exception as e:
    print(e)
```

## 7.7 CHAPTER SUMMARY

This chapter discusses the tools to build interactive applications in Python using Tkinter. It begins by explaining the benefits of GUIs over traditional interfaces. Next, it introduces the core components of a GUI: containers (windows and frames) and widgets (labels, buttons, etc.). This chapter details various widgets and their functionalities, allowing users to interact with the program. Layout managers (pack, grid, and place) are explored to organize these elements effectively. Each offers a unique approach to positioning widgets within the container. Finally, this chapter guides readers through building a GUI for a calculator application and a simple banking system. This hands-on example demonstrates combining widgets and layout managers to create a functioning GUI program.

## 7.8 EXERCISES

### 7.8.1 Test Your Knowledge

1. Explain the role of containers in GUIs and their importance in organizing GUI elements effectively.
2. Describe how to create a label widget in Tkinter.
3. Describe the steps involved in creating an entry widget in Tkinter.
4. Compare and contrast the functionality of entry and text widgets in Tkinter, providing examples of their use cases.
5. How is a checkbutton widget created in Tkinter?
6. How is a Combobox widget created using Tkinter, and what does it offer?
7. Compare modern widgets like Combobox in Tkinter applications with traditional widgets like radiobuttons.
8. Describe the role of layout managers in GUI application development using Tkinter.
9. Describe the functionality of the place layout manager in Tkinter. How does it differ from the pack and grid managers?
10. Reflect on the importance of layout management in GUI application development. How does proper layout management contribute to user experience and interface design?

### 7.8.2 Multiple Choice Questions

1. What is Tkinter?
   a. A programming language
   b. A standard GUI package in Python
   c. A data structure
   d. A database management system
2. What is the purpose of a frame in Tkinter?
   a. To display text
   b. To contain other widgets
   c. To execute Python code
   d. To handle user inputs
3. What are widgets in programming?
   a. Visualization of data
   b. Graphical elements displaying data or facilitating user interaction
   c. Backend algorithms
   d. Data structures
4. Which of the following is not a type of widget?
   a. Button
   b. Label
   c. Array
   d. Entry
5. What is the main purpose of a label widget?
   a. Accepting user input
   b. Displaying information without interaction
   c. Triggering actions
   d. Creating windows
6. What does the entry widget in Tkinter allow users to do?
   a. Display images
   b. Select from a list of choices
   c. Type in text
   d. Execute actions

7. What is the purpose of a button widget in Tkinter?
   a. Display text or images
   b. Accept user input
   c. Trigger actions
   d. Edit text
8. Which method retrieves the selected option from a Combobox widget?
   a. get_value()
   b. retrieve()
   c. get()
   d. fetch()
9. Which parameter of the pack() method specifies the number of pixels surrounding the widget to create padding?
   a. padx
   b. pady
   c. ipadx
   d. ipady
10. The grid geometry manager in Tkinter places widgets in a ___________.
   a. one-dimensional list
   b. one-dimensional array
   c. two-dimensional grid
   d. three-dimensional matrix

### 7.8.3 Short Answer Questions

1. Name two types of containers used in Tkinter other than frames.
2. What invisible containers are used to organize and control their child components?
3. What are the two binary values represented by checkbutton widget in Tkinter?
4. Which widgets are commonly paired together?
5. What do label widgets display without user interaction?
6. Name the non-visible object used to organize the widgets.
7. Name the three built-in layout managers available in Tkinter.
8. Name the method the place layout manager uses to position widgets.
9. Name the method the grid layout manager uses to arrange widgets.
10. How do padx/pady and ipadx/ipady differ in providing padding space for the widget?

### 7.8.4 True or False Questions

1. Frames in Tkinter can have borders.
2. All widgets in Tkinter must be placed directly inside the main window
3. A label widget in Tkinter allows users to input text.
4. Tkinter uses the "pack()" method to set the window title.
5. The Entry widget in Tkinter provides users with a single-line text field.
6. The Button widget in Tkinter allows users to trigger actions.
7. The Combobox widget in Tkinter combines an entry with a list of choices.
8. The radiobuttons in Tkinter are always used together in a set.
9. The "padx" and "pady" parameters with the pack() method in Tkinter specify padding space outside the widget.
10. The mainloop() method displays the main window and keeps it active in Tkinter GUI applications.

### 7.8.5 Fill in the Blanks

1. Fill in the blanks in the code snippet below to create a window GUI using Tkinter the command of the button widget:

```
import tkinter as tk

# Create the main window
main_window = ___.Tk()

# Main loop to display window
main_window.________________()
```

2. Fill in the blanks in the code snippet below to complete the command of the button widget:

```
...
# Create a button widget to display output
welcome_button = tk.Button(self.main_window, text="Display Output", command=self.__________)
welcome_button.pack(pady=10)

# Method to handle welcome button click event
def show_welcome_message(self):
    welcome_label.config(text="Welcome to our application!") # Display welcome message
...
```

3. Fill in the blanks in the code snippet below to use place/pack the entry widget created in the created window:

```
import tkinter as tk

# Create a Tkinter window
window = tk.Tk()
# Create an entry widget
entry = tk.Entry(window)

# Place the entry widget in the window
____________________

# Main loop to display window
window.mainloop()
```

4. Fill in the blanks in the code snippet below to complete the line that helps place the label in the absolute position provided:

```
import tkinter as tk
 # Create a Tkinter window
 window = tk.Tk()
 # Create a label widget
 label = tk.Label(window, text="Hello, tkinter!")
 # Place the label using the place layout manager
 _______________(x=50, y=50)
 # Main loop to display window
 window.mainloop()
```

5. Fill in the blanks in the code snippet below to create a Tkinter window and create a checkbutton within the window:

```
import tkinter as tk

# Create a Tkinter window
window = tk.____()

# Create a checkbutton widget
check_button = tk.Checkbutton(__________, text="Check me!")

# Pack the checkbutton
check_button.pack()

# Main loop to display window
window.mainloop()
```

### 7.8.6 Simple GUI Designs

1. Design a simple GUI application using Tkinter to calculate and display the body mass index (BMI) based on the user input of height and weight. The GUI application should adhere to the following specifications:
   a. Design an interface with Entry widgets for users to input their weight and height, allowing them to choose between SI units (meters and kilograms) or British Imperial units (inches/feet and pounds/lbs.)
   b. Include a “Calculate BMI” button that when clicked, triggers a method to calculate the BMI.
   c. Display the calculated BMI in a label widget.
   d. Provide visual feedback to the user on their BMI result using text messages or images (such as happy/sad face or green/red/orange lights):
      - If BMI is less than 18.5, then the user is underweight.
      - If BMI is between 18.5 and 25, then the user falls in the healthy weight range.
      - If BMI is between 25 and 30, the user falls within the overweight range.
      - If BMI is above 30, then the user falls within the obese range.
      - Provide additional alert signals for BMI values of exactly 18.5 or 25, indicating they are at risk.
   e. Include appropriate error handling to ensure users enter valid numbers for height and weight, displaying error messages if invalid input is detected.
2. Design a simple GUI application using Tkinter to convert values from the International System of Units to the British Imperial System (e.g., length: meters to feet, inches; temperature: Celsius to Fahrenheit; weight: kilograms to pounds, ounces; volume: liters to gallons). The GUI application can be designed as follows:
   a. Design a dropdown menu widget to allow users to select the input from the International System of Units with an “Entry” widget to input the value required to be converted while displaying the IS unit.
   b. Implement the conversion logic for each category in a dedicated method using known formulas.
   c. Include a button to trigger the conversion and display the converted value in a label widget specifying the imperial unit.
   d. Implement input validation to ensure users enter valid numbers using exception handling.

3. Design a simple guessing game application using Tkinter in Python.
   The game should generate a random secret number within a specified range (e.g., 1–50), as follows:
   a. Allow the user to enter a guess.
   b. Validate the user's input to ensure it's a valid number within the specified range.
   c. Inform the user if their guess is correct, too high or too low.
   d. The user will interact with the game through a GUI that includes:
      - A Label widget prompting the user to guess a number.
      - An entry widget for the user to input their guess.
      - A button that the user can press to submit their guess and compare it to the secret number.
      - Another label to display the appropriate feedback based on the user's guess.

### 7.8.7 Coding Problems

1. In this exercise, the primary objective is to develop a GUI using Tkinter that facilitates efficient management of patients for the Dubai Health Authority (DHA). The GUI should include the following features:
   a. Patients' demographic form:
      - Create a form to be used by admin staff to register new patients into the system by adding patients: ID, first name, last name, age, phone number, email, and address.
      - Save into a CSV file the data of the patient.
      - Implement a search feature that enables the staff to search for a patient and display the info on the screen.
   b. Patients' medical information
      - Create a form to be used by nurses and doctors to register ID, first name, last name, age, medical history, current medications, appointments, and visited doctors.
      - Save into a CSV file the data of the patient.
      - Implement a search feature that enables the nurse or doctor to search for a patient and display the information on the screen.
2. Design a Tkinter GUI application that loads the XML file with tourism data for the various destinations in the UAE (as seen in Chapter 6 exercises). Parse the XML file using ElementTree and create instances of the "Destination" class for each destination. The GUI application should:
   a. Contain a Home Page:
      - Display a selected logo of your choice in the middle of the page.
      - Provide two buttons: "View Destinations" and "Add New Destination".
   b. Contain a Window to View Destinations:
      - Open a new window displaying each destination's details (name, location, and description) in separate entries with corresponding labels.
      - Add buttons for:
        - Going back to the "Home Page".
        - Navigating to the "Next" destination if not at the last one.
        - Navigating to the "Previous" destination if not at the first one.
   c. Contain a Window to "Add New Destination":
      - Open a new window to enter the destination's name, location, and description.
      - Save the new destination to the XML file when clicked.

```
<tourism>
    <destination>
        <name>Burj Khalifa</name>
        <location>Dubai</location>
        <description>The iconic Burj Khalifa, the tallest building in the world, offers breathtaking
        views of Dubai's skyline.</description>
    </destination>
    <destination>
        <name>Sheikh Zayed Grand Mosque</name>
        <location>Abu Dhabi</location>
        <description>The Sheikh Zayed Grand Mosque is a stunning architectural marvel and a key
        landmark in Abu Dhabi.</description>
    </destination>
    <destination>
        <name>Al Noor Island</name>
        <location>Sharjah</location>
        <description>Experience the beauty of Al Noor Island, a unique destination in Sharjah featuring
        art, nature, and stunning landscapes.</description>
    </destination>
</tourism>
```

3. Design and implement a Tkinter application for an online course management system. The system should allow users to view courses and their details, register to courses, and view their registered courses. To create the application, do the following:
   a. Home Page:
      - Create a Home Page with a welcome message and the following elements:
        - Three Labels and text entry widgets for "UserID", "First Name", and "Last Name".
        - A button labeled "Enter" that directs the user to the "Course List" page or the "View Registered Courses" based on their input.
   b. Course List Page:
      - Display a list of all available courses using a pack or grid layout. Each course should be organized as the below example:

| **Course Number** | **ICS220** |
|---|---|
| Course name | Programming Fundamentals |
| Instructor name | Adam Smith |
| Days/time | MW/ 10:00AM–11:20AM |

      - Include a button to navigate to the "Course details" page and another one to "Register" to a course.
      - When the "Courses details" page is clicked, it displays detailed information about the selected course.
      - When the "Register" button is clicked, create a CSV file to store the user's ID, first name, last name, and the selected course details.
   c. Course Details Page:
      - Display detailed information about the selected course in a table format. The table should include the following information: course number, course name, description, learning outcomes, main topics, and assessment strategy.
      - Include a button to navigate to the "Course List" page.
   d. Create a "View Registered Courses" Page:
      - Read the CSV file to display all registered courses.

## NOTE

1. In Mac computers the parameter is called "highlightbackground".

# 8 Machine Learning with Python

## 8.1 INTRODUCTION TO MACHINE LEARNING (ML)

Machine learning (ML) is the process where machines learn from vast amounts of data, identify patterns using statistical methods, and make predictions, generating new information. This technology is not just a theoretical concept but a practical tool reshaping every industrial and business domain, aiming to create a better world. Many ML solutions, such as recommendations of online content and products based on past viewership and buying patterns, autocomplete features in search fields or forms, auto-tagging of individuals in pictures, and large language models (LLMs) that can answer most questions, are already a part of our everyday lives. These are just a few examples of the tangible impact of ML.

Artificial intelligence (AI) is a broader term for machines that can simulate human intelligence and work autonomously. ML, a subtopic of AI, is the branch that focuses on machines that can learn from data and make decisions based on past outcomes, such as learning from experience. These two fields are deeply interconnected, with AI providing the framework and ML the tools for machines to learn and evolve. As Arthur Samuel, one of the pioneers in this field, aptly put it in 1967, "Programming computers to learn from experience should eventually eliminate the need for much of this detailed programming effort".[1] Another early definition by Frank Rosenblatt on the potential of AI is: "We are now about to witness the birth of such a machine—a machine capable of perceiving, recognizing and identifying its surroundings without any human training or control".[2]

Although these definitions focus on machines making autonomous decisions, ML specifically focuses on gaining these decision-making abilities by extracting generalized rules or patterns from existing examples and applying these rules to new examples. The ML process can be illustrated by providing an ML algorithm with images of cats. The algorithm can then extract the key characteristics of cats from these images, essentially achieving a state of learning. Subsequently, the algorithm's performance can be evaluated by introducing new images containing cats and dogs and observing its accuracy in predicting cats. This scenario represents a simplified example of the ML process, which will be elaborated on in greater detail in the following sections.

LLMs, like ChatGPT and Gemini, are among the most popular tools in AI today. These tools rely heavily on ML and are trained on massive datasets of text and code. This training process involves complex algorithms that allow them to identify patterns and relationships within the data. These and other chatbots can respond to human cues by analyzing word sequences to understand the context of the data and produce human-like responses. They have been trained on a vast corpus of data from the Internet and other sources to understand the context of the question asked and use their learned experiences to respond. These chatbots disrupt how humans learn and work and mark the beginning of a new era of technological advancement.

## 8.2 THE MACHINE LEARNING PROCESS

An ML project evolves through several stages of development to generate a robust model capable of making accurate predictions. The key stages applied in most projects are preprocessing and analysis, model training, validation, testing, and deployment, as illustrated in Figure 8.1. Most ML projects are iterative. As such, after evaluating a model, you may discover unsatisfactory results. Therefore, you may need to make adjustments, such as transforming features during the pre-processing (Stage 1), applying alternate models (Stage 2), or tuning the existing model(s) and re-testing their performance with the data (Stage 4). Validation (Step 3) is an optional step. Some practitioners split the

 DOI: 10.1201/9781032668321-8

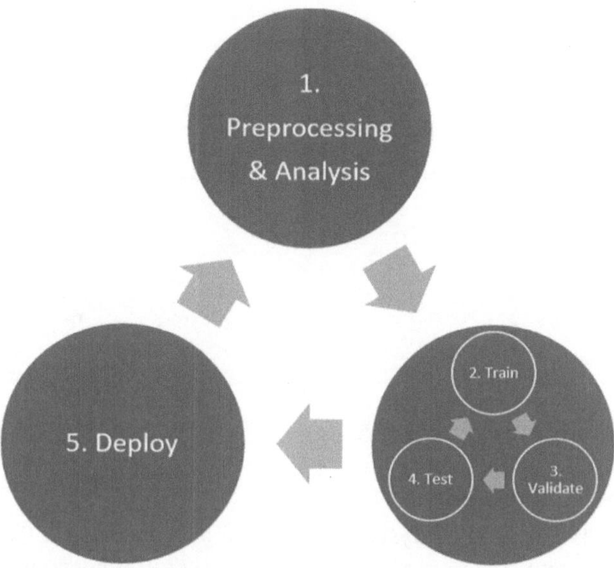


**FIGURE 8.1** The machine learning process.

data into two sets, one for training and the other for testing the model. However, other practitioners split data into three sets: one set of data is for training one or more models, the second set is for fine-tuning the model(s) for better performance, and the third is for testing the finally selected model's performance on unseen data before deploying (Step 5) in a real-world context.

### 8.2.1 Data Preprocessing and Analysis

Data preprocessing is the foundational stage of the project, during which we understand the data and prepare it in a suitable format for training the model. It involves data cleaning tasks such as adding missing values, removing outliers, scaling or normalizing features, and encoding categorical values into numbers. It is widely known in the ML community that pre-processing and data analysis are 80% of ML, while the rest of the stages require 20% of the effort.

#### 8.2.1.1 Python Libraries for Pre-processing

Pandas and NumPy are two essential libraries in the Python ecosystem for preprocessing and analysis. These libraries support tabular data formats, which integrate seamlessly with ML libraries, making them popular for data analysis and ML projects. NumPy is short for Numerical Python. It provides various methods to support pre-processing and advanced mathematical operations such as indexing, slicing, broadcasting, and vectorization. Pandas is built on top of the NumPy library and provides a more intuitive way of manipulating structured data as series and dataframes (a tabular data format). With Pandas, you can perform various operations such as indexing, selection, filtering, grouping, aggregation, and merging data, making it ideal for data preprocessing tasks.

Matplotlib.pyplot and Seaborn are two Python libraries widely used for data visualization and analysis and work well with Pandas and NumPy data formats. Matplotlib.pyplot provides static, interactive, and animated visualizations in Python, making it suitable for various data visualization tasks. With Matplotlib, users can generate various plots, including line plots, scatter plots, bar plots, histograms, and more. Seaborn is built on Matplotlib to easily generate better, aesthetically pleasing, informative statistical graphics. Seaborn excels at producing sophisticated visualizations such as heatmaps, pair plots, violin plots, boxplots, and regression plots with minimal code. Both libraries customize plot elements such as axes, labels, titles, colors, and styles, enabling users to incorporate specific requirements.

Independent variables

Target variable Y

| | Price | Area | Bedrooms | Bathrooms | Basement | Parking |
|---|---|---|---|---|---|---|
| 0 | 13300000 | 7420.0 | 4 | 2.0 | no | 2 |
| 1 | 12250000 | 8960.0 | 4 | 4.0 | no | 3 |
| 2 | 12250000 | 9960.0 | 3 | 2.0 | yes | 2 |
| 3 | 12215000 | 7500.0 | 4 | 2.0 | yes | 3 |
| 4 | 11410000 | 7420.0 | 4 | 1.0 | yes | 2 |

**FIGURE 8.2** First few rows of the housing data.

#### 8.2.1.2 Understanding the Data

The first step for any data-related analysis is to explore and understand the data and the content within the data. To illustrate some pre-processing tasks taken to understand the context, we have selected a sample dataset on housing prices displayed in Figure 8.2, adapted from a dataset on the Kaggle website[3]. In this scenario, "price" is the first column in the dataset, also called the target/dependent variable. The dependent variable represents the outcome of interest within a dataset. It is the variable targeted for explanation or prediction through analysis of other variables in the data. Other features are the area of the house, the number of bedrooms, the number of bathrooms, the basement, and the number of parking spots, which indicate the independent variables that can help predict the target variable, "price".

##### *8.2.1.2.1 Reading Data from Files*

The "read_csv()" method from Pandas reads from a CSV file and generates a dataframe structure, a tabular data representation. Pandas provides a wide range of methods to read data from different file formats to generate a data frame. For example, "read_excel()" parses an Excel file, "read_json()" parses a JSON file, and "read_html()" can parse HTML pages.

##### *8.2.1.2.2 Data Explorations Methods and Description*

Several methods are provided by the Pandas library to explore the data and understand the context. For example, the "head()" method helps retrieve a specified number of rows from the top of the data frame, and the "tail()" method displays a specified number of rows from the bottom of the data frame. These methods help view the contents within the columns. The "info()" method displays basic information about the features (columns) of the data frame, including the total number of columns and rows, the columns' names, the non-null count of each column, and its corresponding data type, as shown in Figure 8.3. This information is useful for understanding the overall structure and integrity of the dataset. As shown in the figure, the result indicates 545 rows and 6 columns are in the dataframe. The column "Price" has 545 non-null values, whereas "Area" has 544 non-null values, indicating one null value. "Bathrooms" have only 541 non-null values. The details also indicate the datatype of each column: integer, float, or object.

The "describe()" method is commonly used to retrieve basic descriptive statistics about the numerical features of the data, including count, mean, standard deviation, quartiles (25th, 50th, and 75th percentiles), minimum, and maximum values in the data, as shown in Figure 8.4.

The "count" is the total count of non-null values in the column. The "mean" is an important measure of central tendency, the average of all the values in the data. The standard deviation (std) provides the spread of the data, in other words, how far the data points are from the central value. A smaller standard deviation value indicates data points are closer to the mean. 25th, 50th, and 75th

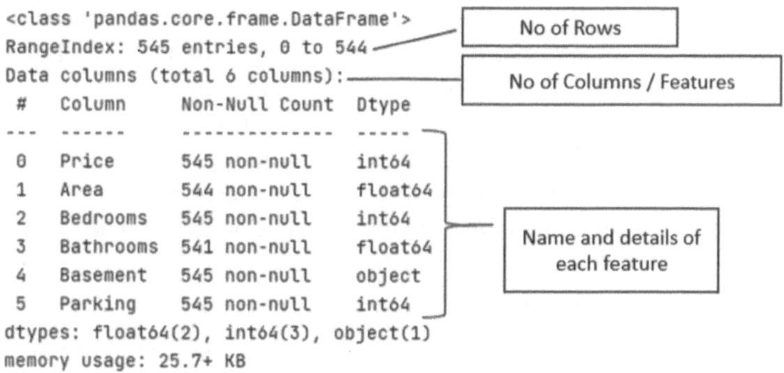

```
<class 'pandas.core.frame.DataFrame'>
RangeIndex: 545 entries, 0 to 544
Data columns (total 6 columns):
 #   Column     Non-Null Count  Dtype
---  ------     --------------  -----
 0   Price      545 non-null    int64
 1   Area       544 non-null    float64
 2   Bedrooms   545 non-null    int64
 3   Bathrooms  541 non-null    float64
 4   Basement   545 non-null    object
 5   Parking    545 non-null    int64
dtypes: float64(2), int64(3), object(1)
memory usage: 25.7+ KB
```



**FIGURE 8.3** Viewing dataframe details using the info() method.

| | Price | Area | Bedrooms | Bathrooms | Parking |
|---|---|---|---|---|---|
| count | 5.450000e+02 | 544.000000 | 545.000000 | 541.000000 | 545.000000 |
| mean | 4.759010e+06 | 5130.229779 | 2.965138 | 1.284658 | 0.693578 |
| std | 1.856814e+06 | 2119.652665 | 0.738064 | 0.502148 | 0.861586 |
| min | 1.750000e+06 | 1650.000000 | 1.000000 | 1.000000 | 0.000000 |
| 25% | 3.430000e+06 | 3596.000000 | 2.000000 | 1.000000 | 0.000000 |
| 50% | 4.340000e+06 | 4580.000000 | 3.000000 | 1.000000 | 0.000000 |
| 75% | 5.740000e+06 | 6360.000000 | 3.000000 | 2.000000 | 1.000000 |
| max | 1.330000e+07 | 15600.000000 | 6.000000 | 4.000000 | 3.000000 |

**FIGURE 8.4** Analyzing basic descriptive statistics using the describe() method.

percentiles are also known as the "Quartiles", Q1, Q2, and Q3. Q1 is the first 25% of the data. Q2, or the median, is 50% of the data or the exact middle value between the lowest and highest values in the data set, which is also a measure of central tendency. Q3 is the 75% of the data. These statistics are also useful in finding outliers in the features. Outliers are data points that fall significantly outside the overall pattern of the data. They can skew the analysis and make it difficult to draw accurate conclusions. Right or left skewness in the data affects the performance of some ML algorithms. In the case of extremely low values, the mean is lower than the median, and the data is considered left-skewed (negative skewness). If the data has very high extreme values, the mean is higher than the median, and the data is considered right-skewed (positive skewness).

On observing the "Price" column, the target variable, Q1, or the 25% of the house "Price" values, are below 3.4 million. Q2 (the median) or 50% of "Price" values are at 4.3 million. In Q3, 75% of the data indicates that the "Price" is 5.74 million. The mean is 4.7 million, indicating right skewness with a few extremely high house prices compared to most house prices in the data set. The standard deviation (std) for "Price" is approximately 1.85 million, indicating how far the data points are from the mean.

Code 8.1 shows the Python code to perform basic data exploration tasks on data read from a CSV file. Line 2 of the code imports the "pandas" library with the commonly used alias "pd", and Line 4 imports the "numpy" library with the alias "np". Lines 7 and 8 are added to ignore warnings generated by the libraries used in the code. The data is then read from a CSV file located at a web link and stored in a Pandas dataframe named "df1" using the "read_csv()" method (Line 11). In Line 14, the "head()" method displays the first "5" rows from the top of the dataframe, allowing for a quick inspection of the data structure and content, as shown in Figure 8.2. Similarly, the “tail()” method (Line 17) displays the last “10” rows of the dataframe. The “info()” method in Line 21 concisely summarizes the dataframe, including the number of rows and features in the dataset, the number of non-null values, and data types in each column. Line 25 illustrates the use of the “describe()” method, which is an initial step toward data analysis, including count, mean, standard deviation, minimum, quartiles (25th, 50th, and 75th percentiles), and maximum values in the data, as shown in Figure 8.4. These statistics provide insights into the distribution and spread of numerical features.

**CODE 8.1 INVESTIGATING THE HOUSING DATA**

```
1. # Import the Pandas library with the acronym 'pd'
2. import pandas as pd
3. # Import the NumPy library with the acronym 'np'
4. import numpy as np
5.
6. # Ignore warning given by libraries
7. import warnings
8. warnings.simplefilter('ignore')
9.
10. # Read the data from the CSV file into a pandas dataframe
11. df1 = pd.read_csv (
        "https://raw.githubusercontent.com/Object-Oriented-Programming-2024/" \
        "Object-Oriented-Programming/main/Chapter8/housing.csv"
    )
12.
13. print("\n******* First 5 Rows of the data *******")
14. print(df1.head(5))
15.
16. print("\n******* Last 10 Rows of the data *******")
17. print(df1.tail(10))
18.
19. # The info method provide details about the features
20. print("\n******* Basic Information about the Data frame *******")
21. print(df1.info())
22.
23. # Display basic summary statistics for numeric features
24. print("\n*******Descriptive statistics for numerical columns*******")
25. print(df1.describe())
```

### *8.2.1.2.3 Cleaning the Data*

Most ML algorithms cannot process null, empty, undefined, or not a number (NaN) values. Therefore, the first data cleaning task is to remove rows with empty values or replace them with an estimate. “*Imputation*” is the term used to handle missing values in a dataset by replacing them with estimates, such as the mean, median, mode, or the highest value. The “fillna()” method of the “pandas” library is used to replace the missing values with any of these estimates. Another important step during data cleaning is to ensure no duplicate rows exist in the data. Many duplicates in the data cause bias in the predictions of the model. The “duplicated()” method automatically finds duplicate rows in the data frame.

The housing data example in Figure 8.3 identified that the features “Area” and “Bathrooms” have null or empty values, which can be replaced with one of the measures of central tendency. To illustrate how to replace these null values, consider Code 8.2, which takes the same data frame as the dataset loaded in Code 8.1.

In Line 2, the “Area” column is accessed and stored in the variable “area”. In Line 3, the “mean()” method is applied to the “area” variable, and the resultant mean is stored in the “average” variable, which is printed in Line 4. In Line 5, the “fillna()” method is applied to the column “Area” taking two parameters: the variable "average" and "inplace=True". This method will fill all the null values in the "Area" column with the "average". The second parameter alters the current dataframe "df1" instead of creating a new dataframe with the updated changes.

In Line 7, the column "Bathrooms" is retrieved and stored in the variable "bathrooms". In Line 8, the most frequently occurring number(s) in the column "Bathrooms" is calculated, also called the mode, and printed in Line 9. In Line 10, the "fillna()" method is used to replace the empty values in the "Bathrooms" column with the mode value, i.e., the most frequent value within a column.

In Line 13, the "isnull()" method is used to find rows with null values, and the "sum()" method is added to count the number of rows with missing values. This ensures that the total number of empty rows is zero and that no more missing values are in the dataframe.

Line 17 uses the sum method to add all the duplicate rows and display their count. This dataset has one duplicate row, which does not impact the model. The "drop_duplicate()" method in Line 18 drops the duplicates from the dataframe.

**CODE 8.2 CLEANING THE DATA BY APPLYING IMPUTATIONS AND REMOVING DUPLICATES**

```
1. # Imputate missing values
2. area = df1['Area']
3. average=area.mean()
4. print(f"Replacing null Area with Average Area : {average}" )
5. df1['Area'].fillna(average, inplace=True)
6.
7. bathrooms = df1['Bathrooms']
8. most_frequent = bathrooms.mode()[0]
9. print(f"\nReplacing null bathrooms values with most frequent occurrence: {most_frequent}")
10. df1['Bathrooms'].fillna(most_frequent, inplace=True)
11.
12. print("Making sure no null values: ")
13. print(df1.isnull().sum())
14.
15. # Check for duplicates
16. print("\n******* Checking for duplicates *******")
17. print(df1.duplicated().sum())
18. df1.drop_duplicates(inplace=True)
19. print(f"Check rows after removing duplicates: {df1.shape}")
```

### 8.2.1.3 Exploratory Data Analysis

Exploratory data analysis (EDA) is crucial in creating an effective model that includes understanding the data distribution, analyzing relationships between features, and identifying potential outliers or biases. Data visualization is vital in EDA, making complex datasets easier to comprehend. Matplotlib and Seaborn libraries provide a variety of charts for visual representation, chosen based on the data type (numerical, categorical, and temporal) as outlined in Table 8.1.

Analyzing each feature separately is called univariate analysis, whereas analyzing two features is called bivariate analysis. Most visualizations, such as histograms, line plots, boxplots, bar charts, area plots, and Kernel Density Estimate (KDE) plots, are appropriate for univariate analysis. Scatter, joint, and violin plots can be used for bivariate analysis. Only a few visualizations, such as heat maps, can support multivariate analysis.

#### *8.2.1.3.1 Univariate Analysis of Numerical Features*

Histograms and boxplots are the most appropriate visualizations to explore continuous numerical features such as "Price" and "Area" variables in our housing data set in Figure 8.2. Histograms

**TABLE 8.1**
**Appropriate charts for each data type**

| | Matplotlib | Seaborn |
|---|---|---|
| Numerical Continuous | Histogram<br>Line plot<br>Scatter plot<br>Box plot<br>Area plot | KDE plot<br>Joint plot<br>Violin plot<br>Heatmap (multi-variant) |
| Numerical Ordinal | Box plot<br>Bar chart | Point plot<br>Bar chart |
| Categorical | Bar chart/ grouped/ stacked<br>Pie chart<br>Dot plot | Bar chart<br>Count plot<br>Boxen plot<br>Strip plot<br>Swarm plot |
| Temporal | Line plot | Time series plot<br>Joint plot with time axis<br>Facet grid with time axis |

group data into ranges or intervals, with the y-axis representing the frequency of data points within each interval, as illustrated in Figure 8.5 (a) for the "Price" variable.

The boxplot in Figure 8.5 (b) helps identify outliers and the number of data points beyond the appropriate range, represented as the inter quartile range (IQR). The IQR covers the central 50% of

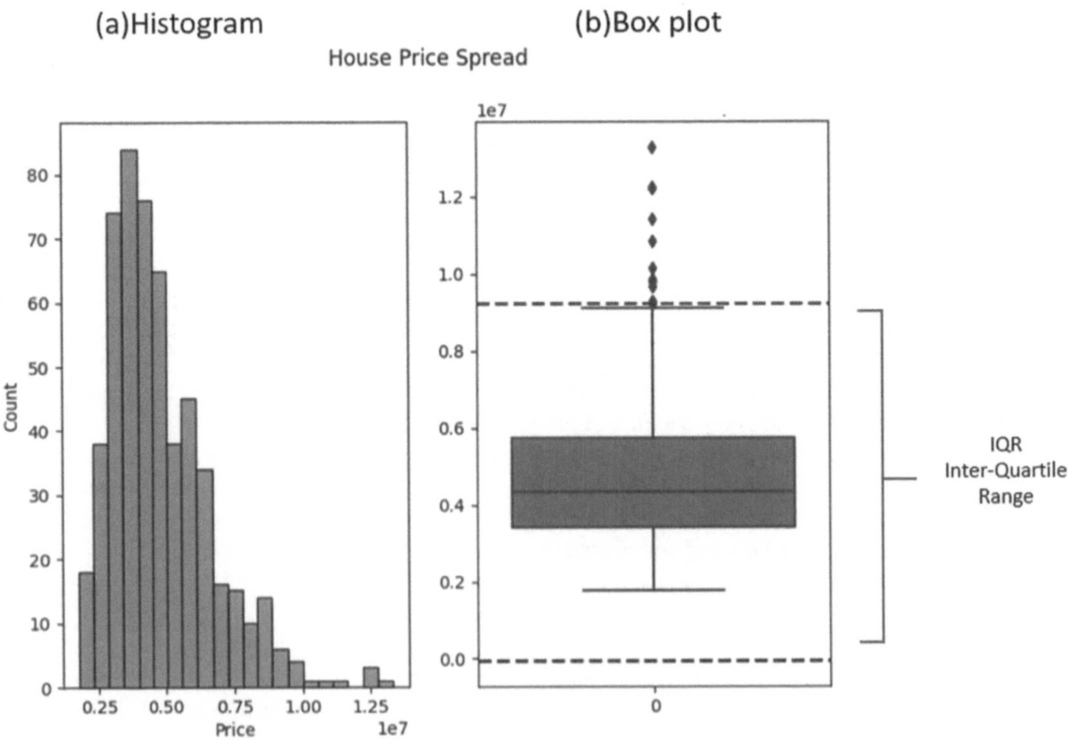


**FIGURE 8.5** Histogram and boxplot representing the price variable, (a) Histogram and (b) Box Plot.

the data (Q3–Q1) and displays the upper and lower bounds, illustrated with green and red lines, respectively. The bounds are calculated as follows:

- Lower bound = Q1 – 1.5 * IQR
- Upper bound = Q3 + 1.5 * IQR

Outliers beyond these boundaries are considered risks and should be addressed, often by capping the values to the upper or lower bound. Alternatively, data transformations discussed later in this chapter can compress the range of values closer to the central data.

Let's compare the distribution of the 'Price' variable in the histogram in Figure 8.6 (a) to a normal distribution in Figure 8.6 (b). A normal distribution has symmetrical data points around the mean, median, and mode, while the histogram of "Price" shows a right skewness with outliers.

Code 8.3 is a continuation of Code 8.2 for predicting house prices. It provides a guideline for creating the histogram and boxplot for the numerical column "Price", as shown in Figure 8.5. Lines 1 and 2 import the libraries "matplotlib" as "plt" and "seaborn" as "sns", which are used for creating graphs in Python. Line 4 uses the "subplots()" method to create a figure, saved in the variable "fig", with two subplots arranged side by side, setting the figure size to "8 by 6" (8 × 6) inches, where "ax1" is assigned to the first subplot and "ax2" is assigned to the second subplot. Line 5 uses the "seaborn" library to create a histogram for the "Price" column using the "histplot()" method on the first subplot, "ax1".

Lines 8 and 9 calculate the first quartile, "Q1" (25%), and the third quartile, "Q3" (75%), for the "Price" column. Line 12 calculates the interquartile range, "IQR". Lines 14 and 15 calculate the upper and lower boundaries of the boxplot for detecting outliers.

Line 16 creates a boxplot for the "Price" column using the seaborn method "boxplot()" and adds it to the second plot "ax2". Lines 18 and 19 use the "axhline()" method to add horizontal dashed lines at the boxplot's calculated upper and lower boundaries. The "y" parameter takes the value at which the line is drawn, which is a lower boundary at Line 18 and an upper boundary at Line 19.

Lines 21 and 22 set the titles for the histogram and boxplot using the subplot objects "ax1" and "ax2", and the graph is displayed in Line 23. Line 26 retrieves all the values beyond the upper bound in the "Price" column, and Line 28 replaces these values with the upper bound value. This step caps the outliers at the upper bound, visible in the boxplot in Figure 8.5. This is an example of a data transformation to improve the performance of our model.

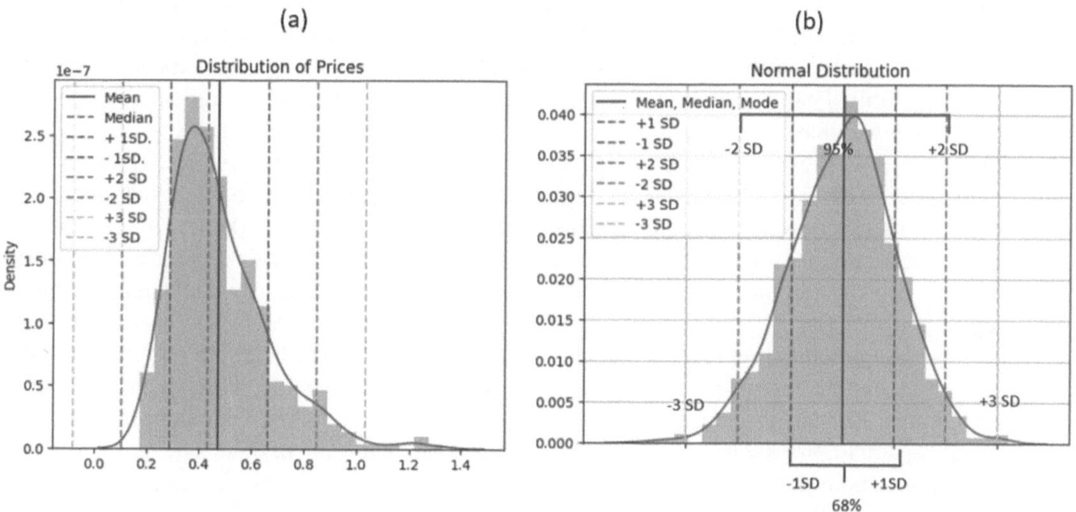


**FIGURE 8.6** (a) Distribution of price values using a histogram; (b) histogram of a normal distribution.

**CODE 8.3 CREATING A HISTOGRAM AND BOXPLOT**

```
import matplotlib.pyplot as plt
import seaborn as sns
# Plot the distribution of target outcome 'Price'
fig, (ax1, ax2) = plt.subplots(1, 2, figsize=(8, 6))
sns.histplot(df1['Price'], ax=ax1)
# Plot the Spread using boxplot
# Calculate the first quartile (Q1) and third quartile (Q3)
Q1 = df1['Price'].quantile(0.25)
Q3 = df1['Price'].quantile(0.75)

# Calculate the interquartile range (IQR)
IQR = Q3 - Q1
# Define the lower and upper bounds to identify outliers
lower_bound = Q1 - 1.5 * IQR
upper_bound = Q3 + 1.5 * IQR
sns.boxplot(df1['Price'], ax=ax2)
# Plot the upper and lower bounds as horizontal lines
plt.axhline(y=lower_bound, color='r', linestyle='--', linewidth=2, label='Lower Bound')
plt.axhline(y=upper_bound, color='g', linestyle='--', linewidth=2, label='Upper Bound')

ax1.set_title('House Prices Histogram')
ax2.set_title("House Price Boxplot")
plt.show()

# Find outlier indices
outlier_indices = (df1["Price"] > upper_bound)
# Replace outliers with the upper bound
df1.loc[outlier_indices,"Price"] = upper_bound
```

#### *8.2.1.3.2 Bivariate Analysis of Numerical Features*

During data analysis, it is important to explore the correlation or impact of independent variables on the dependent variable. When we analyze the relationship between two variables, it is known as bivariate analysis. In our housing dataset, in Figure 8.2, it is essential to understand the relationship between the independent variables and their impact on the dependent variable, "Price". One effective way to analyze the correlation between variables is the heatmap, particularly when dealing with numerous numeric variables, as shown in Figure 8.7.

The heatmap illustrates the correlation between variables, with the green intensity indicating the correlation's strength. The "Area" has the highest correlation with "Price", with a coefficient of 0.58, followed by the number of "Bathrooms", 0.47. Intuitively, these variables are expected to impact house prices notably. Features with high correlation coefficients substantially influence the prediction of the target variable.

Additionally, investigating any interdependence among independent variables, referred to as multicollinearity, is essential. For instance, if the area has a 0.37 correlation coefficient with parking, this suggests a relatively weak correlation. However, values close to 1 or −1 depict a higher interdependence, which can affect the model's performance. In such cases, strategies like feature removal, combination, or other techniques are applied to mitigate multicollinearity and enhance model accuracy.

A heatmap is a graphical representation of a matrix. Each cell in the heatmap represents a value from the matrix, and the intensity of the cell color represents the data value, which is the correlation between features. Typically, higher values are represented by darker colors. For example, the "Area" and "Price" cells in Figure 8.7 have a darker shade of green with a value of 0.58. The matrix displays all the features on both the x-axis and y-axis and the correlation between the features is represented in the cross-section cell.

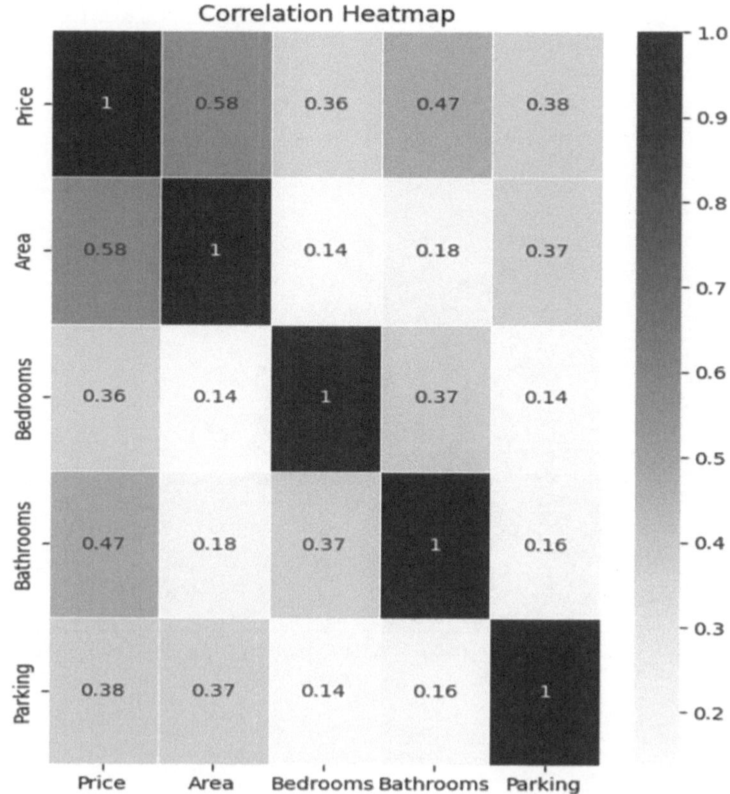


**FIGURE 8.7** Heatmap displaying the correlations between variables in the housing dataset.

Code 8.4 illustrates how to generate the heatmap displayed in Figure 8.7. This code continues from the previous code cells, starting from Code 8.1. Line 2 shows the numerical and nominal features selected from the data frame. The correlation between each feature is calculated using the "corr()" method, and the resulting matrix is stored in the variable "cor_matrix". Line 5 uses "plt" to create a figure of size 6 inches in width and 6 inches in height. In Line 6, Seaborn creates a heatmap based on the "cor_matrix", as the first parameter. The parameter "annot = True" ensures the correlation coefficients are visible on the heatmap. The "cmap" parameter with the value "BuGn" is used to select the color intensity, with shades from light blue to dark green. Line 7 sets the title, and Line 8 displays the heatmap.

## CODE 8.4 VISUALIZING CORRELATIONS IN A HEATMAP

```
1. # Find the correlation between features
2. cor_matrix = df1[['Price','Bedrooms','Bathrooms','Parking']].corr()
3. print(cor_matrix)
4. # Using visualizations for an overview of correlation among features
5. plt.figure(figsize=(6, 6))  # Set the figure size
6. sns.heatmap(cor_matrix, annot=True, cmap='BuGn', linewidths=0.5)
7. plt.title('Correlation Heatmap')
8. plt.show()
9.
10. # Check for correlation between target outcome Price with Area
11. plt.figure(figsize=(6, 6))
12. plt.scatter(x = df1['Area'], y = df1['Price'])
13. plt.ylabel('Price', fontsize=12)
14. plt.xlabel('Area', fontsize=12)
15. plt.show()
```

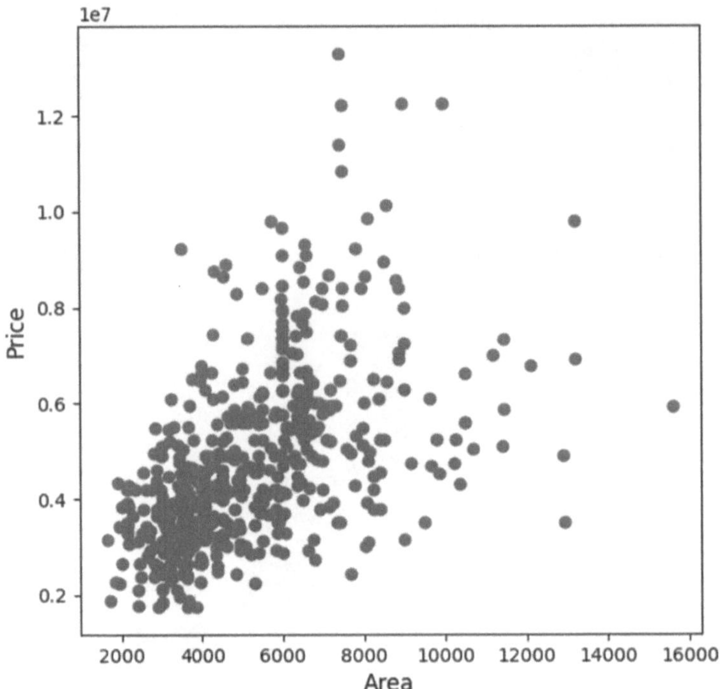


**FIGURE 8.8** Correlation between price and area.

From Lines 11 to 15 of Code 8.4, a scatter plot represents the relationship between "Area" and "Price", as these two variables had the highest correlation, as seen in Figure 8.7. In Line 11, a figure of size 6 by 6 inches is drawn. In Line 12, the "scatter()" method of the Matplotlib library is selected to create a scatterplot with the "Area" on the x-axis and "Price" on the y-axis. The axis is labeled at Lines 13 and 14, and the plot is displayed at Line 15, as illustrated in Figure 8.8.

Observe the relationship between the variables "Area" and "Price" from the scatter plot in Figure 8.8. It can be seen that the price increases whenever the area increases, which makes the scatterplot's direction positive. As the area increases from 8000 to above, the prices are higher against the area size or too low with larger house sizes.

#### *8.2.1.3.3 Univariate Analysis for Categorical Features*

Categorical features are as important as numerical features but do not work well with all types of charts to showcase the distribution of categories. Bar count and cat plots are most commonly used to display categorical data. Although the number of bedrooms, bathrooms, and parking contain numerical values, they are nominal values representing the frequency of each category. If you have categorical columns representing text values, you can use transformations to convert a categorical feature into a nominal data type.

Figure 8.9 displays the bar chart for each nominal feature in our housing dataset and can help highlight more dominant categories. For example, there are more three-bedroom houses and very few houses with one or six bedrooms, as shown in Figure 8.9 (a). Similarly, there are more houses without a parking space and a basement, as shown in Figure 8.9 (c and d).

Code 8.5 creates the bar charts displayed in Figure 8.9. Line 2 creates a figure with four subplots arranged in a 2×2 grid. The figure size is set to 8 by 6 inches. The resulting subplots are stored in the "axes" array. In Line 5, the first subplot (row 0, column 0) is stored in variable "ax". In Line 7, the "value_counts()" method calculates the frequency of each category in the "Bedrooms" feature (e.g., how many apartments have one bedroom, two bedrooms, or three bedrooms?). The frequencies are

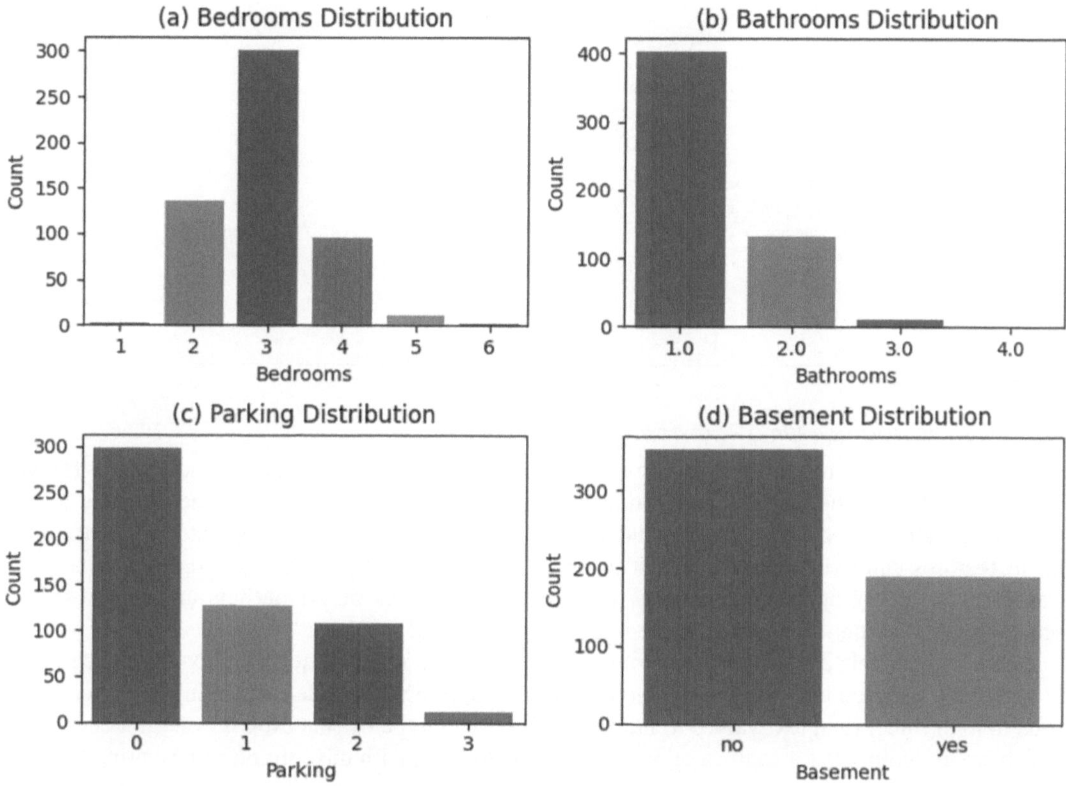


**FIGURE 8.9** Bar charts to visualize categorical data. (a) Bedrooms Distribution, (b) Bathrooms Distributions, (c) Parking Distribution, and (d) Basement Distribution.

then used to create the bar plot using the "plot()" method with "kind=bar", and the subplot is assigned to "ax". Line 8 sets the chart's title, and Line 9 sets the y-axis label. Lines 11 to 15 use a similar code to draw the bar chat for the "Bathrooms". The "Parking" plot is generated from Lines 17 to 21, and the "Basement" plot code is in Lines 23 to 27, each positioned in the respective subplots in the "axes" array. On Line 30, the graph is displayed.

## CODE 8.5 VISUALIZING NOMINAL FEATURES IN BAR CHARTS

```
1. # Create a figure with 2 rows and 2 columns grid of subplots
2. fig, axes = plt.subplots(2, 2, figsize=(8, 6), gridspec_kw={'hspace': 0.4})
3.
4. # At column 0 and row 0 of the plot
5. ax=axes[0,0]
6. # Calculate the frequency of bedrooms per category
7. df1['Bedrooms'].value_counts().plot(kind="bar", ax=ax)
8. ax.set_title('(a) Bedrooms Distribution')
9. ax.set_ylabel("Count")
10.
11. ax=axes[0,1]
12. # Calculate the frequency of Bathrooms per category
13. df1['Bathrooms'].value_counts().plot(kind="bar", ax=ax)
14. ax.set_title('(b) Bathrooms Distribution')
15. ax.set_ylabel("Count")
16.
17. ax=axes[1,0]
```

```
# Calculate the frequency of Parking per category
df1['Parking'].value_counts().plot(kind="bar", ax=ax)
ax.set_title('(c) Parking Distribution')
ax.set_ylabel("Count")

ax=axes[1,1]
# Calculate the frequency of Yes or No in the Basement
df1['Basement'].value_counts().plot(kind="bar", ax=ax)
ax.set_title('(d) Basement Distribution')
ax.set_ylabel("Count")

# Show the plot
plt.show()
```

#### *8.2.1.3.4 Performing Data Transformations*

Data transformations modify existing features to make them more appropriate and simpler for ML, helping to improve the model's performance. Common transformation techniques include mathematical operations, encoding, scaling, and dimensionality reduction. Applying mathematical formulas to features can help reduce outliers in the data set, for example, through log and square root transformations. By applying these transformations, the values of the outliers are compressed or scaled down, making them less extreme relative to the rest of the data.

Some ML models can only process numeric values. As a result, categorical columns need to be dropped or converted into nominal data types before applying the model. Encoding is the technique used to map categorical text values to numeric values. One type of encoding is "one-hot encoding", which creates columns for each category value in the feature, with the column representing the presence or absence of the category in the instance with a value of zero or one. Another technique is "label encoding", which assigns numerical values to each category level but maintains ordinal relationships.

Scaling addresses the problem of having features with different scales, which can affect the model's performance. For example, in our housing dataset, prices are in millions, whereas the frequency of the number of bedrooms can be no more than six. Therefore, a scaling technique can be used to map the price feature to a specific range, such as between zero and one, while preserving the shape of the distribution. The "fit_transform()" method is typically used for applying such transformations.

Another important transformation technique is "dimensionality reduction". This technique helps combine features to avoid multi-collinearity or high-dimensional data. New features can be created by combining existing ones using addition, multiplication, or ratios. These techniques are applied in the case studies at the end of this chapter.

Code 8.6 illustrates the use of some data transformation techniques. During EDA, it is clear that the "price" variable has many outliers and values in millions. In Line 2, the "price" column is first divided by a million, and then the "log()" transformation from "numpy" is applied to achieve a more symmetric distribution (a more normalized distribution).

Lines 5–9 demonstrate one-hot encoding on the "Basement" feature, which contains text-based data ("no" and "yes"). The "get_dummies()" method in Line 5 creates binary columns for each category (where "0" represents the absence of the basement and "1" represents its presence). By using the parameter "drop-first=True", we keep only one column, "Basement_yes", which is renamed to "Basement" in Line 8. The "astype(int)" method ensures the values are integers. The "temp" dataframe is concatenated with the original "df1" dataframe in Line 6, and the original "Basement" column is dropped at Line 7. The new basement column named "Basement_yes" is renamed "Basement" in Line 8. Since we performed the log transformation on our "price" column, additional scaling is unnecessary.

**CODE 8.6 DATA TRANSFORMATIONS**

```
# Apply the log transformation to normalize the spread of the data
df1['price'] = np.log(df1['Price']/1000000)

# Converting categorical boolean feature to one-hot encoding
temp = pd.get_dummies(df1['Basement'], drop_first = True, prefix='Basement').astype(int)
df1 = pd.concat([df1, temp], axis = 1)
df1.drop(['Basement'], axis = 1, inplace = True)
df1.rename(columns={'Basement_yes': 'Basement'}, inplace=True)
df1.head(5)
```

Data transformation is an iterative process. Through experimentation with different options, we can evaluate their impact and choose the options that enhance the model's performance and understanding.

### 8.2.1.4 Model Training, Validate, and Test

Before training a model, it is essential at this stage to choose one or more appropriate models that can be applied to the problem. The main indicator for choosing a model is the data available for training and the problem you are trying to solve. In our housing example, we have training data on house sale prices based on various features of the houses, such as area, number of bedrooms, and so on. The problem we want to solve is to predict the sale price of a house based on these features. This is a supervised ML problem because our target feature, sale price, is found within our dataset. The concept of supervised ML is explained in the next section, Section 8.3, which covers some of the ML problems and appropriate models that can be applied to solve them.

#### *8.2.1.4.1 Machine Learning Python Libraries*

After choosing an appropriate model(s), one must find a Python library that supports the required model. For example, in our housing scenario, we can apply linear regression available in most ML libraries for Python, such as Scikit-learn (sklearn), TensorFlow, Keras, and PyTorch. To accommodate a basic understanding of ML, this chapter focuses on using Scikit-Learn to illustrate the application of some algorithms.

#### *8.2.1.4.2 Splitting the Data*

Before applying a supervised ML model, the first step is to remove the target variable from the dataset.

The second step for modeling is to split the data into three portions for training, validation, and testing. The training set should be the largest to support effective training of the model. The validation set fine-tunes the models and improves their performance. The test set is the unseen data set to confirm the model's performance on unseen data before deployment.

#### *8.2.1.4.3 Model Training*

Training a model aims to generalize from the training data to make near-accurate predictions on the validation and test data. During training, the model learns patterns and relationships in the data and adjusts its internal parameters to minimize the prediction error, known as the loss function.

#### *8.2.1.4.4 Loss Function*

The loss function helps identify the loss of accurate predictions by measuring the difference between predicted and actual values. The most commonly used loss function for linear regression is the mean squared error (MSE). This method measures the average squared difference between predicted and actual values. It penalizes larger prediction errors more, focusing the model on overall accuracy. Another common metric is R-squared, with values ranging from 0 to 1. It is used to evaluate the performance of the regression model by comparing the predictions with actual values, where 1 indicates a perfect model, and 0 indicates that the model does not explain any variability in the target variable.

#### *8.2.1.4.5 Underfitting vs. Overfitting*

Training and selecting an appropriate model for a problem requires careful consideration of the bias vs. variance tradeoff. Bias occurs when the model is too simple to capture the data's complexity, resulting in underfitting (inability to learn true relationships), as illustrated in Figure 8.10 (a). Variance reflects the model's sensitivity to changes in the training data. High variance means that the model captures the complexity and even noise in the data, resulting in overfitting. This means that it will perform well on training data but poorly on unseen data, as illustrated in Figure 8.10(b). The tradeoff between bias and variance arises because decreasing bias increases variance and vice versa. A good fit is a model that can generally give a near-accurate prediction, balancing bias and variance well.

Code 8.7 shows how to train, validate, and test the linear regression model on the housing data.

Lines 1–3 import the necessary libraries and methods for the subsequent code to function; the "train_test_split()" method is imported in Line 1, and "LinearRegression" model in Line 2 from Scikit-learn; the first is used to split the dataset into training and testing sets, and the second to create and train the regression model. In Line 3, "r2_score()" is imported from Scikit-learn to evaluate the performance of the regression model by calculating the R-squared value.

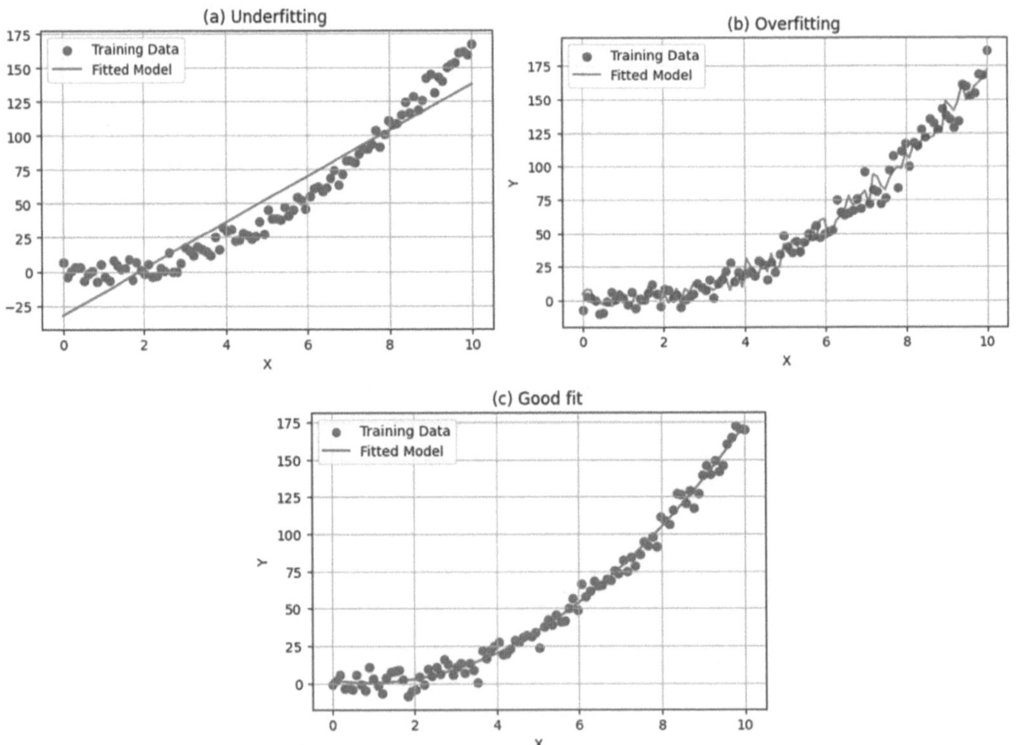


**FIGURE 8.10** (a) Underfit model to support bias. (b) Overfit model with high variance. (c) Sample of a good fit.

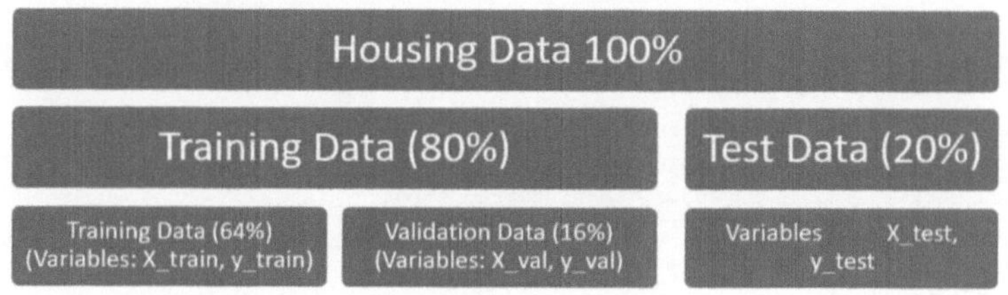


**FIGURE 8.11** Splitting the data into training, validation, and test sets.

In Line 6, the "Price" column is dropped from the data frame, and only the dependent variables are stored in the variable "x". In Line 8, the target variable price is stored in a variable "y". In Lines 11–13, the "train_test_split()" method is used to split the data frames "X" (dependent variables) and "y" (target variable) into an 80% train set and a 20% test set, with "test_size=0.2". The "random state" parameter ensures reproducibility by maintaining the same data split in every run. In Line 16, the training set is further split into 80% for training and 20% for validation, as illustrated in Figure 8.11.

At Line 21, the "LinearRegression()" model is instantiated and stored in a variable called "model". In Line 24, the model is trained using the "fit()" method, with the training data "X_train" and the target variable "y_train". In Line 27, the "predict()" method is called on the model to make predictions on the validation set, where results are stored in "y_val_pred". In Line 30, the "r2_score()" method is used to calculate R-squared by comparing the actual house prices in the validation, "y_val", with the predicted house prices, "y_val_pred".

In Line 34, the "predict()" method is called again with the test data, and the results are stored in "y_test_pred". The mean square error is calculated at Line 37 to understand our predicted results. Lines 41–44 create a chart that compares the actual values with the predictions on a scatter plot (shown in Figure 8.12), with the test set's actual values on the x-axis and the predicted values on the y-axis. The data points are represented in blue. Line 45 displays the scatter plot with a regression line. The regression line is a benchmark representing a perfect correlation between the two variables.

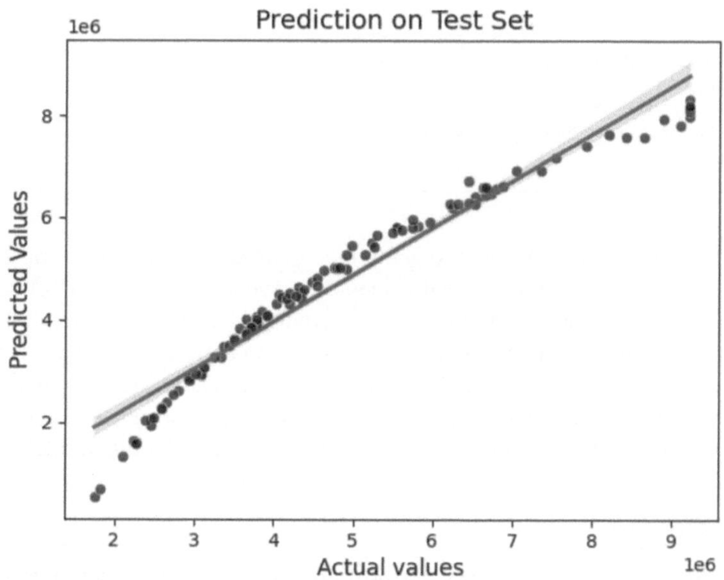


**FIGURE 8.12** Comparing actual values with predicted values.

**CODE 8.7 MODEL TRAINING, VALIDATION, AND TESTING**

```
from sklearn.model_selection import train_test_split
from sklearn.linear_model import LinearRegression
from sklearn.metrics import r2_score, mean_squared_error

# Removing the target variable from the dataset
X = df1.drop("Price", axis=1)
# Target Outcome
y = df1['Price']

# Split the dataset into train and test sets (80% train, 20% test)
X_train, X_test, y_train, y_test = train_test_split(
    X, y, test_size=0.2, random_state=42
)

# Further split the training set into train and validation sets (80% train, 20% validate)
X_train, X_val, y_train, y_val = train_test_split(
    X_train, y_train, test_size=0.2, random_state=42
)

# Initialize the linear regression model
model = LinearRegression()

# Train the model on the training data
model.fit(X_train, y_train)

# Make predictions on the validation set
y_val_pred = model.predict(X_val)

# Evaluate the model on the validation set using R-squared
r2 = r2_score(y_val, y_val_pred)
print("********* Validation R-squared *********\n", r2)

# Make predictions on the test set
y_test_pred = model.predict(X_test)

# Evaluate the model on the test set using R-squared
test_mse = mean_squared_error(y_test, y_test_pred)
print("********* Test  R-squared: *********  \n", test_mse)

# Visualize the difference between predictions and actual values
sns.scatterplot(x=y_test, y=y_test_pred, color='blue')
# Regression line
sns.regplot(x=y_test, y=y_test_pred, scatter=False, color='green')
plt.title('Prediction on Test Set', fontsize=14)
plt.show()
```

In Figure 8.12, the model may be able to generalize to unseen examples, as most predicted values are close to or slightly higher than the actual values. However, some values at both ends indicate that the predicted values are very low compared to the actual price. Therefore, we can try a more complex regression algorithm, such as polynomial regression, to increase our model's variance.

After evaluating a model, we may need to select another model, tune the existing model's hyperparameters, or transform the features to improve the model's performance. Therefore, ML requires iterative development and switching between stages to create an effective model.

### 8.2.2 Model Deployment

The most common task that practitioners perform after validation is to serialize the trained model into a format that can be saved and loaded easily, such as using object serialization with Pickle, as shown in Code 8.8 and discussed earlier in Chapter 6. Line 4 creates the file for the model, and Line 5 stores the model. Lines 8 and 9 load the model from the file.

Another common library for saving and loading ML models is the "joblib" library. This library is an alternative to pickle and is more efficient for storing large datasets of NumPy arrays. Code 8.9 imports this library in Line 1 and illustrates how to store the model at Line 4 using the "dump()" method. Line 7 provides an example of how to read the model from a file and apply it to unseen data in real life.

The trained and saved model can now be integrated into an application to make predictions. This integration can be provided as an API, a microservice, or by embedding the model directly into the application code.

**CODE 8.8 USING PICKLE TO SERIALIZE AND DESERIALIZE A MODEL**

```
1. import pickle
2.
3. # Saving the model to model.pkl file
4. file= open('model.pkl', 'wb')
5. pickle.dump(model, file)
6.
7. # Loading the model from model.pkl file
8. file=open('model.pkl', 'rb')
9. model = pickle.load(file)
```

**CODE 8.9 USING JOBLIB TO STORE AND RETRIEVE A MODEL**

```
1. from joblib import dump, load
2.
3. # Store the model in model.joblib file
4. dump(model, 'model.joblib')
5.
6. # Load the model from model.joblib file
7. loaded_model = load('model.joblib')
```

## 8.3 TYPES OF MACHINE LEARNING MODELS

ML problems can be broadly categorized into supervised, unsupervised, semi-supervised, and reinforcement learning. Figure 8.13 outlines these main categories and sub-categories of ML problems.

- *Supervised learning* problems involve predicting a dependent variable found within the available data. For example, in our house price prediction scenario, the house price needs to be predicted, and the data contains the prices of houses sold previously.
- *Unsupervised learning* problems require finding hidden patterns and structures within data without pre-defined dependent variables. An example is grouping emails into different topics based on their content without predefined categories.
- *Semi-supervised learning* employs a combination of labeled examples and a larger pool of unlabeled data to infer relationships and generalize effectively with minimal reliance on labeled data. An example is classifying images of handwritten digits using a small set of labeled images and a larger set of unlabeled images. The labeled images provide guidance, while the unlabeled images help refine the model's understanding.

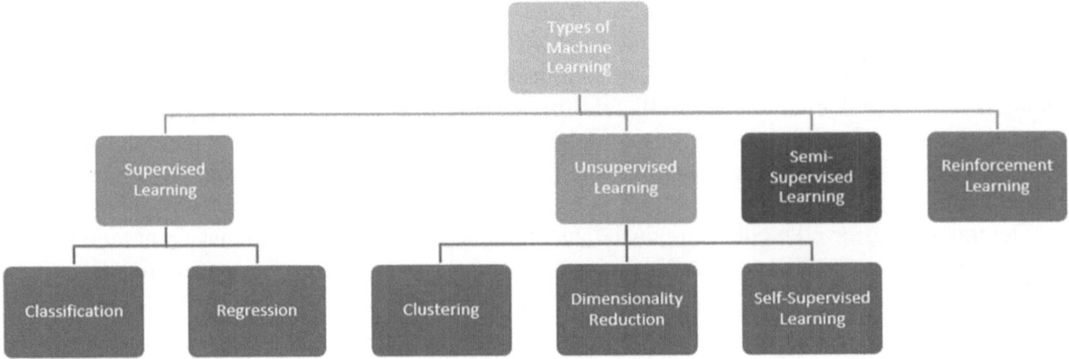


**FIGURE 8.13** Types of machine learning problems.

- *Reinforcement learning* problems require models to interact with an environment and learn through trial and error. An example is training a robot to navigate a maze through trial and error.

The following sections cover some algorithms applied within these broad categories. Over time, you will gain a deeper insight into the flexible nature of the algorithms, their strengths, and their application to various categories and stages of ML.

## 8.4 SUPERVISED LEARNING

Supervised learning involves training a model based on past examples to recommend future outcomes. We consider a problem under this category if each data point (record/row) is associated with a labeled outcome or target variable. For example, consider a dataset containing various health parameters such as body mass index (BMI), blood glucose levels, blood pressure, and age, as illustrated in Figure 8.14. Based on the patient's health parameters, a human diagnosis is present for each patient in the dataset, labeled as diabetic or non-diabetic. Each patient record is commonly referred to in ML as an "instance". The model is trained on a large, labeled dataset to find relationships between the independent variables, the health parameters, and the target labeled variable, diabetic or non-diabetic. This enables the model to predict new patients and determine whether they are diabetic or non-diabetic.

Supervised learning tasks can be further divided into two subtypes: classification and regression tasks. If the target variable is categorical, we apply classification. If it is numerical, we apply

| | Independent Variables | | | | Target Variable |
|---|---|---|---|---|---|
| Records | BMI | Blood Glucose Level (mg/dL) | Blood Pressure | Age | Labels (Diabetic= 1 Non-Diabetic=0) |
| 1 | 23.3 | 183 | 64 | 32 | 1 |
| 2 | 33.6 | 148 | 72 | 50 | 1 |
| 3 | 26.6 | 85 | 66 | 31 | 0 |
| 4 | 28.1 | 89 | 66 | 21 | 0 |
| 5 | 43.1 | 137 | 40 | 33 | 1 |

**FIGURE 8.14** Labelled data to predict diabetes.

**TABLE 8.2**
**Machine learning models based on target variables**

| | Algorithms | | | | | |
|---|---|---|---|---|---|---|
| **Classification (Categorical Target)** | Logistic Regression | Support Vector Classifier (SVC) | K-Neighbors Classifier | Decision Tree Classifier | Random Forest Classifier | Multilayer Classifier |
| **Regression (numerical target)** | Linear regression | Support Vector regressor (SVR) | K-Neighbors regressor | Decision tree regressor | Random Forest regressor | Multilayer regressor |

regression. Table 8.2 lists a few ML models available by Scikit-learn for classification and regression. These models are discussed in detail in the following sections.

## 8.4.1 Classification

A classification problem requires a model to predict the class or category to which a new observation belongs. Therefore, the target variable is categorical, such as in the case of predicting diabetes. The labeled data has two classes, diabetic or non-diabetic, as shown in Figure 8.15 (a). Another example is sentiment analysis, which determines whether a post on social media is positive, negative, or neutral – three classes (Figure 8.15 (b)). If the target outcome has two classes, it is considered a binary classification problem. If there are more than two classes, it is considered a multi-class classification problem. For example, classifying whether a picture contains a cat and whether an email is spam is a binary classification problem. However, identifying different animals in an image and categorizing important, promotional, or personal emails are examples of multi-class problems.

### 8.4.1.1 Evaluation Metrics

The training process for an ML algorithm seeks to develop a generalized model with a good fit, resulting in accurate predictions for most cases. While the model's predictions will be close to the actual values, a perfect match is not always achievable. Evaluation metrics assess the model's performance by measuring the discrepancy between predicted and actual values within the validation and test sets. This evaluation helps determine whether the model's performance is satisfactory, if further refinement is necessary, or if an alternative model should be explored.

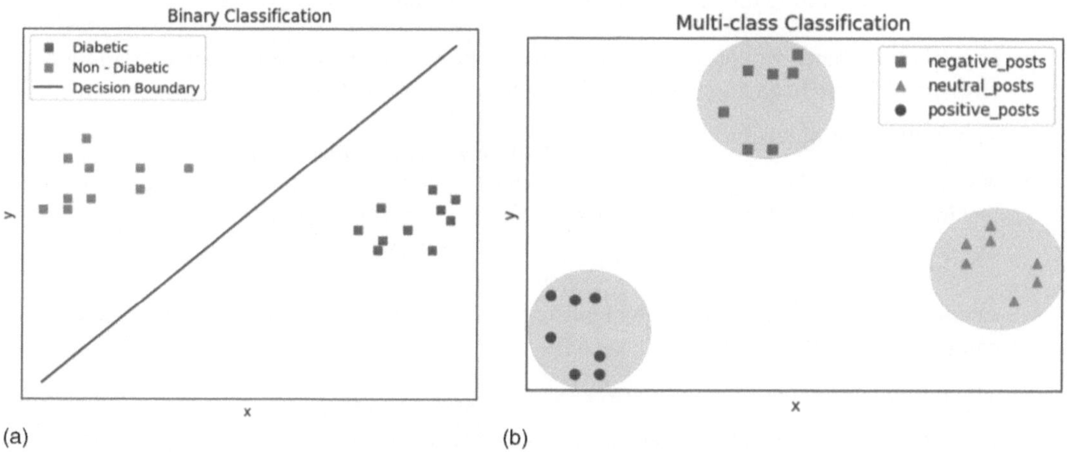


**FIGURE 8.15** Types of Classification. (a) Binary classification problem. (b) Multi-class classification problem.

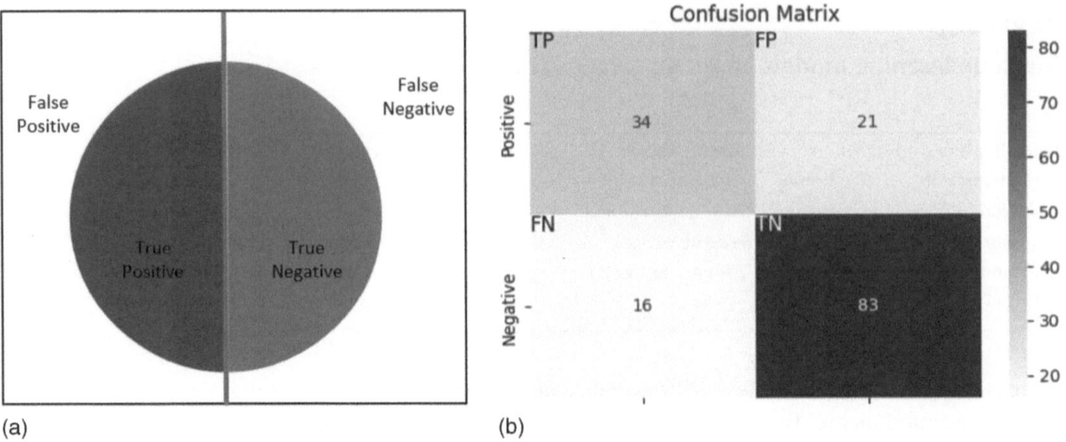


**FIGURE 8.16** Conceptual representation of the Confusion Matrix. (a) Circular representation with correct prediction (TP, TN) inside the circle. (b) Matrix representation with correct predictions(TP, TN) shown diagonally.

#### *8.4.1.1.1 Evaluation Metrics for Classification Problems*

The "confusion matrix" contains four types of measurements that determine the difference between the predicted and actual values: True Positive (TP), False Positive (FP), True Negative (TN), and False Negative (FN), as shown in Figure 8.16.

- **TP:** Represents correct positive predictions. For example, in identifying whether an image contains a cat or not, TP is an instance that is correctly identified as having a cat.
- **FP:** Represents incorrect positive predictions. For example, all images identified as having a cat but did not contain cats are FP.
- **TN:** Represents correct negative predictions. For example, the number of images correctly identified as not having a cat.
- **FN:** Represents positive instances that are incorrectly classified as negative. For example, cat images are identified as not having cats in them.

These four measurements are used to calculate further metrics for evaluating classification algorithms. "Sklearn.metrics" is a module in the Scikit-learn library of Python that facilitates various measures for evaluating ML models. The common evaluation metrics are precision, recall, f1-score, accuracy, receiver operating characteristic (ROC), and area under the ROC curve (AUC).

- *Precision* measures the proportion of correct predictions against all positive predictions. High precision indicates few negative instances are classified as positive.

  The formula for precision is: $Precision = \frac{True\ Positives}{True\ Positives + False\ Positive}$
- *Recall* measures the proportion of correct positive predictions among all actual positive instances. High recall indicates the model effectively identifies positive cases.

  The formula for the recall is: $Recall = \frac{True\ Positives}{True\ Positives + False\ Negative}$
- *Accuracy* measures the proportion of correctly classified instances in the dataset. Accuracy can be misleading in imbalanced datasets.

  The formula for accuracy is: $Accuracy = \frac{True\ Positives + True\ Negatives}{Total\ Predicitons}$
- *F1 Score* balances measure considering both precision and recall, especially useful in imbalanced datasets. A score close to 1 indicates good performance.

  The formula for F1 score is: $F1Score = 0.2 * \frac{Precision * Recall}{TPrecision + Recall}$

- *ROC* is a chart that plots the trade-off between TP rate (sensitivity) and FP rate (specificity) at various classification thresholds.
- *AUC* measures the model's ability to distinguish between the positive and negative classes. AUC ranges from 0 to 1, and higher values indicate better performance.

In Figure 8.17, the shape of the ROC bulges out from the red diagonal line, representing reasonably good performance (AUC of 0.84).

Scikit-learn provides various classification models, such as logistic regression, k-nearest neighbors, Random Forests, support vector machines (SVMs), decision trees, and more.

Code 8.10 applies the logistic regression algorithm to the diabetes dataset from the Kaggle website[4]. Lines 1–5 import the necessary libraries for ML and data processing. Lines 7 and 8 are used to avoid displaying warnings generated by the libraries used. Line 11 loads the dataset into the dataframe "df1". In Line 14, the target feature is stored in the variable "y"; in Line 15, all other features are stored in "X". In Line 18, the X and Y data frames are split into train and test datasets, 80% for training and 20% for testing. Lines 21 and 22 normalize the features using a standard scaler. Line 25 further splits the training set into a training set (80%) and a validation set (20%).

Lines 28–31 apply the logistic regression model from Scikit-learn. Line 28 imports "LogisticRegression", Line 29 initializes the logistic regression model, and Line 31 creates a list of six values for the hyperparameter "C", ranging from very small to large values, such as 0.001 and 100. This parameter influences the behavior and performance of a model and is discussed in more detail in the model's description in the following section.

At Line 34, a validation technique called k-fold cross-validation is applied. It splits the data into k subsets using k-1 subsets for training and 1 subset for validation, iterating until each subset has been tested. In Scikit-learn, the "GridSearchCV" class performs cross-validation and tests different hyperparameter combinations to find the best-performing model. It is initialized with the logistic regression model, hyperparameters dictionary, five-folds (cv=5), and accuracy as the evaluation metric. "Accuracy" is the loss function used by the grid search to determine the model's performance with various hyperparameter settings.

In Line 37, the "fit()" method trains logistic regression on the data "X_train" and "y_train", iterating over different subsets and hyperparameter settings each time to find the best-performing model.

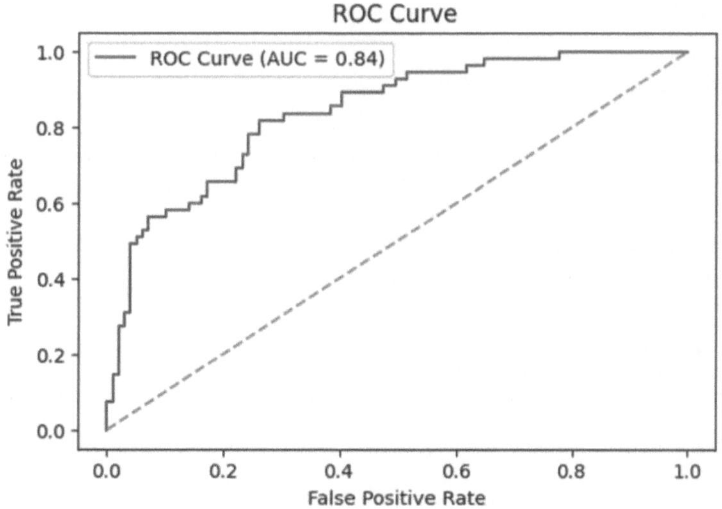


**FIGURE 8.17** Receiver operating characteristic (ROC) and the area under the curve (AUC).

In Line 38, the "best_estimator_()" method is used on the "grid_search" variable to return the best-performing model. In Line 39, to validate the results of training, the "predict()" method is used on the "X_val" validation set.

Accuracy, F1 score, and confusion matrix are calculated and printed on Lines 42, 43, 46, 47, and 50, 51, respectively. Lines 54–74 visualize the ROC and AUC using "matplotlib". Line 71 prints the best model, and Line 72 stores it in a file using "joblib".

### CODE 8.10 APPLYING A CLASSIFICATION MODEL TO THE DIABETES DATASET

```
1. import pandas as pd, seaborn as sns, matplotlib.pyplot as plt
2. from sklearn.model_selection import train_test_split, GridSearchCV
3. from sklearn.preprocessing import StandardScaler
4. from sklearn.metrics import accuracy_score, f1_score, confusion_matrix, roc_curve,roc_auc_score
5. from joblib import dump
6. # Code to avoid the display of warnings from the libraries used
7. import warnings
8. warnings.simplefilter('ignore')
9.
10. # Loading the Diabetes dataset
11. df1 = pd.read_csv (
        "https://raw.githubusercontent.com/Object-Oriented-Programming-2024/" \
        "Object-Oriented-Programming/main/Chapter8/diabetes.csv"
        )
12.
13. # Splitting the dataframe into features and target variables
14. y = df1["Outcome"]
15. X = df1.drop(columns=['Outcome'])
16.
17. # Splitting the dataset into training and testing sets (80% train, 20% test)
18. X_train, X_test, y_train, y_test = train_test_split(X, y, test_size=0.2, random_state=42)
19.
20. # Applying a standard scaler to normalize the features
21. scaler = StandardScaler()
22. X_train = scaler.fit_transform(X_train)
23.
24. # Further split the training set into train and validation sets (80% train, 20% validate)
25. X_train,X_val,y_train,y_val = train_test_split(X_train, y_train, test_size=0.2,random_state=42)
26.
27. # Applying the logistic regression
28. from sklearn.linear_model import LogisticRegression
29. model=LogisticRegression()
30. # Various settings for the hyperparameter C
31. hyperparams = {'C': [0.001, 0.01, 0.1, 1, 10, 100]}
32.
33. # Initialize GridSearchCV with five folds and find the best hyperparameter settings
34. grid_search = GridSearchCV(model, hyperparams, cv=5, scoring='accuracy')
35.
36. print(f"Training the model…")
37. grid_search.fit(X_train, y_train)
38. model = grid_search.best_estimator_
39. y_pred = model.predict(X_val)
40.
41. # Calculating and printing the accuracy
42. accuracy = accuracy_score(y_pred, y_val)
43. print(f"Accuracy: {accuracy:.2f}")
44.
45. # Calculating and printing the F1 score
46. f1 = f1_score(y_pred, y_val)
47. print(f"F1 Score: {f1:.2f}")
48.
```

```
49. # Calculating and printing the confusion matrix
50. cm = confusion_matrix(y_val, y_pred)
51. print(f"Confusion Matrix:\n{cm}")
52.
53. # Predicting probabilities for ROC and AUC
54. y_prob = model.predict_proba(X_val)[:, 1]
55.
56. # Calculating ROC curve
57. fpr, tpr, thresholds = roc_curve(y_val, y_prob)
58.
59. # Calculating AUC score
60. auc_score = roc_auc_score(y_val, y_prob)
61.
62. # Creating the ROC curve chart
63. plt.figure(figsize=(6, 4))
64. plt.plot(fpr, tpr, label=f'ROC Curve (AUC = {auc_score:.2f})')
65. plt.plot([0, 1], [0, 1], 'r--')
66. plt.title(f'ROC Curve ')
67. plt.legend()
68. plt.show()
69.
70. # Printing and storing the best model
71. print(model)
72. dump(model, f"best_model.joblib")
```

When the confusion matrix is printed at code Line 51, the following output will be displayed: [[74 7] [19 23]] as illustrated in Box 8.1. The first row represents the actual positive values (TP and FN). These values indicate that the model can identify 74 cases correctly as positive. However, seven cases were positive but were identified as negative. The second row represents the actual negatives (FP and TN). Nineteen cases were negative but were identified as positive, whereas 23 cases were correctly identified as negative. In this case, 19 cases are incorrectly identified as positive. Based on the scenario, this FP value can be a concern. For example, in the case of diabetes or other diseases, this number of FPs means incorrect diagnosis. The FNs are also a concern for the same reasons.

## BOX 8.1 OUTPUT OF APPLYING LOGISTIC REGRESSION

**Output**

```
Training the model…
Accuracy: 0.79
F1 Score: 0.64
Confusion Matrix:
[[74  7]
 [19 23]]
LogisticRegression(C=1)
```

What are some ways to improve the performance of this model? First, we did not perform any data analysis. Our dataset's features are numeric, as shown in Table 8.2. Plotting each feature on a histogram, such as "Glucose" and "Blood Pressure", reveals outliers, as shown in Figure 8.18. You can apply scaling, normalization, or transformation to the features to ensure that features are on a similar scale and have similar ranges. For example, log transformation was applied to the price feature earlier in Code 8.6.

Another step to improve the model's performance would be to apply other classification models to our data. Since the ML process remains the same, in Code 8.10, Lines 28–31 can be replaced with another model's import statement, model initialization, and hyperparameter settings. In many cases, more data is required to improve performance. The following sections describe some of the models used for solving classification problems.

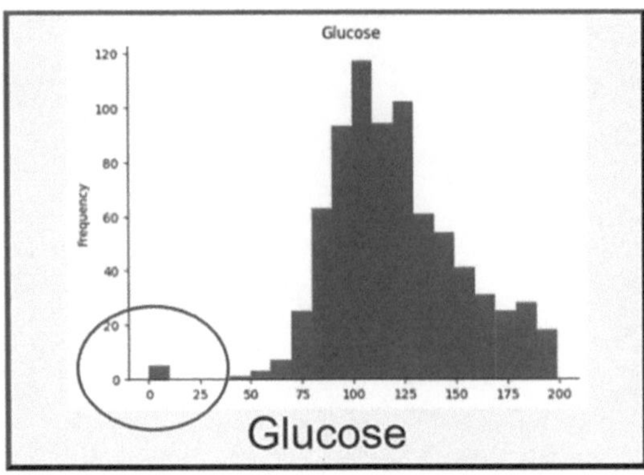


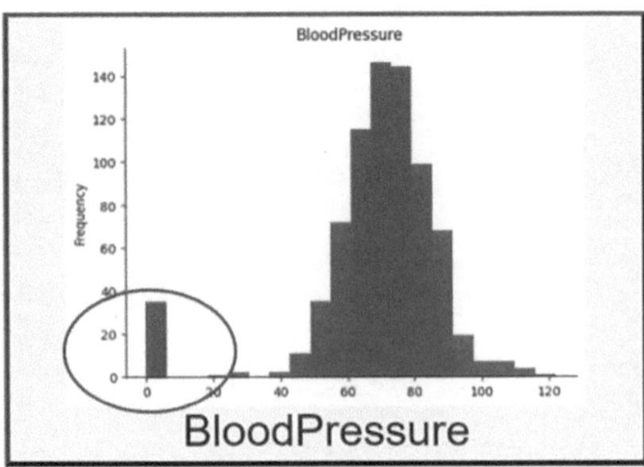


**FIGURE 8.18** Outliers in the features.

### 8.4.1.2 Logistic Regression

Logistic regression can be used for categorical target variables. It predicts the class an instance belongs to, based on existing labeled data. It is a linear model, meaning the relationship between the independent features directly affects the target outcome.

The logistic regression finds the probability of an instance being classified as 0 or 1 (e.g., diabetic or non-diabetic) by adding the features linearly with assigned weights to predict the outcome. The formula is as follows:

$$\text{Probability} = \beta_0 + \beta_1 x_1 + \beta_2 x_2 + \ldots + \beta_n x_n$$

where $\beta_0$ is the slope of the line, $\beta_1$ to $\beta_n$ are the weights assigned to the features, and $x_1$ to $x_n$ are the feature values.

A sigmoid or logistic function is applied to limit the probabilities between 0 and 1. The probabilities are compared to a threshold value (Figure 8.19) to predict the class outcome. For example, if the probability is greater than 0.5, then a patient is classified as diabetic, otherwise non-diabetic.

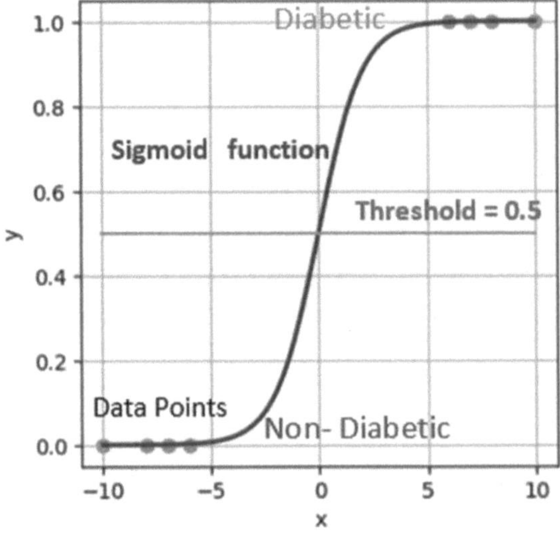


**FIGURE 8.19** Logistic regression representation.

Hyperparameters are parameters set before the training process begins, influencing the behavior and performance of a model. In logistic regression, regularization strength is the hyperparameter called "C". It helps prevent overfitting by penalizing models for assigning larger weights to features, values of $\beta_1$ to $\beta_n$. A smaller value of C increases the regularization strength and penalizes the algorithm more. In comparison, a larger value decreases regularization strength and may cause overfitting.

The maximum iterations or "max_iter" hyperparameter specifies the maximum number of iterations the algorithm allows to tune its parameters and reduce the loss function. If the algorithm does not find the best solution within the specified maximum number of iterations, it stops and returns the current parameter values.

Logistic regression is mainly used to understand a classification problem, especially binary classification. It is beneficial in identifying the most important features that impact predictions. While it can handle large datasets, its performance can be affected by high-dimensional data. Outliers can also affect the model's performance, but their impact can be reduced by applying pre-processing techniques.

#### 8.4.1.3 K-Nearest Neighbors (KNN)

The K-nearest neighbor (KNN) relies on the simple concept that data points with similar feature values will have a similar outcome. For example, consider a scatter plot where data points represent people with diabetes and others without diabetes, as illustrated in Figure 8.20. When a new data point is added for prediction, the KNN algorithm identifies neighbors with feature values similar to those of the data point. Then, the algorithm predicts the class (diabetes or no diabetes) based on the majority class of its neighbors. KNN can be applied to both classification and regression problems.

"K" is the hyperparameter that determines how many neighbors the algorithm considers when making a prediction. The term "N Neighbors" describes the set of "K" nearest neighbors identified for a specific data point. A small value for K focuses on a few close neighbors, which increases the risk of overfitting. A larger "K" value creates a more general model that reduces overfitting but may blur the distinction between classes and result in underfitting.

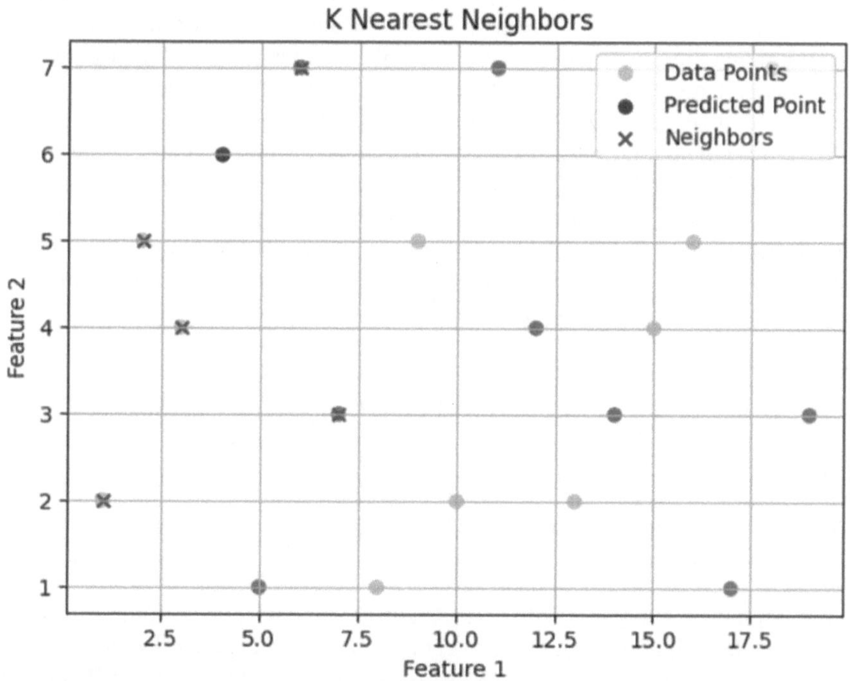


**FIGURE 8.20** K-nearest neighbor conceptual representation.

Lines 28–31 can be replaced in Code 8.10 to apply the k-nearest neighbor model to the diabetes dataset. In Line 31, "n_neighbours" is the list of numeric values to consider as neighbors when training the model.

**CODE 8.11 APPLYING K-NEAREST NEIGHBOR CLASSIFIER**

```
28. from sklearn.neighbors import KNeighborsClassifier
29. model= KNeighborsClassifier()
30. # Various settings for the hyperparameter n_neighbors
31. hyperparams = {'n_neighbors': [3, 5, 7, 9, 11]}
```

KNN is a simple model that does not rely on the relationships between features. Therefore, it is very suitable for complex and non-linear classification problems. It adapts to changes in training data, making it very suitable for evolving and noisy datasets. However, it requires a lot of computation power when processing large datasets, as it computes the distance of each data point to the nearest neighbors to make predictions.

#### 8.4.1.4 Support Vector Machine (SVM)

SVM can be applied to classification and regression problems. A support vector classifier (SVC) tries to find a hyperplane or decision boundary that can segregate data into multiple classes. The hyperplane with the maximum margin between the classes is selected. Figure 8.21 (a) shows that many decision boundaries can separate the classes. However, the one with the greatest distance from the data points of both classes is selected.

Another important feature of SVM is the kernel method, which can transform linearly non-separable classes into a higher-dimensional space to find clear distinctions between classes. Figure 8.21 (b) shows data points in two-dimensional space. Data points can be linearly separable in a higher dimension, as shown in Figure 8.21 (c).

The SVC model in Code 8.12 can be applied to the diabetes dataset in Code 8.10, Lines 28–31. The hyperparameter "C" in Line 31 controls the trade-off between maximizing the margin between classes and minimizing the classification error. A smaller "C" value leads to a larger margin between classes. In contrast, a larger "C" value results in a narrower margin. Finding the right value for "C" is important to generalize well to unseen data. Line 31 contains another hyperparameter, the kernel method, which helps map features and decision boundaries to a higher dimensional space. There are three types of kernel methods: linear, polynomial, and radial basis function (RBF). The linear method creates a linear decision boundary suitable for linearly separable data. In contrast, a polynomial transforms the original features into polynomial features and is suitable for data that requires curved boundaries for separation. RBF transforms the data into an infinite higher dimensional space suitable for complex non-linear decision boundaries.

**CODE 8.12 APPLYING SUPPORT VECTOR CLASSIFIER**

```
28. from sklearn.svm import SVC
29. model=SVC(probability=True)
30. # Various Hyperparameter settings for Support Vector Classifier
31. hyperparams = {'C': [0.001, 0.01, 0.1, 1, 10, 100], 'kernel': ['linear','polynomial','rbf']}
```

SVM is very suitable for datasets with many features and can support complex non-linear datasets with the help of the polynomial and RBF kernel methods. It can generalize well for classification tasks and is less likely to overfit. However, SVM is less interpretable, as it is difficult to understand the decision boundary with non-linear kernels.

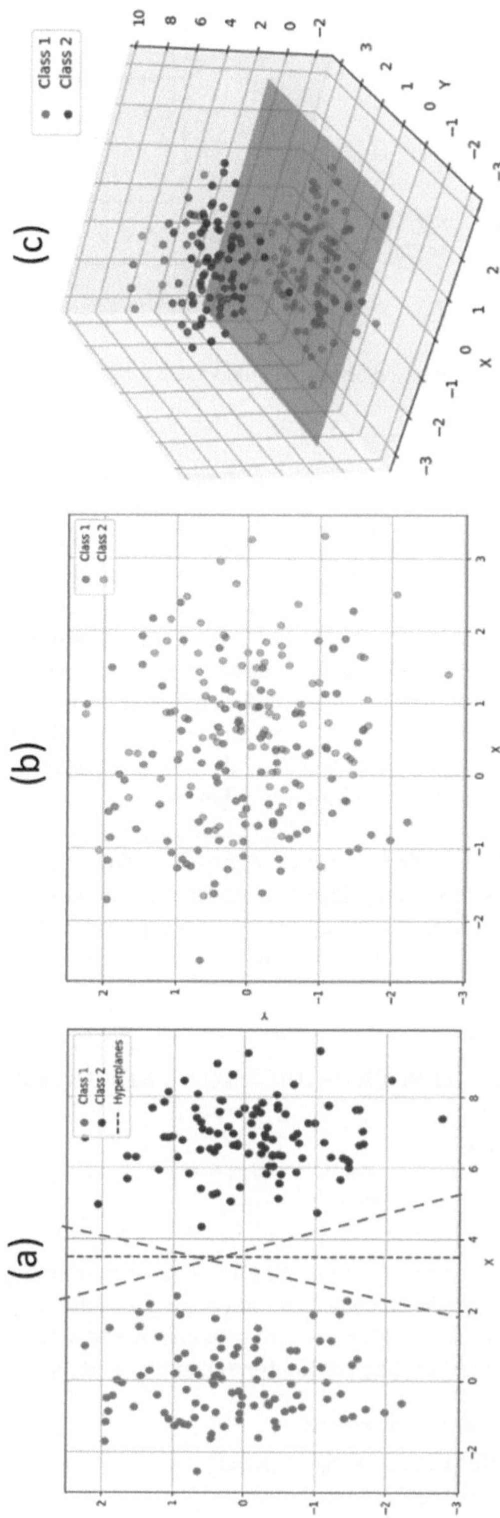


**FIGURE 8.21** (a) Multiple hyperplanes. (b) Linearly inseparable classes. (c) Linearly separable classes in a higher dimension.

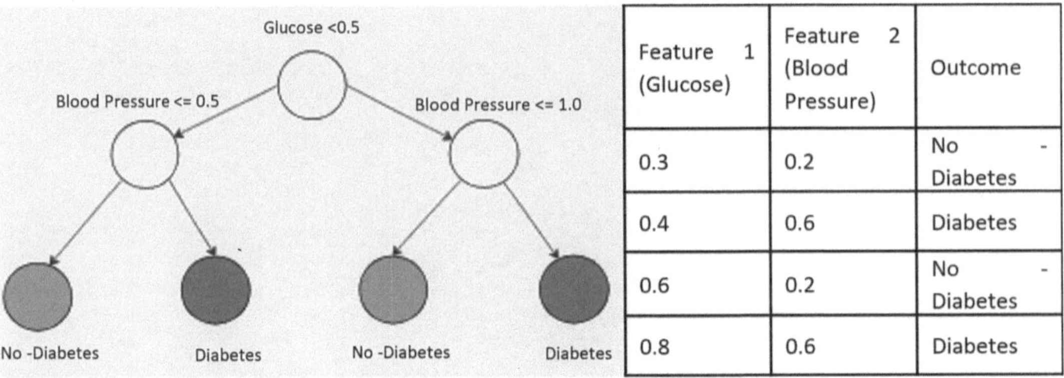


| Feature 1 (Glucose) | Feature 2 (Blood Pressure) | Outcome |
|---|---|---|
| 0.3 | 0.2 | No -Diabetes |
| 0.4 | 0.6 | Diabetes |
| 0.6 | 0.2 | No -Diabetes |
| 0.8 | 0.6 | Diabetes |

**FIGURE 8.22** A simple decision tree based on normalized values for glucose and blood pressure.

#### 8.4.1.5 Decision Tree

A decision tree can be used for both classification and regression tasks. It creates a tree-like structure based on feature values, where each internal node represents a decision based on a feature value, and each child node represents another decision based on another feature value. The decisions continue until the leaf node represents the target feature, resulting in a non-linear path to the prediction.

Let's consider the diabetes dataset with only 4 rows. Using two features, glucose and blood pressure, a decision tree can be drawn based on simple decisions, as shown in Figure 8.22. The decision tree begins with a decision separating data points based on glucose values less than 0.5 (2 instances on the left side) and greater than 0.5 (2 instances on the right side). Child nodes are created based on the decisions for blood pressure. If the blood pressure is less than 0.5, then the number of non-diabetes outcomes is recorded (1st row). However, if the blood pressure exceeds 0.5, the number of diabetic outcomes is recorded (2nd row). Similarly, other branches are recorded.

The following decision tree model in Code 8.13 can be applied to the diabetes dataset in Code 8.10 (Lines 28–31) instead of the code for logistic regression. The "max_depth" hyperparameter at Line 31 of Code 8.13 restricts how many levels of nodes a decision tree can have. Increasing the maximum depth allows the decision tree to capture more complex relationships in the training data. However, deeper trees can lead to overfitting, where the model memorizes the data rather than generalizing well to unseen data.

**CODE 8.13 APPLYING DECISION TREE CLASSIFIER**

```
28. from sklearn.tree import DecisionTreeClassifier
29. model=DecisionTreeClassifier()
30. # Various Hyperparameter settings for depth of the trees
31. hyperparams ={'max_depth': [3, 5, 7, None]}
```

Decision trees are widely used due to their simplicity and transparency. They can capture non-linear relationships between features and target outcomes and are useful in feature selection. However, decision trees are sensitive to small variations in datasets, can overfit, and are biased with imbalanced datasets. They are mostly used as building blocks for ensemble models, such as Random Forest.

#### 8.4.1.6 Ensemble Model – Random Forest

"Ensemble" combines multiple weak models to generate stronger and more accurate predictions. Random Forest is an ensemble combining independent decision trees to generate an average of their predictions for more accurate results. Therefore, it randomly creates subsets of data points and features to train multiple decision trees. Consider one data point on three different decision trees, as shown in Figure 8.23. Random Forest can select the outcome based on the prediction from most trees.

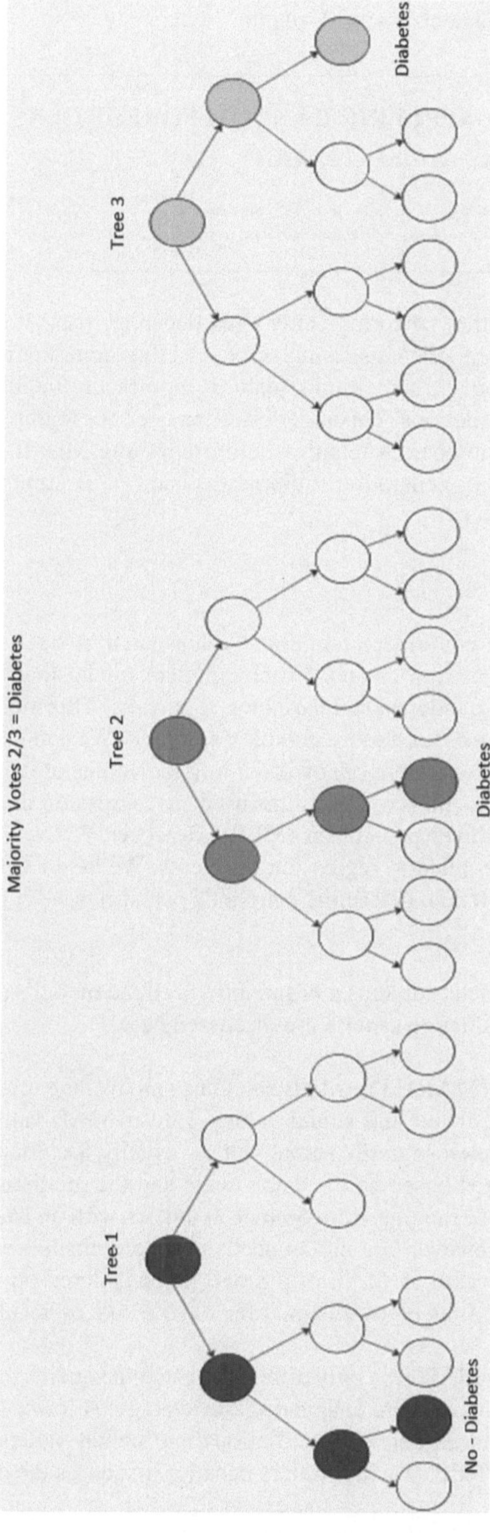


**FIGURE 8.23** Random Forest prediction based on three decision trees.

Code 8.1 can be used in Code 8.10 (28–31) to apply the Random Forest to the diabetes dataset. The hyperparameter "n_estimators" refers to the number of decision trees to create. "Max_depth" is the same hyperparameter as the decision tree's depth.

**CODE 8.14 APPLYING RANDOM FOREST CLASSIFIER**

```
28. from sklearn.ensemble import RandomForestClassifier
29. model=RandomForestClassifier()
30. # Various Hyperparameter settings for linear regression
31. hyperparams = {'n_estimators': [50, 100, 200], 'max_depth': [3, 5, 7, None]}
```

The Random Forest algorithm performs better than decision trees. It averages the predictions from individual trees and incorporates randomness by selecting combinations of features and samples to create the trees. Moreover, it measures feature importance, indicating which features are more influential in making predictions. This model is often used for feature selection during exploratory analysis to reduce the number of features before applying ML. It is also very suitable for datasets with outliers. However, generating multiple decision trees simultaneously requires more memory and computation power.

## 8.4.2 Regression

Regression algorithms predict continuous numerical values such as stock prices, credit risk, sales, and weather predictions. Regression creates a mathematical model that expresses the dependent variable, `y`, as a function of the independent variables, $x_1$ to $x_n$. This model can be a simple equation like a straight line, `y=mx+b`, or a more complex equation. We can use these equations to predict the values of the dependent variable (`y`) by providing the values of the independent variables.

Under this topic, we will cover a few algorithms used for regression like linear regression, polynomial regression, and multi-layer perceptron (MLP). However, KNN, SVM, decision trees, and Random Forests can also be applied to regression problems. We begin with the evaluation metrics for regression problems, which help determine a model's performance on the available data.

### 8.4.2.1 Evaluation Metrics

Evaluation metrics for regression problems measure how well the model's predictions align with the actual values. Some of the evaluation criteria are discussed here:

- **Mean Absolute Error (MAE)**: The MAE calculates an average after finding the absolute differences between predicted and actual values. Lower MAE values generally indicate that predicted values are close to the actual values, usually less than 0.5. However, MAE values must be considered based on the data's scale and the prediction type. For example, predicting taxi fare prices ranging from 50 to 100 dollars with an MAE of 0.8 can be considered a small error. However, if a model needs to predict insulin dosage based on blood glucose levels, a model with an MAE of 0.8 will be considered poor-performing and not applicable. Underestimation or overestimation by 0.8 mg of insulin can cause serious health concerns.
- **MSE**: MSE is similar to MAE. The only difference is that it squares the differences between actual and predicted values before calculating the average. A lower MSE signifies that the average squared differences between predictions and actual values are small. However, MSE is not very useful for data with outliers because it squares the errors.
- **R Squared Score ($R^2$)**: $R^2$ measures the extent to which the variability in the dependent variable (y) is predictable by the independent variables ($x_1$ to $x_n$). $R^2$ values range from

0 to 1, with higher values indicating a better fit. For example, a model with an $R^2$ value of 0.7 for predicting airline prices based on distance, class, and rating can be considered a good reference for a baseline fare. However, improvement can be made with hyperparameter tuning or considering additional features.

### 8.4.2.2 Linear Regression

Linear regression is one of the simplest and most widely used regression techniques. It helps determine the relationship between the independent and dependent variables using a linear equation (y=mx+b), where "y" is the dependent variable, "x" represents the independent variable, and "m" represents the slope. After fitting the best linear line to the data, the model can make future predictions for "y".

For multiple independent variables, multiple linear regression is used with the formula given:

$$Y = \beta_0 + \beta_1 x_1 + \beta_2 x_2 + \ldots + \beta_n x_n$$

where y is the predicted value, $\beta_0$ is the intercept, and $\beta_1$ to $\beta_n$ are the coefficients representing the change in the dependent variable when there is a one-unit change in the independent variable.

#### *8.4.2.2.1 Loss Function*

The loss/cost function measures how well the model's predicted values match the actual values in the training data. The most common loss function for linear regression is the MSE. The goal of the model is to find the best line that minimizes this error, ensuring the model's predictions are close to the actual values.

#### *8.4.2.2.2 Gradient Descent*

Gradient descent is an optimization technique used to adjust the model's coefficients ($\beta_1$ to $\beta_n$) iteratively in the direction that reduces the loss function. Starting with random initial values, the algorithm calculates the gradient (slope) of the loss function concerning each coefficient and updates them in the opposite direction of the gradient. The learning rate determines the size of these updates. The process continues until the loss method cannot be reduced further, achieving convergence.

Code 8.15 illustrates applying simple linear regression to a sales dataset from Kaggle[5]. The dataset contains four features: TV, radio, and newspaper advertisement budgets and the corresponding sales (dependent variable).

**CODE 8.15 APPLYING LINEAR REGRESSION TO PREDICT SALES BASED ON ADVERTISING BUDGET**

```
import pandas as pd, seaborn as sns, matplotlib.pyplot as plt
from sklearn.linear_model import LinearRegression
from sklearn.model_selection import train_test_split
from sklearn.metrics import r2_score, mean_squared_error, mean_absolute_error
df = pd.read_csv(
    "https://raw.githubusercontent.com/Object-Oriented-Programming-2024/" \
    "Object-Oriented-Programming/main/Chapter8/advertising_budget_sales.csv"
)

# Storing the independent features in X
X = df.drop('Sales', axis=1)
# Target Outcome / dependent feature in Y
y = df['Sales']

```

```
# Splitting the dataset into train and test sets (80% train, 20% test)
X_train, X_test, y_train, y_test = train_test_split(X, y, test_size=0.2, random_state=42)
# Further splitting the training set into train and validation sets (80% train, 20% validate)
X_train, X_val, y_train, y_val = train_test_split(X_train, y_train, test_size=0.2, random_state=42)

# Initializing the linear regression model
model = LinearRegression()

# Training the model
model.fit(X_train, y_train)

# Making predictions on the validation set
y_val_pred = model.predict(X_val)

# Evaluate the model on the validation set
val_MAE = mean_absolute_error(y_val, y_val_pred)
val_MSE = mean_squared_error(y_val, y_val_pred)
r2 = r2_score(y_val, y_val_pred)
print("********* Validation Results *********\n")
print(f"Mean Absolute Error: {val_MAE:.4f}")
print(f"Mean Square Error: {val_MSE:.4f}")
print(f"R Square: {r2:.4f}")

# Visualize the difference between predictions and actual values
sns.scatterplot(x=y_val, y=y_val_pred, color='blue')
# Regression line
sns.regplot(x=y_val, y=y_val_pred, scatter=False, color='green')
plt.title('Prediction on Validation Set', fontsize=14)
plt.show()

# Make predictions on the test set
y_test_pred = model.predict(X_test)

# Evaluate the model on the test set
test_MAE = mean_absolute_error(y_test, y_test_pred)
test_MSE = mean_squared_error(y_test, y_test_pred)
test_r2 = r2_score(y_test, y_test_pred)
print("********* Test Results *********\n")
print(f"Mean Absolute Error:{test_MAE:.4f}")
print(f"Mean Square Error:{test_MSE:.4f}")
print(f"R Square:{test_r2:.4f}")

# Visualize the difference between predictions and actual values
sns.scatterplot(x=y_test, y=y_test_pred, color='blue')
# Regression line
sns.regplot(x=y_test, y=y_test_pred, scatter=False, color='green')
plt.title('Prediction on Test Set', fontsize=14)
plt.show()
```

#### *8.4.2.2.3 Feature Selection*

After performing linear regression, you can use the ".coef_" attribute on the model to identify features that have a higher impact on the dependent variable. This helps in feature selection and dimensionality reduction.

Code 8.16 shows how to find the significant features after fitting a linear regression model. Line 2 retrieves the coefficient values from the model, and Line 3 retrieves the names. Line 6 iterates through the features and prints each with its corresponding coefficient value.

**CODE 8.16 PRINTING CORRELATION COEFFICIENTS**

```
# Extract coefficients
coef_values = model.coef_
feature_names = X.columns

# Print coefficients and corresponding feature names
for i in range(len(feature_names)):
    print(f"{feature_names[i]}: {coef_values[i]:.4f}")
```

### 8.4.2.2.4 *Regularization of Linear Regression*

Regularization is a technique for creating a more generalized model and balancing between under and overfitting. Ridge and Lasso regression are two popular regularization techniques that add a penalty term to the cost/loss function.

Ridge regression adds an L2 penalty, the square of the coefficients, to the loss function, reducing the coefficients' magnitude and shrinking them toward zero. This reduces the model's sensitivity to noise and improves its generalization, but L2 does not provide feature selection. The Ridge regression in Code 8.17 can be applied to the advertising budget example in Code 8.15 by updating Lines 18–22 as follows:

**CODE 8.17 APPLYING RIDGE REGULARIZATION**

```
18. from sklearn.linear_model import Ridge
19. model = Ridge(alpha=1.0)  # Regularization strength (alpha)
20. model.fit(X_train, y_train)
```

Lasso Regression adds an L1 penalty to the loss function. L1 refers to the L1 norm (also called L1 regularization or sparsity penalty). Introducing a penalty term into the model's loss function is a mathematical technique. L1 provides feature selection, and it is an absolute value of coefficients, shrinking them, and in certain cases, to exactly zero. L1, the Lasso regression in Code 8.18 can be applied to the advertising budget example in Code 8.15 by updating Lines 18–22 as follows:

**CODE 8.18 APPLYING LASSO REGULARIZATION**

```
18. from sklearn.linear_model import Lasso
19. model = Lasso(alpha=1.0)  # Regularization strength (alpha)
20. model.fit(X_train, y_train)
```

### 8.4.2.2.5 *Polynomial Regression*

Polynomial regression is a form of linear regression that captures non-linear relationships by converting independent features into polynomial terms of a specified degree. Polynomials or higher degrees can facilitate a flexible model and a closer fit to the training data but can lead to overfitting. We can prevent overfitting by applying cross-validation and regularization techniques. The polynomial regression in Code 8.19 can be applied to the advertising budget example in Code 8.15 by updating Lines 18–22. Code 8.19 converts all features to polynomial features with degree 2 before applying linear regression.

**CODE 8.19 APPLYING POLYNOMIAL REGRESSION**

```
from sklearn.preprocessing import PolynomialFeatures
poly_features = PolynomialFeatures(degree=2)
X_train_poly = poly_features.fit_transform(X_train)
model = LinearRegression()
model.fit(X_train_poly, y_train)
```

Linear and polynomial regression are simple and interpretable models that provide coefficients representing the relationship between independent and dependent variables. However, they may not capture complex relationships, potentially leading to underfitting.

#### 8.4.2.3 Multi-layer Perceptron (MLP)

MLP is a type of artificial neural network (ANN) provided by the Scikit-learn library to address classification, regression, and pattern recognition problems. The functions of the human brain inspire ANNs. They consist of layers of interconnected nodes representing neurons in the human brain. They consist of three main layers: an input layer, one or more hidden layers, and an output layer, as illustrated in Figure 8.24. Nodes in each layer are connected to all nodes in the following layer. The connections between the nodes are given weights to represent the strength of the connection. The weights are adjusted during the training process to minimize the loss function.

Each node in the input layer represents one feature of the data. The input layer processes the input data and passes it to the hidden layers. Nodes in the hidden layers perform computations and assign weights to the connections. Each node in the output layer generates predictions based on the assigned weights received from the last hidden layer.

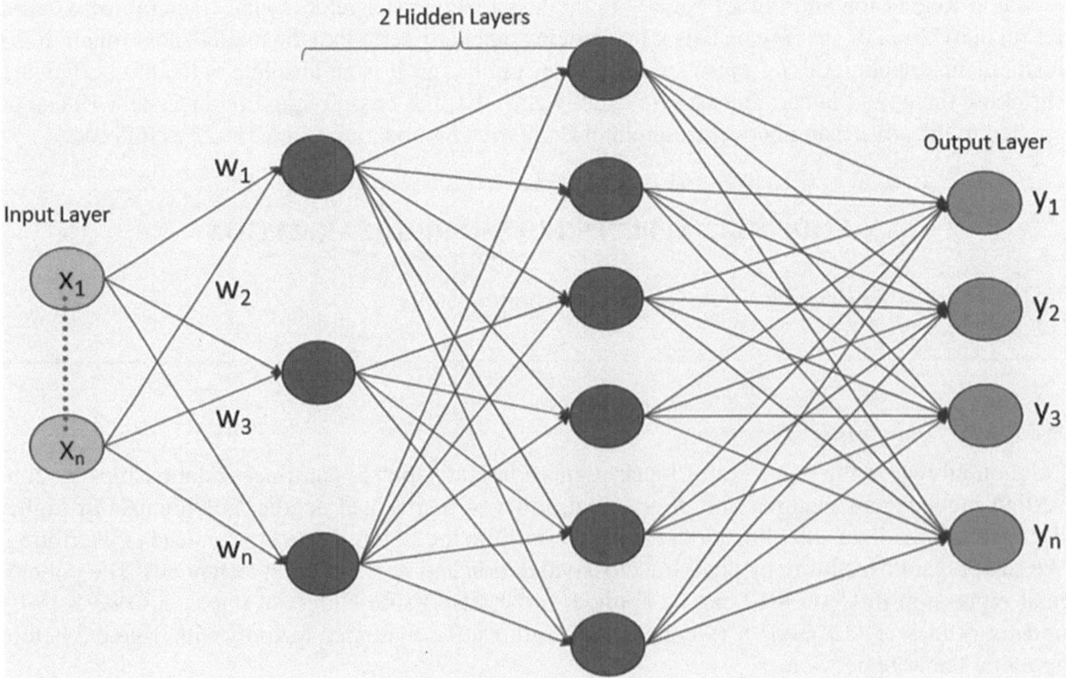


**FIGURE 8.24** Multi-layer perceptron (MLP) representation.

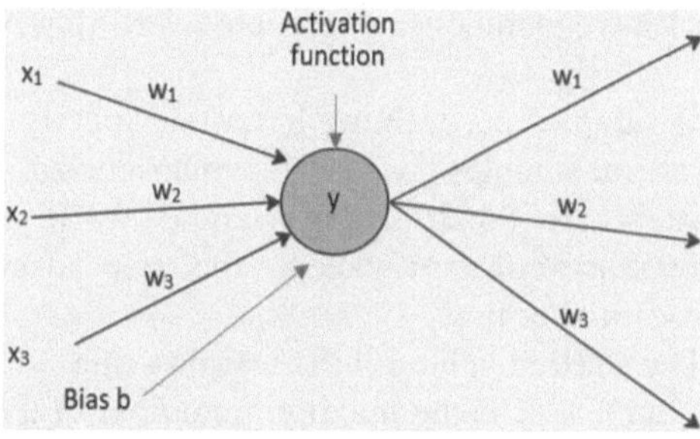


**FIGURE 8.25** A node in a hidden layer of MLP.

Each node in the hidden layer applies an activation function on the sum of the weights received from the previous layers, including a bias term, as shown in Figure 8.25. This function introduces non-linearity and enables the model to learn complex patterns.

The number of output nodes depends on the problem being solved: multiple nodes for classification (one per class) or a single node for regression (predicted value). The MLP regressor in Code 8.20 can be applied to the advertising budget example in Code 8.15 by updating Lines 18–22. In Line 19, the MLPRegressor is initialized with various hyperparameter settings discussed below.

**CODE 8.20 APPLYING MULTI-LAYER PERCEPTRON REGRESSOR**

```
18. from sklearn.neural_network import MLPRegressor
19. model = MLPRegressor(hidden_layer_sizes=(100, 50),  # Two hidden layers with 100 & 50 nodes
                    activation='relu',  # ReLu activation method
                    solver='adam',      # Adam Optimization algorithm
                    alpha=0.01,         # L2 regularization parameter
                    batch_size='auto',  # Size of minibatches for stochastic optimizers
                    learning_rate='constant',  # Learning rate schedule for weight updates
                    learning_rate_init=0.001,  # Initial learning rate
                    max_iter=200,       # Maximum number of iterations
                    random_state=42)    # Random seed for reproducibility
20. model.fit(X_train, y_train)
```

MLP has many hyperparameter settings to consider, and tuning them for the best performance is important. The following are the settings to consider:

- **Number of hidden layers**: The number of hidden layers and nodes in each hidden layer are hyperparameters that can be adjusted during training based on the problem's complexity and the dataset's size. For example, (100, 50) for two hidden layers, with the first layer having 100 nodes and the second layer having 50 neurons.
- **Activation function**: This function introduces non-linearity to the output, allowing the model to learn complex patterns. Common activation functions include Rectifier Linear Unit (ReLU), which converts all negative values of x to zero and maps all positive values of x as a straight line with the slope 1, sigmoid, which maps input values x to a range between 0 and 1, resulting in a probability, and hyperbolic tangent (tanh), a symmetric method that maps input values of x to the range between -1 and 1.
- **Optimizer algorithm**: This algorithm updates network weights during training. Common algorithms include Stochastic Gradient Descent (SGD, which updates parameters using the gradient computed from randomly selected data point(s)) and Adam (which adds momentum and adaptive learning rates).

- **Regularization**: L1 or L2 are techniques to prevent overfitting by penalizing features with large weights.
- **Batch size**: Number of samples per training iteration. Specifying batch_size="auto" lets the algorithm choose several samples based on the dataset size.
- **Learning rate**: The step size for adjusting weights while optimizing each iteration. Selecting the learning_rate="constant" keeps the step size the same, if learning_rate="adaptive" decreases it over time.
- **Learning_rate_init**: The starting value for the learning rate.
- **Max iterations (max_iter)**: This is the maximum number of training iterations or epochs to prevent indefinite running if there is no convergence.

MLP can automatically learn relevant features from raw input data through the hidden layers, which can be advantageous when dealing with high-dimensional or complex datasets. Moreover, they scale well to large datasets by increasing the number of hidden layers and neurons, although this requires more computational resources. However, MLP can cause overfitting, especially when the model is complex and the training data is limited. There are regularization techniques that can be used to avoid overfitting. MLP requires tuning many hyperparameters to achieve optimal performance, demanding extensive experimentation, and time. Moreover, it is considered a black-box model, meaning it is difficult to understand how it arrives at its predictions.

#### 8.4.2.4 Deep Neural Networks

Deep neural networks (DNNs) are also a type of ANN, like MLPs, but they have more hidden layers than MLPs. These hidden layers can capture more complex relationships such as hierarchies. LLMs, such as Gemini and Chat GPT, use DNNs to train vast amounts of text data to predict the next word in the sequence. The input is usually a sequence of words or tokens, and the output is a sequence of words or tokens. The model is trained to predict the subsequent word in a sequence given the prior context.

## 8.5 UNSUPERVISED MACHINE LEARNING

Unsupervised learning models discover patterns and relationships within the data without any labeled target variable. Unlike supervised learning, it is difficult to determine the success of unsupervised learning as no labeled data is available for comparison. These algorithms cannot help in making predictions. However, they help understand complex datasets and perform clustering and dimensionality reduction.

### 8.5.1 Clustering Problems

Clustering algorithms group similar data points. For example, they can group customers by purchasing behavior or recommend movies based on past viewership. K-means is a popular clustering algorithm that partitions data into k clusters. It finds centroids (center points) within groups of related data points and assigns new data points to the nearest cluster. Figure 8.26 displays three clusters with their corresponding centroids represented with stars.

The iris dataset provided by the Scikit-learn library is a classic and well-known dataset used for ML tasks, particularly in classification problems. It represents a collection of data points that describe different species of Iris flowers. Code 8.21 applies the K-means clustering algorithm to the iris dataset. Lines 1–4 in the code import the necessary libraries: NumPy, Pandas, and Matplotlib. After importing the necessary libraries from code Lines 1 to 4, the iris dataset is accessed from Scikit-learn using the method load_iris() at Line 7. The features are stored in variable "X" at Line 8.

K means clustering is sensitive to differences in the scales of the features. Therefore, at code Line 10, a standard scaler is applied to ensure all features are at similar scales. In Line 13, the KMeans algorithm is initiated with the hyper-parameter "n_clusters=3", specifying the number of clusters.

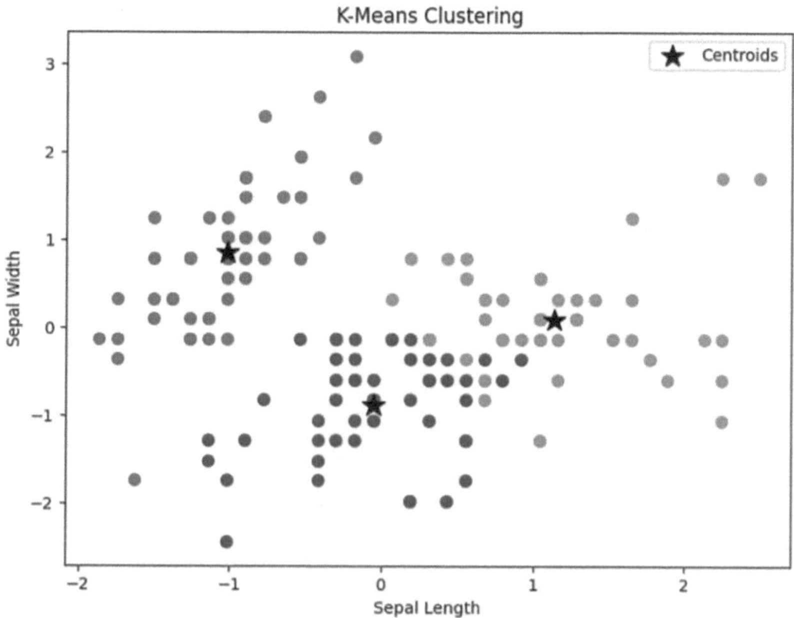


**FIGURE 8.26** K-means clustering.

Since the iris dataset has three species, we have specified three. The "init=k-means++" parameter ensures that the initial values for the centroids are far apart, improving the model's efficiency and performance.

In Line 16, the K-means is applied to the features using the "fit()" method. The centroids of the three clusters are stored in a variable at Line 19, and the three class labels assigned to each datapoint are retrieved at Line 20. A graph using Matplotlib is created at Line 23. The three clusters are added to the graph separately as scatter plots at Lines 25, 26, and 27. In Line 29, the centroids are added to the graph as stars. Lines 31–35 add a title, "x" and "y" labels, a legend, and display the graph.

### CODE 8.21 APPLYING THE K-MEANS CLUSTERING ALGORITHM

```
1. import numpy as np, pandas as pd, matplotlib.pyplot as plt
2. from sklearn.cluster import KMeans
3. from sklearn.datasets import load_iris
4. from sklearn.preprocessing import StandardScaler
5.
6. # Loading the Iris dataset
7. iris = load_iris()
8. X = iris.data
9.
10. X=StandardScaler().fit_transform(X)
11.
12. # Initializing KMeans with 3 clusters
13. kmeans = KMeans(n_clusters=3, init="k-means++")
14.
15. # Fitting the model
16. kmeans.fit(X)
17.
18. # Getting the cluster centroids and labels
19. centroids = kmeans.cluster_centers_
20. labels = kmeans.labels_
21.
```

```
# Plotting the clusters
plt.figure(figsize=(8, 6))

plt.scatter(X[labels == 0, 0], X[labels == 0, 1], s=50)
plt.scatter(X[labels == 1, 0], X[labels == 1, 1], s=50)
plt.scatter(X[labels == 2, 0], X[labels == 2, 1], s=50)

plt.scatter(centroids[:, 0], centroids[:, 1], s=200, c='black', marker='*', label='Centroids')

plt.title('K-Means Clustering')
plt.xlabel('Sepal Length')
plt.ylabel('Sepal Width')
plt.legend()
plt.show()
```

Unlike supervised learning, evaluating the quality of K-means clustering is subjective and requires manual interpretation of the clusters.

### 8.5.2 Dimensionality Reduction

Dimensionality reduction techniques are used to reduce the number of features in the dataset while preserving the relationships between the features. Principal component analysis (PCA) is a common technique that reduces features by combining the original features into principal components that maintain the data's original variance with minimal loss. PCA is a very useful technique for visualizing high-dimensional data in lower dimensions. For example, the iris dataset from Scikit-learn contains four features: sepal length, sepal width, petal length, and petal width. We can apply PCA to reduce the features to two and visualize the new features on a scatter plot, as shown in Figure 8.27.

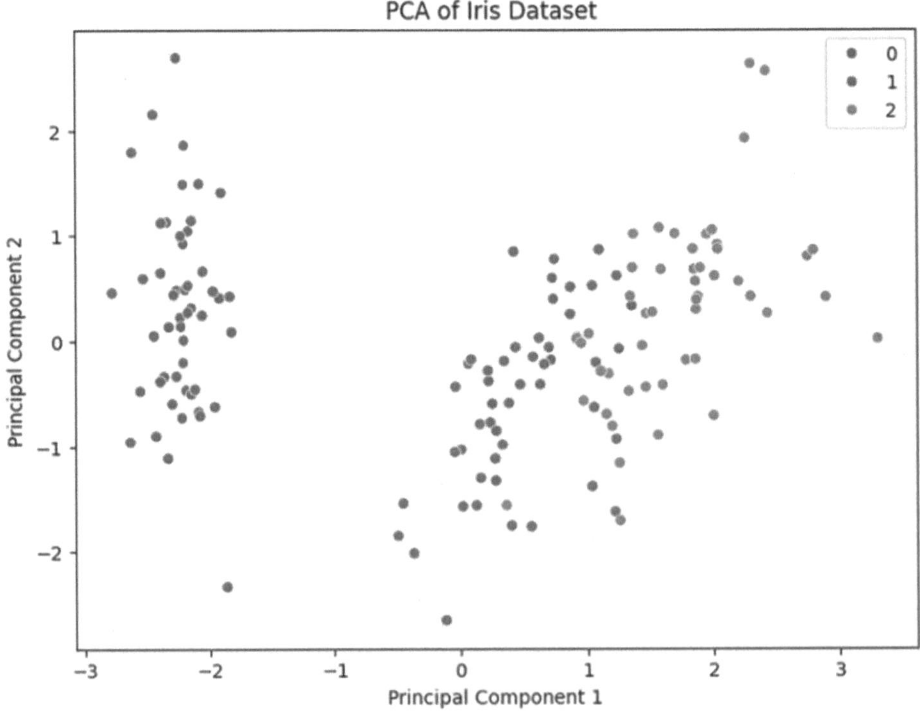


**FIGURE 8.27** Principal component analysis of the Iris dataset.

Code 8.22 shows how to apply PCA to the iris dataset and reduce the dimensions to two. Lines 1–4 import the necessary libraries. Line 7 imports the iris dataset from Scikit-learn and Line 8 stores the features in the variable "X". At Line 12, the PCA algorithm is initialized with "n_components = 2", setting the number of principal components.

The "fit_transform()" method at Line 13 applies PCA to the data. In Line 16, "df_pca", is a dataframe created from the NumPy array "X". In Line 19, the number of columns of the new dataframe is printed. In Line 21, a chart is initialized to display the seaborn scatter plot created with the "df_pca" data frame at Line 22. PCA component 1 is displayed on the x-axis, and component 2 is displayed on the y-axis. To showcase that the new dataset with two components can still maintain three clusters of flowers, the datapoints are colored based on the three types of iris flowers using iris.target column for the hue parameter. Lines 23–26 present the title, display the captions for the x and y axis, and show the graph. The resulting figure is illustrated in Figure 8.27, clearly demonstrating the three types of iris flowers in different clusters. This signifies that principal components can effectively maintain the shape of the data after dimensionality reduction and can be used to train an ML algorithm.

**CODE 8.22 APPLYING UNSUPERVISED MACHINE LEARNING TECHNIQUE – PRINCIPLE COMPONENT ANALYSIS**

```
1. import pandas as pd, matplotlib.pyplot as plt, seaborn as sns
2. from sklearn.preprocessing import StandardScaler
3. from sklearn.datasets import load_iris
4. from sklearn.decomposition import PCA
5.
6. # Load the Iris dataset
7. iris = load_iris()
8. X = iris.data
9. X = StandardScaler().fit_transform(X)
10.
11. # Performing PCA
12. pca = PCA(n_components=2)  # Reduce the data to 2 principal components
13. X_pca = pca.fit_transform(X)
14.
15. # Create a dataframe with the principal components
16. df_pca = pd.DataFrame(data = X_pca, columns = ['PCA 1', 'PCA 2'])
17.
18. # The number of dimensions after PCA
19. print("Number of dimensions after PCA:", df_pca.shape[1])
20.
21. plt.figure(figsize=(8, 4))
22. sns.scatterplot(x='PCA 1', y='PCA 2', data=df_pca, palette='Set1', hue=iris.target)
23. plt.title('PCA of Iris Dataset')
24. plt.xlabel('Principal Component 1')
25. plt.ylabel('Principal Component 2')
26. plt.show()
```

### 8.5.3 Self-Supervised Learning

Self-supervised learning uses unlabeled data by training models on correct and corrupted examples. The model learns patterns from the correct instances and tries to fix the corrupted instances. For example, consider a dataset with employee information (age, experience, department, and salary). To learn meaningful representations of employees without labels, we introduce noise or randomly mask certain features as corrupted versions. A model such as a neural network learns to reconstruct the original employee data from the corrupted data, known as auto-encoding. These learned patterns can improve both supervised and unsupervised learning tasks.

## 8.6 SEMI-SUPERVISED LEARNING

Semi-supervised learning combines supervised learning techniques with unlabeled data to make predictions or find patterns. Labeled data is often expensive or time-consuming, while unlabeled data is abundant. These algorithms learn from a combination of small amounts of labeled data and a large amount of unlabeled data to make predictions and infer the structure of the data. One common approach is to train a model on labeled data and then use it to predict and infer structure in unlabeled data. These techniques are applied to problems in natural language processing, image recognition, and speech recognition.

## 8.7 REINFORCEMENT LEARNING

Reinforcement learning trains an algorithm or agent through trial and error. The agent receives feedback from the environment, getting rewards for accurate actions and penalties for inaccurate ones. Through practice, the agent learns a set of rules that maximizes rewards. For example, an agent learning to build a house with Lego pieces starts by randomly assembling pieces and refining its strategy based on feedback, either a penalty or reward. This type of learning is well-suited for robotics control, where robots can learn to navigate and perform tasks through exploration and reward-based feedback, helping them adapt to the changing environment.

## 8.8 MODEL SELECTION

ML algorithms are versatile and can be adapted for various categories and stages of ML. For example, classification algorithms, primarily used for supervised ML problems, can also be adapted for anomaly detection. Another example is regression algorithms, commonly used for predicting numerical target variables within the supervised learning category, which can assist in feature selection by finding the most important features. Dimensionality reduction algorithms, typically used in unsupervised learning, can help in supervised learning problems by identifying and removing multi-collinearity.

Based on the given problem and the dataset, we also need to consider other factors that influence the selection of ML models:

- **Interpretability**: Simple models like logistic regression and decision trees are easy to understand and interpret. Complex models such as SVMs and MLPs are difficult to interpret.
- **Performance**: Models like Random Forests and MLPs handle complex and large datasets efficiently, while simpler models like decision trees and K-means clustering may perform inefficiently on such datasets.
- **Linearity vs. complexity**: Algorithms like linear and logistic regression capture linear relationships, whereas SVM and MLP can capture complex relationships.
- **Computational resources**: Simple linear and logistic regression algorithms require fewer computational resources. In contrast, SVM and MLP need significant computational resources, especially with large datasets.

## 8.9 CASE STUDY 1 – PREDICTING CUSTOMER BEHAVIOR

This case study investigates bank customers who are more likely to take personal loans, enabling the bank to target customers more effectively and boost profit margins. The bank customer data was taken from the Kaggle website,[6] which includes demographic and financial details such as age, experience, income (in thousands), zip code, number of family members, education level (1 for undergraduates, 2 for graduates, and 3 for advanced professionals), annual income, credit card spending per month, mortgage (in thousands), and account types. The target variable in this dataset is personal loan, a categorical column with yes or no values. Therefore, this is a binary classification problem.

Code 8.23 applies methods that help gain basic insights into the data. For example, the "head(5)" method at Line 5 displays the first five rows of the data, and the "info()" method provides data types and memory usage details at Line 9. In Line 13, the "describe()" method provides descriptive statistics for the numerical columns. For example, for the age column, the mean age is 45.34 years, and the standard deviation is 11.46, indicating a moderate spread from the mean and a minimum age of 23 and a maximum of 67 years. Similarly, the mean experience is approximately 20.10 years. However, the minimum experience value is –3, which seems to be a data error. To investigate this concern further, in Line 16, the "nunique()" method displays the total number of unique values in the experience column. The result is 47 unique values in the column, with three negative values: –1, –2, and –3. In Line 18, the "value_counts()" method is used to print each unique value with its frequency of occurrence. The "clip()" method is used in Line 20 to replace these negative values with a zero.

### CODE 8.23 UNDERSTANDING THE BASIC CONTEXT OF THE DATASET

```
1. import pandas as pd
2. import matplotlib as plt
3. df = pd.read_csv(
       "https://raw.githubusercontent.com/Object-Oriented-Programming-2024/"
       "Object-Oriented-Programming/main/Chapter8/bank_personal_loan.csv"
   )
4. print("\n******* First 5 Rows of the data *******")
5. print(df.head(5))
6.
7. #Displying basic details about the features
8. print("\n******* Basic Information about the Data frame *******")
9. print(df.info())
10.
11. # Display basic summary statistics for numeric features
12. print("\n*******Descriptive statistics for numerical columns*******")
13. print(df.describe())
14.
15. # Displaying the number of unique values in the Experience column
16. print(f"No of unique categories: {df['Experience'].nunique()}")
17. # Displaying all unique values with their frequency
18. print(df['Experience'].value_counts())
19. # Replacing unique values with a zero
20. df['Experience'] = df['Experience'].clip(lower=0)
```

The next step in pre-processing is to perform EDA using visualizations to understand our data further. We begin with univariate analysis, exploring each column separately. Code 8.24 creates a histogram and boxplot for each continuous numerical column. The for loop at Line 6 iterates through the list of these numerical columns: age, experience, income, CCAvg, and mortgage. At Line 8, a Matplotlib chart is created with two subplots, one to contain the histogram and the other to contain the boxplot of the same column.

In Line 9, the histogram for the column is created. Lines 13–21 calculate the boxplot's quartiles, IQR, and upper and lower bounds. Line 24 identifies outliers beyond these bounds. Line 26 creates the boxplot, and Lines 28 and 29 add the lower and upper bounds to the boxplots. Titles for the histogram and boxplot are added in Lines 31 and 32, and the chart is displayed in Line 33. To handle outliers, Line 36 replaces them with the upper bound value.

### CODE 8.24 EXPLORATORY DATA ANALYSIS OF NUMERICAL COLUMNS

```
import seaborn as sns
import matplotlib.pyplot as plt
import numpy as np

numerical_cols= ["Age","Experience","Income","CCAvg","Mortgage"]
for column in numerical_cols:
    # Plot the histogram
    fig, (ax1, ax2) = plt.subplots(1, 2, figsize=(8, 6))
    sns.histplot(data=df[column], ax=ax1, kde=True)

    # Plot the Spread using boxplot
    # Calculate the first quartile (Q1) and third quartile (Q3)
    Q1 = df[column].quantile(0.25)
    Q3 = df[column].quantile(0.75)

    # Calculate the interquartile range (IQR)
    IQR = Q3 - Q1

    # Define the lower and upper bounds to identify outliers
    lower_bound = Q1 - 1.5 * IQR
    upper_bound = Q3 + 1.5 * IQR

    # Find outlier indices
    outlier_indices = ((df[column] < lower_bound) | (df[column] > upper_bound))

    sns.boxplot(y=df[column], ax=ax2)
    # Plot the upper and lower bounds as horizontal lines
    plt.axhline(y=lower_bound, color='r', linestyle='--', linewidth=2, label='Lower Bound')
    plt.axhline(y=upper_bound, color='g', linestyle='--', linewidth=2, label='Upper Bound')

    ax1.set_title(f'{column} Histogram')
    ax2.set_title(f'{column} Boxplot')
    plt.show()

    # Replace outliers with the upper bound
    df.loc[outlier_indices, column] = upper_bound
```

Now, we can visualize the categorical columns with Code 8.25. These columns contain numerical values representing categories. For example, the target variable, "Personal Loan", uses "0" to mean no personal loan was taken and "1" to mean yes personal loan was taken. Line 2 creates a chart with four rows and two columns to display eight-count plots. Line 4 selects the first subplot. In Line 6, a dictionary maps numerical values in "Personal Loan" to labels ("No" and "Yes"), replacing the values in a temporary dataframe called "df_temp" in Line 8. The "seaborn" library is used in Line 9 to create a count plot for "Personal Loan" with a color palette and sets the title in Line 10. The same process is repeated for other categorical columns.

### CODE 8.25 EXPLORATORY DATA ANALYSIS OF CATEGORICAL COLUMNS

```
1. # Create a figure with four rows and two columns grid of subplots
2. fig, axes = plt.subplots(4, 2, figsize=(10, 15))
3.
4. ax=axes[0,0]
5. # Mapping of numeric values to labels
6. mapping = {0: "No", 1: "Yes"}
7. # Temporarily replacing numeric values with labels in the column
8. df_temp = df.replace({"Personal Loan": mapping})
```

```
sns.countplot(x='Personal Loan', data=df_temp, ax=ax, palette="Set2")
ax.set_title(f"(a) Personal Loan")

ax=axes[0,1]
sns.countplot(x='Family', data=df, ax=ax, palette="Set2")
ax.set_title('(b) Family Distribution')


ax=axes[1,0]
# Mapping of numeric values to labels
mapping = {1: "Undergraduate", 2: "Graduate", 3: "Professional"}
# Temporarily replacing numeric values with labels in the column
df_temp = df.replace({"Education": mapping})
sns.countplot(x='Education', data=df_temp, ax=ax, palette="Set2")
ax.set_title(f"(c) Education Distribution")

ax=axes[1,1]
# Mapping of numeric values to labels
mapping = {0: "No", 1: "Yes"}
# Temporarily replacing numeric values with labels in the column
df_temp = df.replace({"Securities Account": mapping})
sns.countplot(x='Securities Account', data=df_temp, ax=ax, palette="Set2")
ax.set_title(f"(d) Securities Account Distribution")

ax=axes[2,0]
# Mapping of numeric values to labels
mapping = {0: "No", 1: "Yes"}
# Temporarily replacing numeric values with labels in the column
df_temp = df.replace({"CD Account": mapping})
sns.countplot(x='CD Account', data=df_temp, ax=ax, palette="Set2")
ax.set_title(f"(e) CD Account Distribution")

ax=axes[2,1]
# Mapping of numeric values to labels
mapping = {0: "No", 1: "Yes"}
# Temporarily replacing numeric values with labels in the column
df_temp = df.replace({"Online": mapping})
sns.countplot(x='Online', data=df_temp, ax=ax, palette="Set2")
ax.set_title(f"(f) Online Banking Distribution")

ax=axes[3,0]
# Mapping of numeric values to labels
mapping = {0: "No", 1: "Yes"}
# Temporarily replacing numeric values with labels in the column
df_temp = df.replace({"CreditCard": mapping})
sns.countplot(x='CreditCard', data=df_temp, ax=ax, palette="Set2")
ax.set_title(f"(f) Bank's CreditCard")

ax=axes[3,1]
sns.countplot(x='ZIP Code', data=df, ax=ax, palette="Set2")
ax.set_title('(h)  Distribution')

# Adjust spacing between subplots
plt.subplots_adjust(hspace=1)
plt.xticks([])

# Show the plot
plt.show()
```

We can now visualize relationships between columns of interest. For example, clients with higher income, more credit card spending, or more family members might be more likely to take a personal loan. Code 8.26 generates charts that compare two variables, a bi-variate analysis.

Lines 1–7 generate a boxplot comparing clients' income levels with personal loan status. Lines 10–15 generate another boxplot for credit card spending (CCAvg) and personal loan status. Lines 17–23 generate a count plot to compare the number of family members with personal loan status.

The boxplot for income shows that clients who take loans tend to have higher median incomes. Similarly, clients with higher credit spending are more likely to take loans. However, the count plot shows that clients who take loans have varying numbers of family members.

### CODE 8.26 BI-VARIATE ANALYSIS

```
1. # Plotting the relationship between "Income" and "Personal Loan"
2. plt.figure(figsize=(6, 6))
3. sns.boxplot(x='Personal Loan', y='Income', data=df)
4. plt.title('Relationship between Income and Personal Loan')
5. plt.xlabel('Personal Loan (0: No, 1: Yes)')
6. plt.ylabel('Income')
7. plt.show()
8.
9. # Plotting the relationship between Credit Card Spending -"CCAvg" and "Personal Loan"
10. plt.figure(figsize=(8, 6))
11. sns.boxplot(x='Personal Loan', y='CCAvg', data=df)
12. plt.title('Relationship between Credit Card Spending and Personal Loan')
13. plt.xlabel('Personal Loan (0: No, 1: Yes)')
14. plt.ylabel('Credit Card Spending (CCAvg)')
15. plt.show()
16.
17. # Plotting the relationship between "Family" and "Personal Loan"
18. plt.figure(figsize=(6, 6))
19. sns.countplot(x='Family', hue='Personal Loan', data=df, palette='Set2')
20. plt.title('Relationship between Family and Personal Loan')
21. plt.xlabel('Number of Family Members')
22. plt.ylabel('Count')
23. plt.legend(title='Personal Loan', loc='upper right')
24. plt.show()
```

Code 8.27 performs a multi-variate analysis using a heatmap to investigate multiple variables. In Line 1, all numerical columns are selected from the data frame. Line 4 uses the "corr()" method to create a correlation matrix between these variables. Line 6 creates a 10 by 8 (10 X 8) chart. In Line 8, Seaborn creates a heatmap from the correlation matrix, ensuring the chart displays the correlation coefficient between variables within the matrix with the "BuGn" color map. The chart's title is set at Line 8 and displayed at Line 9. The resulting chart is displayed in Figure 8.28.

Upon reviewing the correlations, two important observations are made: the zip code correlates very poorly with the target variable. Therefore, at Line 11, the column is dropped from the database. Second, age and experience have a high correlation of 0.99. This multi-collinearity can impact the ML model. Therefore, at Line 12, the average of both columns is taken to create a new column representing both. In Lines 13 and 14, age and experience are dropped from the data frame.

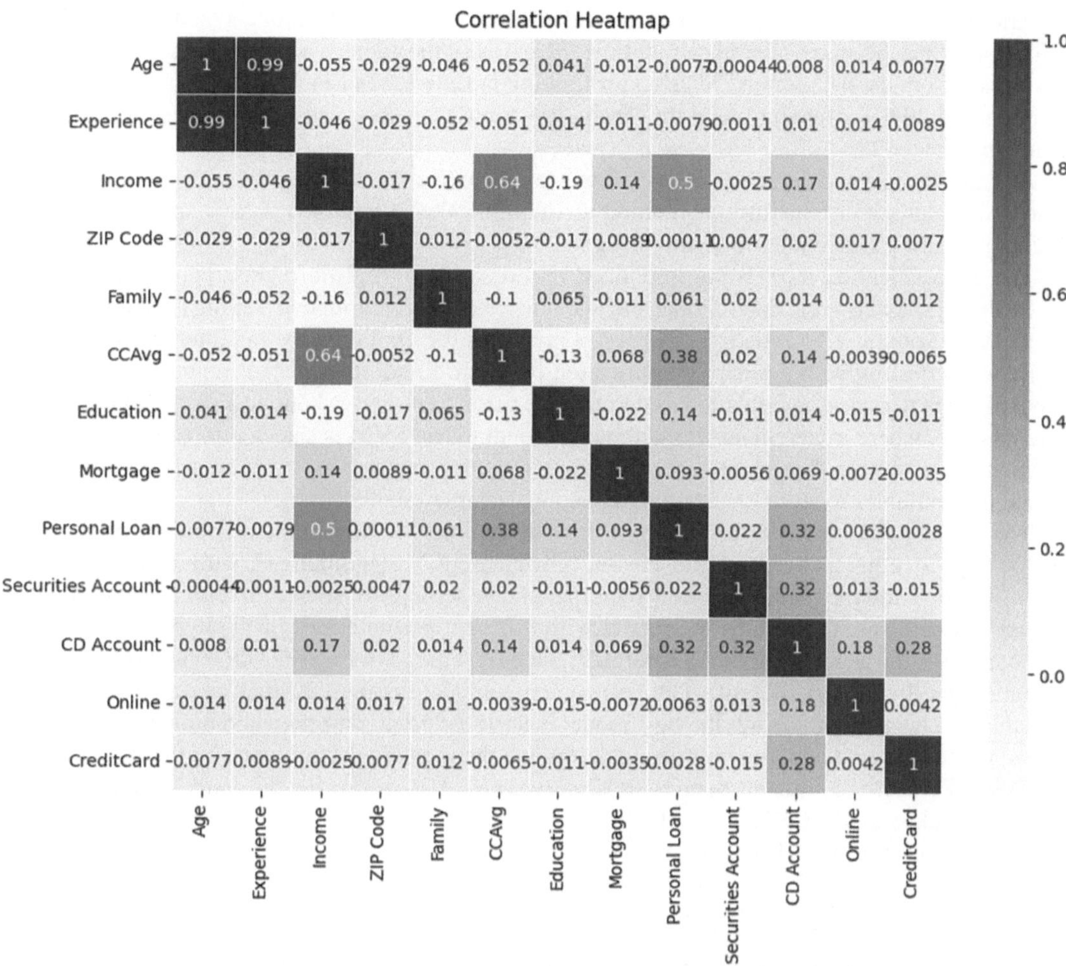


**FIGURE 8.28** Correlation heatmap of all numerical features.

### CODE 8.27 MULTI-VARIANT ANALYSIS

```
numerical_cols= df.select_dtypes(include=['int64', 'float64']).columns

# Find the correlation between features
cor_matrix = df[numerical_cols].corr()
# Using visualizations for an overview of correlation among features
plt.figure(figsize=(10, 8))
sns.heatmap(cor_matrix, annot=True, cmap='BuGn', linewidths=0.5)
plt.title('Correlation Heatmap')
plt.show()

df.drop('ZIP Code', axis=1, inplace=True)
df['Age_Experience'] = (df['Age'] + df['Experience']) / 2
df.drop('Age', axis=1, inplace=True)
df.drop('Experience', axis=1, inplace=True)
```

After basic pre-processing, we continue with Code 8.28 to train ML models on the data. It begins with importing the necessary libraries. Line 12 extracts the target variable "Personal Loan" and removes it from the data frame "X" used for training in Line 13. In Line 16, the data is split 80% for training and 20% for testing using the "train_test_split()" method, ensuring consistent splits with "random_state=42".

At Line 19, a standard scaler is initialized to normalize the training data "X_train" using the standard scaler's "fit_transform()" method at Line 20. The normalized data is split into 80% for training and 20% for validation testing in Line 23.

Three classification models are initialized in a dictionary: logistic regression, Random Forest, and SVM (Lines 26–31). Each model includes different hyperparameter settings for optimization using "GridSearchCV". Logistic regression is preferred for its simplicity and interpretability, especially when dealing with a binary classification problem. Random Forests is an ensemble learning method for better accuracy. It is based on many decision trees and is beneficial for imbalanced data like this one, where more clients are without personal loans. SVM is selected to capture non-linear relationships between data if they exist. Following model training and evaluation, Lines 33 and 34 create two variables to store the best model and its accuracy score.

The for loop at Line 37 iterates through each of the three models in the dictionary and picks the various hyperparameter settings. At Line 39, "GridSearchCV" is initialized, with the current model in the iteration and the model's various hyperparameter settings to perform fivefold cross-validation "cv=5". "GridSearchCV" trains the model with different permutations of the hyperparameter settings and uses the accuracy score to determine the best settings. Using the "fit()" method, the model is trained on the data at Line 40. The best hyperparameter settings for that model are stored in a variable at Line 41. In Line 42, the best model is used for predicting the validation set. Lines 45 to 63 calculate various evaluation metrics to determine the model's performance on the validation set: accuracy, F1 score, confusion matrix, ROC curve, and AUC score for the model. Lines 66 to 73 visualize each model's ROC curve with the AUC score. Lines 75–77 store the model most accurately in the variables created at Lines 33 and 34.

### CODE 8.28 MODEL TRAINING AND VALIDATION

```
1. from sklearn.model_selection import train_test_split, GridSearchCV
2. from sklearn.preprocessing import StandardScaler
3. from sklearn.metrics import accuracy_score, precision_score,recall_score,f1_score, confusion_
   matrix, roc_auc_score, roc_curve
4. from joblib import dump
5. from sklearn.linear_model import LogisticRegression
6. from sklearn.ensemble import RandomForestClassifier
7. from sklearn.svm import SVC
8. import warnings
9.
10.
11. # Splitting the dataframe into features and target variables
12. y = df['Personal Loan']
13. X = df.drop(columns=['Personal Loan'])
14.
15. # Splitting the dataset into training and testing sets (80% train, 20% test)
16. X_train, X_test, y_train, y_test = train_test_split(X, y, test_size=0.2, random_state=42)
17.
18. # Applying a standard scaler to normalize the features
19. scaler = StandardScaler()
20. X_train = scaler.fit_transform(X_train)
21.
22. # Further split the training set into train and validation sets (80% train, 20% validate)
23. X_train, X_val, y_train, y_val = train_test_split(X_train, y_train, test_size=0.2, random_state=42)
24.
```

```
# Initialize models and their hyperparameters
models = {
    'Logistic Regression': (LogisticRegression(), {'C': [0.001, 0.01, 0.1, 1, 10, 100]}),
    'Random Forest': (RandomForestClassifier(), {'n_estimators': [50, 100, 200],
    'max_depth': [5, 10, None]}),'SVM': (SVC(probability=True), {'C': [0.1, 1, 10],
    'kernel': ['linear', 'rbf']})
}

best_accuracy = 0
best_model = None

# Training the models on all hyper parameter setting combinations
for name, (model, hyperparams) in models.items():
    print(f"Training {name}…")
    grid_search = GridSearchCV(model, hyperparams, cv=5, scoring='accuracy')
    grid_search.fit(X_train, y_train)
    model = grid_search.best_estimator_
    y_pred = model.predict(X_val)

    # Calculating and printing the accuracy
    accuracy = accuracy_score(y_pred, y_val)
    print(f"Accuracy: {accuracy:.2f}")

    # Calculating and printing the F1 score
    f1 = f1_score(y_pred, y_val)
    print(f"F1 Score: {f1:.2f}")

    # Calculating and printing the confusion matrix
    cm = confusion_matrix(y_val, y_pred)
    print(f"Confusion Matrix:\n{cm}")

    # Predicting probabilities for ROC and AUC
    y_prob = model.predict_proba(X_val)[:, 1]

    # Calculating ROC curve
    fpr, tpr, thresholds = roc_curve(y_val, y_prob)

    # Calculating AUC score
    auc_score = roc_auc_score(y_val, y_prob)

    # Creating the ROC curve chart
    plt.figure(figsize=(6, 4))
    plt.plot(fpr, tpr, label=f'ROC Curve (AUC = {auc_score:.2f})')
    plt.plot([0, 1], [0, 1], 'r--')
    # plt.xlabel('False Positive Rate')
    # plt.ylabel('True Positive Rate')
    plt.title(f'ROC Curve ')
    plt.legend()
    plt.show()

    if accuracy > best_accuracy:
        best_accuracy = accuracy
        best_model = model
```

The Random Forest classifier indicates the best performance. The results are displayed in Box 8.2. The accuracy of 0.97 and F1 score of 0.88 are the highest among the other models and indicate good performance. The model correctly predicted 702 instances as positive and identified 77 as negative. However, 21 instances were predicted as negative, whereas they were positive. In our scenario, if we do not target these 21 clients for personal loans, it will not cause harm to the bank or clients. However, in the case of health-related data, these incorrect predictions would have been a concern and inapplicability of the model in the real world. Moreover, there are no instances incorrectly predicted as positive. This information indicates that our Random Forest model is good for final testing.

## BOX 8.2 OUTPUT FOR THE RANDOM FOREST CLASSIFIER

**Output**

```
Accuracy: 0.97
F1 Score: 0.88
Confusion Matrix:
[[702   0]
 [ 21  77]]
```

Once we have found a model that performs well on the validation set, we are now ready to evaluate it again on the test set before deployment in the real world. Code 8.29 evaluates the model on the test set. Lines 2 and 3 normalize the test set, as our model was trained on normalized data. When we use this model with real-world data, we must ensure that it is also normalized before prediction. In Line 6, the model predicts the "X_test" set. The predictions are stored in "y_pred_test". From Lines 9 to 27, the evaluation metrics, accuracy, F1 score, confusion matrix, ROC curve, and AUC score are calculated. Lines 30–35 visualize the ROC Curve with the AUC score. Lines 37 and 38, print and store the model in a file called "best_model.joblib".

## CODE 8.29 MODEL TESTING

```
1. # Applying a standard scaler to normalize the features
2. scaler = StandardScaler()
3. X_test = scaler.fit_transform(X_test)
4.
5. # Predicting on the test set using the best model
6. y_pred_test = best_model.predict(X_test)
7.
8. # Calculating and printing the accuracy
9. test_accuracy = accuracy_score(y_pred_test, y_test)
10. print(f"Accuracy: {test_accuracy:.2f}")
11.
12. # Calculating and printing the F1 score
13. f1 = f1_score(y_test, y_pred_test)
14. print(f"F1 Score: {f1:.2f}")
15.
16. # Calculating and printing the confusion matrix
17. cm = confusion_matrix(y_test, y_pred_test)
18. print(f"Confusion Matrix:\n{cm}")
19.
20. # Predicting probabilities for ROC and AUC
21. y_prob = best_model.predict_proba(X_test)[:, 1]
22.
23. # Calculating ROC curve
24. fpr, tpr, thresholds = roc_curve(y_test, y_prob)
25.
26. # Calculating AUC score
27. auc_score = roc_auc_score(y_test, y_prob)
28.
29. # Creating the ROC curve chart
30. plt.figure(figsize=(6, 4))
31. plt.plot(fpr, tpr, label=f'ROC Curve (AUC = {auc_score:.2f})')
32. plt.plot([0, 1], [0, 1], 'r--')
33. plt.title(f'ROC Curve ')
34. plt.legend()
35. plt.show()
36. # Printing and storing the best model
37. print(f"Best Model: {best_model}")
38. dump(best_model, "best_model.joblib")
```

**BOX 8.3 OUTPUT AFTER APPLYING THE MODEL ON THE TEST SET**

**Output**

```
Accuracy: 0.99
F1 Score: 0.95
Confusion Matrix:
[[894   1]
 [  9  96]]
```

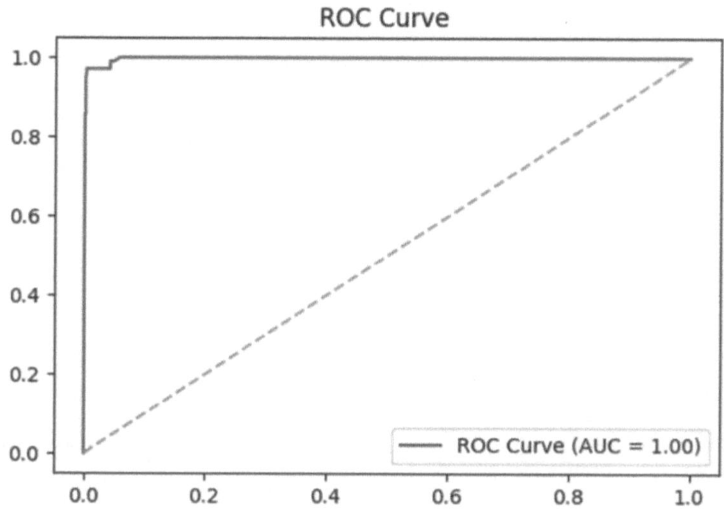


**FIGURE 8.29** ROC curve and the AUC score for the Random Forest classifier.

The results of the evaluation metrics are illustrated in Box 8.3. The accuracy of 0.99 and F1 score of 0.95 indicates very good performance. The model correctly predicted 894 instances as positive and 96 instances as negative. However, nine instances were predicted as negative, whereas they were positive, which is a very low number of data points. Moreover, only one instance was incorrectly predicted as positive. This information indicates that our Random Forest model is good for deployment.

Figure 8.29 displays the ROC curve and the AUC score of 1, which indicates perfect performance.

## 8.10 CASE STUDY 2 – PREDICTING EMPLOYEE SALARIES

This case study applies ML algorithms to employee data from the Kaggle website.[7] The purpose is to train a model to predict salaries based on employee information such as age, gender, education level, job title, years of experience, country, and race. Code 8.30 begins by performing basic pre-processing tasks on the dataframe by calling: "head()", "info()", and "describe()" methods (Lines 5, 9, and 13). It also performs basic imputations, such as in Lines 19 and 24, where missing age and years of experience values are replaced by the average of the columns because these columns contain continuous numerical values. Missing values for gender and education level are replaced by the most frequent occurrence at code Lines 29 and 34, as they are categorical columns. Null salary values are replaced by the minimum salary at Line 42. Code Line 45 checks to ensure no more null values exist in the data frame.

## CODE 8.30 BASIC PRE-PROCESSING

```
import pandas as pd
import matplotlib.pyplot as plt, seaborn as sns
df = pd.read_csv("https://raw.githubusercontent.com/Object-Oriented-Programming-2024/"
                 "Object-Oriented-Programming/main/Chapter8/salary_data.csv"
)
print("\n******* First 5 Rows of the data *******")
print(df.head(5))

#Displying basic details about the features
print("\n******* Basic Information about the Data frame *******")
print(df.info())

# Display basic summary statistics for numeric features
print("\n*******Descriptive statistics for numerical columns*******")
print(df.describe())

# Imputate missing values
age = df['Age']
average=age.mean()
print(f"\n Replacing null Age values with Average Age : {average}" )
df['Age'].fillna(average, inplace=True)

gender = df['Gender']
most_frequent = gender.mode()[0]
print(f"\n Replacing null gender values with most frequent occurrence : {most_frequent}")
df['Gender'].fillna(most_frequent, inplace=True)

education = df['Education Level']
most_frequent = education.mode()[0]
print(f"\n Replacing null Education Level with most frequent occurrence : {most_frequent}")
df['Education Level'].fillna(most_frequent, inplace=True)

title = df['Job Title']
most_frequent = title.mode()[0]
print(f"\n Replacing null Job Title  values with most frequent occurrence : {most_frequent}")
df['Job Title'].fillna(most_frequent, inplace=True)

experience = df['Years of Experience']
average=experience.mean()
print(f"\n Replacing null Years of Experience values with Average  : {average}" )
df['Years of Experience'].fillna(average, inplace=True)

min_salary = df['Salary'].min()
df['Salary'].fillna(min_salary, inplace=True)

print("Making sure there are no null values: ")
print(df.isnull().sum())
```

Categorical columns, such as education level, job title, gender, country, and race, contain text labels for each category in the column. Therefore, in Code 8.31, Lines 1 and 2 import two types of encoders: "OrdinalEncoder" for ordered columns such as education level and "LabelEncoder" for labels without an order. Although job titles can be ordered, the number of job titles in the data set is 193, and assigning an ordered label to each title will be inefficient. Code Line 5 retrieves all the dataset's column names with non-numeric values. The for loop from Lines 8 to 12 loops through each non-numeric column and prints the number of unique categories in each column using the "nunique" method at Line 10. All the unique categories in each column are printed at Line 11. Lines 15–17 clean up the education level column to ensure each logical category is represented by one unique label. Line 20 creates a list of labels for the education column. In Line 26, a label encoder replaces these text-based labels with ordered numerical values using the ordinal encoder. Job title, gender, country, and race column categories are also converted to numeric values using the label encoder at code Lines 30, 33, 35, and 37, respectively. In Line 38, the "head" method displays the data frame's first five rows and sees the transformations.

### CODE 8.31 DATA TRANSFORMATIONS – ENCODING CATEGORICAL FEATURES

```
1. from sklearn.preprocessing import OrdinalEncoder
2. from sklearn.preprocessing import LabelEncoder
3.
4. # Getting categorical columns
5. categories = df.select_dtypes(include=['object']).columns
6.
7. # Displaying categories for each categorical column
8. for col in categories:
9.     print(f"Categories for '{col}':")
10.    print(f"No of unique categories: {df[col].nunique()}")
11.    print(df[col].unique())
12.    print()
13.
14. #Replacing duplicated categories within Education Level
15. df['Education Level'] = df['Education Level'].replace("Bachelor's Degree","Bachelor's")
16. df['Education Level'] = df['Education Level'].replace("Master's Degree","Master's")
17. df['Education Level'] = df['Education Level'].replace("phD","PhD")
18.
19. # Performing ordinal encoding for Education Level and providing the order
20. education_levels = ["High School", "Bachelor's", "Master's", "PhD"]
21.
22. # Initialize the OrdinalEncoder with education levels
23. ordinal_encoder = OrdinalEncoder(categories=[education_levels])
24.
25. # Performing ordinal encoding for the 'Education' column
26. df['Education Level'] = ordinal_encoder.fit_transform(df[['Education Level']])
27.
28. # Performing Label Encoding for 'Job Title'
29. label_encoder = LabelEncoder()
30. df['Job Title'] = label_encoder.fit_transform(df['Job Title'])
31.
32. # Performing Label Encoding for 'Gender'
33. df['Gender'] = label_encoder.fit_transform(df['Gender'])
34.
35. df['Country'] = label_encoder.fit_transform(df['Country'])
36.
37. df['Race'] = label_encoder.fit_transform(df['Race'])
38. print(df.head())
```

Now that all the columns in the dataset are converted to numerical values, data analysis using visualizations can be performed. Code 8.32 creates a heatmap to understand the correlations between the columns. At code Line 1, all the numerical columns are retrieved. In Line 3, correlations between all variables are calculated and stored in a variable called "cor_matrix". Lines 6–9 display the matrix as a heatmap. The heatmap results indicate that columns age, years of experience, and education level correlate highly with the target variable, salary.

### CODE 8.32 GENERATING A HEATMAP FOR CORRELATIONS

```
numerical_cols= df.select_dtypes(include=['int', 'float']).columns
# Find the correlation between features
cor_matrix = df[numerical_cols].corr()
print(cor_matrix)
# Using visualizations for an overview of correlation among features
plt.figure(figsize=(6, 6))
sns.heatmap(cor_matrix, annot=True, cmap='BuGn', linewidths=0.5)
plt.title('Correlation Heatmap')
plt.show()
```

Upon observing the correlation heatmap, you will notice that race and country have a low correlation with salary. As a result, in Code 8.33, the race and country values are scaled using the standard scaler at Line 3 and then combined to represent a new column, Race_Country, at Line 6. The original columns for Race and Country are dropped at Line 9. Another column of interest is job title, which has a very low correlation with the target variable, salary.

### CODE 8.33 FEATURE ENGINEERING

```
1. from sklearn.preprocessing import StandardScaler
2. scaler = StandardScaler()
3. age_experience = scaler.fit_transform(df[['Race', 'Country']])
4.
5. # Combine the normalized features
6. df['Race_Country'] = (df['Race'] + df['Country'])
7.
8. # Drop the original 'Age' and 'Years of Experience' columns
9. df.drop(['Race', 'Country'], axis=1, inplace=True)
```

Code 8.34 uses the unsupervised learning technique, the K means clustering algorithm, to generate a new column of clusters based on age and years of experience, identified at code Lines 4 and 5. A standard scaler is used at Line 8 to normalize these features before performing K means clustering. The number of clusters k is selected as 4 (Line 10). The K means algorithm is initialized with k clusters, and the initialization method is "k-means++" (Line 12). The algorithm is applied to the two features in the dataset at Line 15. Line 18 stores the centroids of the clusters, and Line 19 stores the new column containing the clusters. A chart is initialized at Line 22. The for loop at Lines 25 and 26 iterates through each point in the data frame and displays it on the scatter plot. Lines 29 and 30, add the centroids to the scatter plot. The chart's title and legend are set at Lines 32 and 33, and the chart is shown at Line 34. Code Line 37 adds the clusters to the data frame as a new column. The scatter plot generated from this code is displayed in Figure 8.30.

### CODE 8.34 APPLYING UNSUPERVISED LEARNING ALGORITHM – K-MEANS CLUSTERING

```
1. from sklearn.cluster import KMeans
2. from sklearn.preprocessing import StandardScaler
3.
4. feature1 = "Age"
5. feature2 = "Years of Experience"
6.
```

```
# Preprocessing: Normalize the data
X = StandardScaler().fit_transform(df[[feature1, feature2]])

k=4
# Initializing KMeans with k clusters
kmeans = KMeans(n_clusters=k, init="k-means++")

# Fitting the model
kmeans.fit(X)

# Getting the cluster centroids and labels
centroids = kmeans.cluster_centers_
labels = kmeans.labels_

# Plotting the clusters
plt.figure(figsize=(8, 6))

# Plot data points for each cluster
for i in range(k):
 plt.scatter(X[labels == i, 0], X[labels == i, 1], s=50, label=f'Cluster {i+1}')

# Plot centroids
plt.scatter(centroids[:, 0], centroids[:, 1], s=200, c='black',
            marker='*', label='Centroids')

plt.title('K-Means Clustering')
plt.legend()
plt.show()

# Adding the cluster labels to the dataframe
df['Age_Experience'] = labels
```

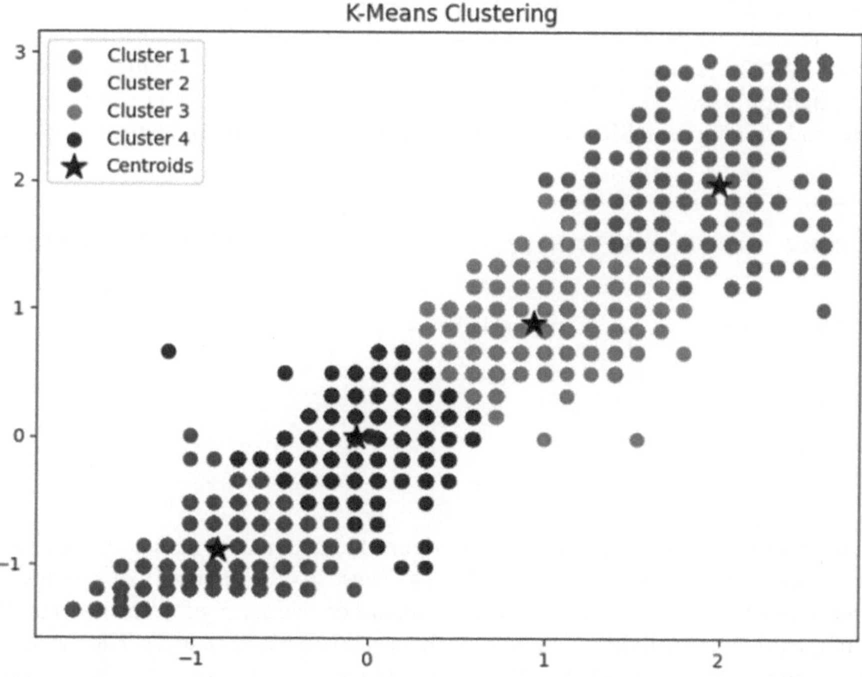


**FIGURE 8.30** K-means clusters on the age and years of experience features.

The results of the K-means displayed in Figure 8.29 are very useful in identifying four major clusters based on years of experience and age. The cluster column can now be added to the data frame as an additional feature.

Code 8.35 applies various regression models to the data set to predict salaries. Linear regression is used with ridge regularization to penalize higher weights and improve performance. Decision tree regression is applied because it is capable of capturing non-linear relationships. Random Forest is used as an ensemble method and is more powerful due to the aggregation of decision trees. An MLP regressor is used to apply neural networks to the problem in cases where complex relationships exist between the features.

Code 8.35 begins with importing the necessary libraries. Code Line 10 drops the target column from the data frame for training and stores input features in variable "X". In Line 14, the target variable, salary, is stored in "y". At code Line 17, a standard scaler is applied to the input features. The training data is split into train, validate, and test sets at code Lines 23 to 27. A dictionary is created from Lines 31 to 36, with various regression models. Hyperparameter settings for each model are added to another dictionary, param_grid (Lines 39–52). In Line 54, the "best_performance_model" list is created to store the R2 value of the best-performing model at index 0 and the model itself at index 1.

The for loop (Lines 56–80) iterates through each model in the dictionary for training and validation. In Line 57, GridSearchCV is initialized with a model in each iteration, with various hyperparameter settings from the param_grids for that model. The GridSearchCV will apply k-fold cross-validation and find the best hyperparameter settings for the model based on the evaluation metric MAE.

At code Line 59, the model is trained on the training data. Line 62 saves the model with the best hyperparameter settings in a variable. At Line 65, the model makes predictions on the validation set. Lines 68–70 display the evaluation metrics, MAE, MSE, and R2 score. These metrics are printed from Lines 72 to 76. Lines 78–80 compare the current model's performance in the iteration with the previously best-performing model. If the R2 score of the current model in the iteration is better, then it is stored in the best-performing model list at index 1 and its R2 score at index 0.

At Line 82, outside the for loop, the results of the best-performing model are displayed. These results will indicate that the Random Forest algorithm performed better than all other models, as shown in Box 8.4. The results may vary slightly, and the execution time for the MLP Regressor will take around an hour or two, as the grid search CV has many hyperparameter combinations for MLP to fine-tune, and training each MLP itself requires multiple iterations and layers.

### BOX 8.4 OUTPUT AFTER APPLYING RANDOM FOREST REGRESSION

**Output**

```
****** Validation Results for Random Forest Regression ******

Best Hyperparameters: {'max_depth': None, 'n_estimators': 300}
Mean Absolute Error: 3899.1059
Mean Square Error: 95275612.2776
R Square Error: 0.9668
```

Before the results in Box 8.4 are interpreted, we need to understand the salary column has a mean of 115326.96 and a standard deviation of 52786.18. These results indicate that salaries vary widely from the mean of about 52 thousand for the 1st standard deviation. Therefore, based on the context, the MAE of the predicted salaries of about four thousand is not a huge error. The mean square error is always squared, which is even larger than MAE. However, the R square score of about 0.966 is close to 1, showing that the model performs well on the validation set.

## CODE 8.35 TRAINING AND VALIDATING REGRESSION MODELS

```
from sklearn.preprocessing import StandardScaler
from sklearn.model_selection import train_test_split, GridSearchCV
from sklearn.metrics import r2_score, mean_squared_error, mean_absolute_error
from sklearn.linear_model import Ridge
from sklearn.tree import DecisionTreeRegressor
from sklearn.ensemble import RandomForestRegressor
from sklearn.neural_network import MLPRegressor

# Storing the independent features in X
X = df.drop('Salary', axis=1)


# Target Outcome / dependent feature in Y
y = df['Salary']

# Initializing the StandardScaler
scaler = StandardScaler()

# Fit and transform the features
X_scaled = scaler.fit_transform(X)

# Splitting the dataset into train and test sets (80% train, 20% test)
X_train, X_test, y_train, y_test = train_test_split(X_scaled, y, test_size=0.2,
                                                    random_state=42)

# Further splitting the training set into train and validation sets (80% train, 20% validate)
X_train, X_val, y_train, y_val = train_test_split(X_train, y_train, test_size=0.2,
                                                    random_state=42)

# Initializing the models with default hyperparameters
models = {
 'Linear Regression': Ridge(),
 'Decision Tree Regression': DecisionTreeRegressor(random_state=42),
 'Random Forest Regression': RandomForestRegressor(random_state=42),
 'MLP Regression': MLPRegressor(random_state=42)
}

# Defining hyperparameter dictionary for each model
param_grids = {
 'Linear Regression': {'alpha': [0.01, 0.1, 1.0, 10.0]},
 'Decision Tree Regression': {'max_depth': [None, 5, 10, 15]},
 'Random Forest Regression': {'n_estimators': [100, 200, 300],
                              'max_depth': [None, 5, 10,15]},
 'MLP Regression': {
     'hidden_layer_sizes': [(100, 50)],
     'activation': ['relu'],
     'solver': ['adam'],
     'alpha': [0.001, 0.01],
     'learning_rate': ['constant', 'adaptive'],
     'max_iter': [500, 1000]
 }
}

best_performance_model= [0,None] # Varaible to store the best performing model

for name, model in models.items():# Train and evaluate each model using GridSearchCV
 grid_search = GridSearchCV(model, param_grids[name], scoring='neg_mean_absolute_error',
                            cv=5)
 grid_search.fit(X_train, y_train)

 # Saving the model with the best hyperparameter settings
 best_model = grid_search.best_estimator_

```

```
64.  # Predicting on the validation set using the model
65.  y_val_pred = best_model.predict(X_val)
66.
67.  # Evaluating the model on the validation set
68.  val_MAE = mean_absolute_error(y_val, y_val_pred)
69.  val_MSE = mean_squared_error(y_val, y_val_pred)
70.  r2 = r2_score(y_val, y_val_pred)
71.
72.  print(f"********* Validation Results for {name} *********\n")
73.  print(f'Best Hyperparameters: {grid_search.best_params_}')
74.  print(f'Mean Absolute Error: {val_MAE:.4f}')
75.  print(f'Mean Square Error: {val_MSE:.4f}')
76.  print(f'R Square: {r2:.4f}\n')
77.
78.  if best_performance_model[0]< r2: #comparing the R2 score of the best performing model in the
     previous iterations with the current model
79.      best_performance_model[0]=r2
80.      best_performance_model[1]=best_model
81.
82. print(f"Best Model: {best_performance_model}") # Printing the best performing model
```

To ensure that our selected Random Forest model performs well on the unseen test set, Code 8.35 segregated the test set at Line 23. This test set is used in the following Code 8.36. The "best_performance_model" from Code 8.35 is used for prediction on the test set at Line 2 of Code 8.36. MAE, MSE, and R2 score are calculated to measure the model's performance at code Lines 5, 6, and 7, respectively. A scatterplot is created from code Lines 15 to 22 to plot the predicted values against the actual y values in the test set. A line is drawn to show the linear trend learned by the model and the tension between the actual values and the predicted values, as shown in Figure 8.31. The "best_performance_model" is dumped into a file using the Pickle library (Lines 25–27).

### CODE 8.36 TESTING THE MODEL BEFORE DEPLOYMENT

```
 1. # Make predictions using the best performing model on the test set
 2. y_test_pred = best_performance_model[1].predict(X_test)
 3.
 4. # Evaluate the model on the test set
 5. test_MAE = mean_absolute_error(y_test, y_test_pred)
 6. test_MSE = mean_squared_error(y_test, y_test_pred)
 7. r2_test = r2_score(y_test, y_test_pred)
 8.
 9. print("********* Test Results *********\n")
10. print(f'Mean Absolute Error: {test_MAE:.4f}')
11. print(f'Mean Square Error: {test_MSE:.4f}')
12. print(f'R Square: {r2_test:.4f}')
13.
14. # Visualize the difference between predictions and actual values on the test set
15. plt.figure()
16. sns.scatterplot(x=y_test, y=y_test_pred, color='blue')
17. # Regression line
18. sns.regplot(x=y_test, y=y_test_pred, scatter=False, color='green')
19. plt.title('Prediction on Test Set (Random Forest Regression)', fontsize=14)
20. plt.xlabel('Actual Salary')
21. plt.ylabel('Predicted Salary')
22. plt.show()
23.
24. # Store the best model
25. import pickle
26. file= open('best_model.pkl', 'wb')
27. pickle.dump(best_performance_model[1], file)
```

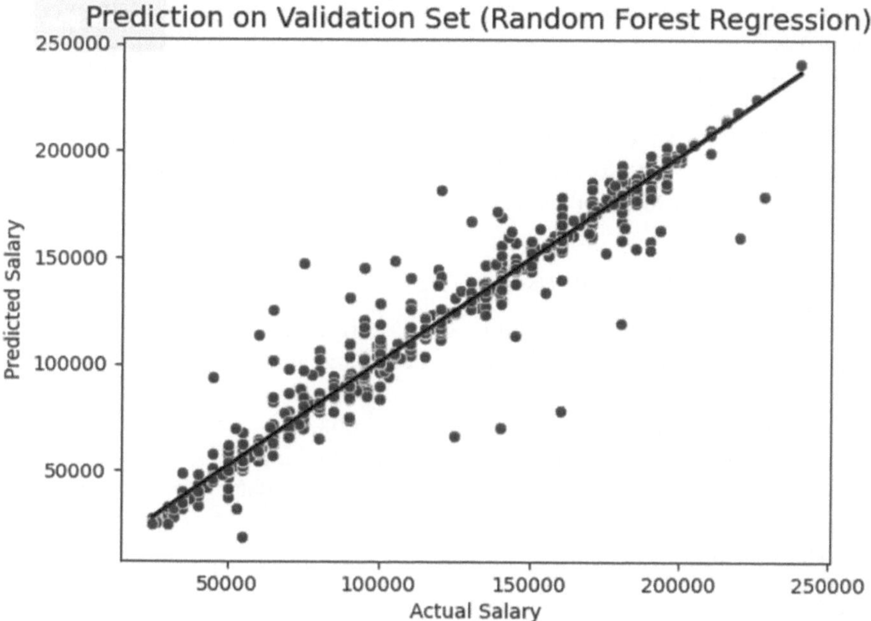


**FIGURE 8.31** Visualizing the results of the predictions on the test set.

The scatterplot of the predicted vs. actual values in Figure 8.30 indicates that the model performs well on the test set as well, as most data points fall close to the trendline, indicating that predicted values are close to the actual values. However, a few outliers in predictions are far from the trendline.

## 8.11 CHAPTER SUMMARY

This chapter provided an overview of the ML landscape. It covered some of the fundamentals of how to apply ML techniques to a variety of datasets. This chapter covered basic ML algorithms for classification, regression, and unsupervised learning problems. Moreover, with the help of the ML process that involves pre-processing, data transformations, training, validating, and testing, you can draw conclusions, identify patterns, and make predictions. This chapter is a stepping stone toward a deeper understanding of ML and how to apply it in a real-world context. There are many more analysis, modeling, training, and testing techniques to help predict or structure messy data, big data, or multi-model data problems. These new models and techniques can help you stay at the forefront of AI innovation due to its capacity to exceed human limitations. However, it is also necessary to consider any ethical implications of using these algorithms to support human activities and the possible drawbacks of machines making independent decisions without human involvement.

## 8.12 EXERCISES

### 8.12.1 Test Your Knowledge

1. Explain the importance of data preprocessing in ML.
2. Describe the stages involved in the ML process.
3. Compare the concepts of overfitting and underfitting in ML models.
4. Describe common data cleaning techniques for handling missing values and duplicates in a dataset.
5. Explain the difference between classification and regression tasks in supervised learning.

6. What are the hyperparameters in ML models? How does tuning them impact model performance?
7. Compare logistic regression with SVM and decision tree classification models.
8. Discuss the applications, advantages, and limitations of linear regression in ML.
9. Compare and contrast supervised learning with unsupervised learning.
10. Analyze how factors of performance, linearity vs. complexity, and computational resources affect the selection of ML models for a given problem.

### 8.12.2 Multiple Choice Questions

1. Which of the following statements about ML is correct?
   a. ML focuses on simulating human intelligence and autonomy.
   b. ML uses data to identify patterns and make predictions.
   c. ML is limited to programming explicit instructions.
   d. ML solely relies on human control for decision-making.
2. Which stage of the ML process involves transforming data into a format suitable for training?
   a. Model training.
   b. Validation.
   c. Data preprocessing.
   d. Deployment.
3. What is the primary purpose of EDA in the ML project?
   a. Training the model.
   b. Preprocessing data.
   c. Understanding data characteristics.
   d. Model deployment.
4. Which technique adjusts the scale of numerical features in ML models?
   a. One-hot encoding.
   b. Feature scaling.
   c. Imputation.
   d. Dimensionality reduction.
5. What evaluation metric measures the proportion of correctly classified instances in a dataset?
   a. Precision.
   b. Recall.
   c. F1 score.
   d. Accuracy.
6. Which ML model most suits datasets with high-dimensional, non-linearly separable features?
   a. Logistic regression.
   b. Decision trees.
   c. KNN.
   d. SVM.
7. Which ML approach requires models to learn through interactions with an environment and trial-and-error-based learning?
   a. Supervised learning.
   b. Unsupervised learning.
   c. Semi-supervised learning.
   d. Reinforcement learning.

8. Which algorithm is commonly used for clustering in unsupervised learning?
   a. Decision trees.
   b. K-means.
   c. SVM.
   d. Random Forest.
9. What is the main purpose of PCA in ML?
   a. To increase the number of features.
   b. To reduce overfitting.
   c. To visualize high-dimensional data.
   d. To classify data into categories.
10. Which factor is not typically considered when selecting an ML model?
   a. Performance.
   b. Popularity.
   c. Computational resources.
   d. Interpretability.

### 8.12.3 Short Answer Questions

1. Provide two examples of everyday applications that have applied ML.
2. To assess the model's performance on unseen data, what are the three portions the data is split into?
3. What is the name of the common loss function used to measure the average squared difference between predicted and actual values?
4. Which type of ML problem is the one that contains the target feature as a part of the dataset?
5. Name different types of ML problems.
6. What does ROC stand for in the context of evaluation metrics?
7. Which method is applied by logistic regression to squish the probability between 0 and 1?
8. What is the importance of k in the K-means clustering algorithm?
9. Which ML model discovers patterns and relationships within data without any labeled target variable?
10. Which algorithms are commonly used for predicting numerical target variables within supervised learning and can assist in feature selection by analyzing feature importance?

### 8.12.4 True or False Questions

1. Pandas and NumPy are Python libraries for data preprocessing and analysis in ML projects.
2. Feature scaling is necessary to ensure all numerical features in a dataset have the same scale or range of values.
3. The primary purpose of validation in ML is to fine-tune models and improve their performance before deployment.
4. Semi-supervised learning relies solely on labeled data to train models effectively.
5. Grid search helps identify the optimal values of hyperparameters by testing every possible combination within a specified range.
6. Precision measures the proportion of true predictions among all actual positive instances.
7. Ridge regression adds a penalty term to the loss function that can set some coefficients exactly to zero, effectively performing feature selection.
8. Self-supervised learning involves training models on correct and corrupted examples to learn meaningful representations.

9. Random Forest models are known for their high interpretability compared to decision trees.
10. Dimensionality reduction algorithms can only be used in unsupervised learning problems.

### 8.12.5 Fill in the Blanks

1. A dataset on countries' wealth loaded in dataframe "df1" has a feature called "gdp_per_capita" with "NaN" values. Fill in the appropriate function to fill these "NaN" values with the average of the other values:

```
df1['gdp_per_capita'].fillna(___________, ___________)
```

2. Fill in the blanks in the code snippet below to complete the command that will visualize the histogram of the column "dgp_per_capita" of the dataframe "df1" described above:

```
import matplotlib.pyplot as plt
import seaborn as sns

# Plot the distribution of gdp_per_capita
sns.__________(___________)
plt.title('GDP per Capita Histogram')
plt.show()

```

3. Fill in the blanks in the code snippet below to call the appropriate function for the SVC model:

```
from sklearn.svm import SVC
model=______________________

hyperparams = {'C': [0.001, 0.01, 0.1, 1, 10, 100], 'kernel': ['linear', 'rbf']}
```

4. Fill in the blanks in the code snippet below to call the appropriate function for the Random Forest model:

```
from sklearn.ensemble import RandomForestClassifier
model=_______________________________

hyperparams = {'n_estimators': [50, 100, 200], 'max_depth': [3, 5, 7, None]}
```

### 8.12.6 Coding Problems for Data Preprocessing and Analysis

1. This exercise will help you practice analyzing and visualizing data from the dataset "cities_lifestyle_measures".[8] This dataset contains various features related to the lifestyle measures of different cities. The specific tasks to be performed are as follows:
   a. Import the dataset "cities_lifestyle_measures" into a dataframe.
   b. Use appropriate methods to display the data types of the features in the dataset.
   c. Compute and display summary statistics for the feature "sunshine hours".

   d. Create a histogram to visualize the distribution of the "sunshine hours" feature.
   e. Create a bar chart to visualize the "street_sidewalks" feature.

2. Use the same dataset from problem 1: "cities_lifestyle_measures" to answer the below questions.
   a. Identify and list the number of NaN values for each column in the dataset.
   b. Replace the dataset's NaN values with the respective column's mean value.
   c. Perform scaling on the "obesity_levels" feature to normalize its values.
   d. Create a figure with two parts as follows:
      - Part 1: Plot a boxplot for the feature "outdoor_activities".
      - Part 2: Plot a scatter plot to visualize the relationship between "annual_hours_worked" and "obesity_levels" and compute their correlation.
   e. Apply one-hot encoding to the "street_sidewalks" feature to convert categorical values into numerical format.

### 8.12.7 Coding Problems on Supervised and Unsupervised Machine Learning Problems

3. In this exercise, you are tasked to predict a person's weight based on parameters that represent their daily exercise routine using the "Linnerud" dataset. The "Linnerud" dataset is one of the sample datasets from sklearn.

   Here is the code to load the "Linnerud" dataset into a dataframe called "df1":

```
import pandas as pd
from sklearn.datasets import load_linnerud

# Load the dataset with features and targets in separate dataframes
features_df, targets_df = load_linnerud(return_X_y=True, as_frame=True)

# Combine features and targets into a single dataframe
df1 = pd.concat([features_df, targets_df], axis=1)

# Print the first few rows of the dataframe
print(df1.head(10))
```

   a. Explore the features in the data frame using appropriate Pandas functions.
   b. Identify which features contain null values and impute them with appropriate values.
   c. Perform EDA by visualizing every numerical feature using a histogram. Include your histograms in a single figure with different subplots to compare them.
   d. In case of any outliers identified on your resultant histograms, handle them by replacing them with the lower or upper bound value.
   e. Now, let's look for the correlation between weight and each exercise column separately (Weight vs. Chins, Weight vs. Situps, and Weight vs. Jumps). Write your conclusions about the presence or absence of a correlation.
   f. Apply a standard scaling method to normalize the features.
   g. Train Random Forest and logistic regression to predict the target weight feature.
   h. Measure and evaluate the performance of both models on the validation set. Consider metrics such as MAE, MSE, and R-squared.
   i. Select the best-performing model based on the evaluation metrics.

4. In this exercise, you are tasked to predict a person's glucose level based on various medical parameters using the "Diabetes" dataset from sklearn. The "Diabetes" dataset is a well-known dataset used for regression tasks in ML.

The following code is to load the "Diabetes" dataset into a dataframe called "df1":

```
import pandas as pd
from sklearn.datasets import load_diabetes

# Load the dataset with features and targets in separate dataframes
diabetes = load_diabetes(as_frame=True)
df1 = pd.concat([diabetes.data, diabetes.target], axis=1)

# Rename the target column for clarity
df1.rename(columns={0: 'glucose_level'}, inplace=True)

# Print the first few rows of the dataframe
print(df1.head(10))
```

a. Explore the features in the dataframe using appropriate Pandas functions.
b. Identify which features contain null values and impute them with appropriate values.
c. Perform EDA by visualizing each feature.
d. In case of any outliers identified on your resultant histograms, handle them by replacing them with the lower or upper bound value.
e. Analyze the correlation between glucose level with other columns: blood pressure, thyroid stimulating hormone, and total serum cholesterol. Write your conclusions about the presence or absence of a correlation.
f. Apply a standard scaling method to normalize the features.
g. Select suitable regression models to predict the glucose level (s6). Suggested models include linear regression, Random Forest regression, MLP regression, and decision tree regression.
h. Train the models and evaluate their performance on the validation set. Consider metrics such as MAE, MSE, and R-squared.
i. Select the best-performing model based on the evaluation metrics.

5. In this exercise, you are required to work on the iris dataset from sklearn used earlier in this chapter for unsupervised learning. The iris.target variable of the dataset contains three species of iris flowers. You are required to predict the species of the iris flowers based on sepal length, sepal width, petal length, and petal width. The code below stores the independent and the target variables in the dataframe df1.

```
import pandas as pd
from sklearn.datasets import load_iris

# Loading the dataset
iris = load_iris()
df1 = pd.DataFrame(data=iris.data, columns=iris.feature_names)
df1['species'] = iris.target
```

a. Explore the features in the dataframe using appropriate Pandas functions.
b. Create visualizations for each feature to understand their distributions.
c. Analyze the correlation between features using a heatmap.
d. Apply a standard scaling method to normalize the features.
e. Select suitable classification models to predict the target variable. Suggested models include logistic regression, Random Forest classifier, MLP classifier, and decision tree classifier.
f. Split the data into training, validation, and test sets.

g. Train the selected models and evaluate their performance on the validation set. Consider metrics such as accuracy, precision, recall, and F1-score.
h. Create the ROC curves to visualize the performance.
i. Consider each model's interpretability, complexity, consistency in performance, and computational efficiency, and select the most appropriate model for the given problem.
j. Evaluate the selected model on the test set to ensure its performance on unseen data.

## NOTES

1. A. L. Samuel, "Some Studies in Machine Learning Using the Game of Checkers. II—Recent Progress", in IBM Journal of Research and Development, vol. 11, no. 6, pp. 601–617, Nov. 1967, doi: 10.1147/rd.116.0601.
2. F. Rosenblatt, "The Design of an Intelligent Automaton", Research Trends, vol. 6, no. 2, pp. 1, Summer 1958.
3. https://www.kaggle.com/datasets/yasserh/housing-prices-dataset.
4. https://www.kaggle.com/datasets/mathchi/diabetes-data-set.
5. https://www.kaggle.com/datasets/yasserh/advertising-sales-dataset/data.
6. https://www.kaggle.com/datasets/itsmesunil/bank-loan-modelling/data.
7. https://www.kaggle.com/datasets/amirmahdiabbootalebi/salary-by-job-title-and-country/data.
8. https://raw.githubusercontent.com/Object-Oriented-Programming-2024/Object-Oriented-Programming/main/Chapter8/cities_lifestyle_measures.csv.

# 9 Natural Language Processing and Text Mining with Python

## 9.1 TEXT PREPROCESSING

NLP involves a series of operations to prepare raw text data for analysis or further processing. Text data can be messy, unstructured, and filled with noise, so preprocessing is essential to simplify it into a more manageable form for machines. The aim is to improve the quality and consistency of textual data for applying machine learning models, which significantly enhances their performance. The most common text preprocessing steps include Tokenizing, Filtering Stop Words, stemming, Tagging Parts of Speech, Lemmatizing, Chunking, and Chinking. In the following subsections, we will discuss how to perform these various text preprocessing operations.

### 9.1.1 Tokenizing

Tokenization breaks text into words or sentences, making it easier to work with manageable segments. This helps transform unstructured data into a structured format for analysis. Word-level tokenization allows you to identify frequently occurring terms, while sentence-level tokenization enables examination of the relationships between words in a broader context. To set up your Python environment for working on NLP tasks using the Natural Language Toolkit (NLTK), which is a widely used NLP library in Python, follow the guidelines provided in the NLTK installation.[1] Line 2 of Code 9.1 imports the NLTK library into the script, enabling its diverse features and modules to be used. After the import, Line 5 uses the command "nltk.download('punkt')" to fetch and install the "punkt" tokenizer models from NLTK's repository. These models are essential for accurately splitting text into sentences and words, a fundamental step in many NLP tasks.

This setup ensures that all the necessary components are in place for conducting text analysis and processing within the Python environment.

**CODE 9.1 IMPORTING NLTK AND DOWNLOADING PUNKT**

```
1. # Import the Natural Language Toolkit library
2. import nltk
3.
4. # Download the Punkt sentence tokenizer (if not already downloaded)
5. nltk.download('punkt')
```

Code 9.2 uses NLTK to demonstrate basic text tokenization, including word and sentence tokenization. Initially, specific methods for these tokenization tasks are imported from NLTK: "word_tokenize()" and "sent_tokenize()" are brought into the scope with the import statements in Lines 2 and 3, respectively. These methods are designed to split text into individual words and sentences. The variable text is assigned a sample string in Line 6, serving as the input for tokenization. This string is then processed by "word_tokenize()" in Line 9, which breaks the text into a list of words, storing the result in the variable "words". To show the outcome, Line 12 prints the heading "Word Tokens:", followed by the list of word tokens printed in Line 13. Subsequently, the "sent_tokenize()" method is applied to the same text in Line 16, segmenting it into sentences, with the results in the


DOI: 10.1201/9781032668321-9

variable sentences. Finally, Lines 17 and 18 echo "Sentence Tokens:" and the list of sentence tokens to the console. This sequence of operations showcases the fundamental NLP capability of segmenting text into meaningful components, facilitating further linguistic analysis or processing tasks

**CODE 9.2 WORD TOKENIZATION**

```
1. # Import necessary libraries from NLTK for tokenization
2. from nltk.tokenize import word_tokenize
3. from nltk.tokenize import sent_tokenize
4.
5. # Define the text data
6. text = "Hello there! Welcome to the world of NLP."
7.
8. # Tokenize the text into words
9. words = word_tokenize(text)
10.
11. # Print word tokens
12. print("Word Tokens:")
13. print(words)
14.
15. # Print sentence tokens
16. sentences = sent_tokenize(text)
17. print("Sentence Tokens:")
18. print(sentences)
```

### 9.1.2 Text Normalization

Text normalization is a vital procedure that involves transforming text into a standard format, simplifying processing, and analyzing natural language tasks. This process ensures consistency throughout the text by converting all characters to lowercase, stripping punctuation, and Tokenizing it into individual elements for a better analysis. Text normalization also involves removing common, less meaningful words and applying stemming and lemmatization techniques to reduce words to their base or root forms, thus enhancing the effectiveness of computational text processing. Implementing these steps reduces the text's complexity and uniformly treats the same word's variations, improving computational text processing outcomes.

Let's begin the text normalization with the simple step of converting the text to lowercase. Code 9.3 depicts a basic text normalization technique that converts the text into a uniform case. Line 2 initializes a string variable "text". Subsequently, Line 5 employs the "lower()" method to convert the entire string to lowercase, which is then stored in the "text_lower" variable. Finally, Line 6 prints the lowercase string.

**CODE 9.3 CONVERTING TEXT TO LOWERCASE FORMAT**

```
1. # Sample text data
2. text = "This is an Example Sentence."
3.
4. # Convert the text to lowercase for further processing, then print it
5. text_lower = text.lower()
6. print(text_lower)
```

A corpus is a large set of text data that can support the development of NLP models. The corpus contains a collection of written texts specific to a genre or domain that can be analyzed to study the application of language and common patterns in a specific context. NLTK contains a corpus of stop words. These words contribute to the sentence's grammatical structure but add no meaning, such as "is", "of", "the", and "but". The stop words corpus contains 179 stop words in the English language. Code 9.4 uses the NLTK library to filter out stop words during text processing. Code Lines 2 to 4 import the necessary libraries: "nltk", "word_tokenize", and "stopwords". Line 7 downloads the stop words corpus from NLTK. Line 10 assigns the string "This is an Example Sentence" to a variable called text. In Line 13, a set of English stop words is created for future use in filtering. The "word_tokenize()" method, imported from NLTK, is applied in Line 16 to tokenize the sentence into individual words. Line 19 uses a list comprehension to filter out the stop words, resulting in a new list called "filtered_text", which contains only the meaningful words of the sentence. Lastly, Line 21 prints the "filtered_text" list, showcasing the transformation of the original text by excluding common, less informative words.

**CODE 9.4 REMOVING STOP WORDS FROM TEXT**

```
1. # Import nltk library, work_tokenize and stopwords
2. import nltk
3. from nltk.tokenize import word_tokenize
4. from nltk.corpus import stopwords
5.
6. # Download the 'stopwords' data package from nltk, if not already downloaded
7. nltk.download('stopwords')
8.
9. # Define a sample text
10. text = "This is an Example Sentence"
11.
12. # Create a set of English stopwords
13. stop_words = set(stopwords.words('english'))
14.
15. # Tokenize the text into individual words
16. tokens = word_tokenize(text)
17.
18. # Filter out the stopwords from the tokenized words
19. filtered_text = [word for word in tokens if not word.lower() in stop_words]
20.
21. # Print the filtered text, which excludes the stopwords
22. print(filtered_text)
```

Now we move onto stemming, the process of reducing words to their word stem or root form. Stemming is a set of rules to remove suffixes from a word to obtain its root. For example, the root form of the words; "works", "working", and "worked" is the stem word "work". The suffixes "s", "ing", and "ed" are removed to get the root, "work". It is important to note that the resulting root word in some cases is not accurate, such as when the stem for the word "workable" is "workabl". However, this is a computationally efficient method of converting words to a standard form, especially for large amounts of data.

Code 9.5 depicts an example of using stemming. First, the "word_tokenize" and "PorterStemmer" classes are imported from the NLTK library (Lines 2 and 3). "PorterStemmer" is a widely used stemming algorithm in text processing. After that, an instance of the class is created and assigned to the variable stemmer (Line 6). We then define the "Reading reads read" string and assign it to the variable "text" (Line 9). The "word_tokenize()" method is applied to tokenize the text into individual words or tokens (Line 12). Then, we apply a stemming process to each token and its root forms, showcasing a basic text normalization technique. This is done through a list comprehension, where "stemmer.stem(word)" is called for every word in tokens, resulting in a list of stemmed words stored in the "stemmed_words" variable (Line 15). Finally, Line 18 prints out this list of stemmed words, demonstrating how the original words have been reduced.

### CODE 9.5 STEMMING A SENTENCE

```
# Import the necessary libraries and modules
from nltk.tokenize import word_tokenize
from nltk.stem import PorterStemmer

# Create an instance of the PorterStemmer
stemmer = PorterStemmer()

# Define a sample text
text = "Reading reads read"

# Tokenize the text into individual words
tokens = word_tokenize(text)

# Stem each tokenized word using the PorterStemmer
stemmed_words = [stemmer.stem(word) for word in tokens]

# Print the list of stemmed words
print(stemmed_words)
```

We will now explore the concept of lemmatization, which is like stemming but reduces the word to its root form based on the context and meaning of the word. Lemmatization aims to remove inflectional endings and return a word's base or dictionary form. The key difference between stemming and lemmatization is that while stemming merely chops common suffixes from word tokens, lemmatization ensures that the resulting word is an existing, normalized form of the word (i.e., lemma) that can be in the dictionary. For example, if we perform stemming on the word "flies" the suffix "ies" is chopped off, and the result is "fl" which isn't a real word. If we perform lemmatization, it identifies "flies" as the plural form and returns "fly".

Code 9.6 illustrates lemmatization. First, the NLTK, "word_tokenize", and "WordNetLemmatizer" modules from NLTK are imported at Lines 2, 3, and 4, respectively. In Line 5, the "wordnet" corpus is downloaded. This corpus contains English words, their synonyms, and semantic relationships, which are useful for performing text-processing tasks such as lemmatization. Then, in Line 10, a "lemmatizer" is instantiated using the module. In Line 13, the test phrase "I'm reading a reading read by a reader" is defined as a string. This phrase is tokenized in Line 16. In Line 19, lemmatization is applied to each token, with the POS specified as verb ("v"), to ensure that verbs are reduced to their base form. Finally, Line 22 prints the lemmatized words, demonstrating how the original verbs have been normalized to their root forms.

### CODE 9.6 LEMMATIZING A SENTENCE

```
1. # Import the necessary modules from the nltk library
2. import nltk
3. from nltk.tokenize import word_tokenize
4. from nltk.stem import WordNetLemmatizer
5.
6. # Download the 'wordnet' resource from nltk
7. nltk.download('wordnet')
8.
9. # Create an instance of the WordNetLemmatizer
10. lemmatizer = WordNetLemmatizer()
11.
12. # Define a sample text
13. text = "I'm reading a reading read by a reader"
14.
```

```
# Tokenize the text into individual words
tokens = word_tokenize(text)

# Lemmatize each tokenized word with WordNetLemmatizer and part of speech set as 'verb'
lemmatized_words = [lemmatizer.lemmatize(word, pos='v') for word in tokens]

# Print the list of lemmatized words
print(lemmatized_words)
```

## 9.2 TEXT ANALYSIS

### 9.2.1 PARTS-OF-SPEECH (POS) TAGGING

POS tagging is a critical component of NLP. It involves the assignment of a specific tag to each word in a given text, which represents its respective parts of speech, such as nouns, verbs, adjectives, and so on. The primary objective of POS tagging is to facilitate our comprehension of the syntax and meaning conveyed by a sentence by identifying its grammatical structure. As a foundational step in many NLP applications, POS tagging is essential in parsing, named entity recognition, and machine translation. Automated POS tagging is typically accomplished through the utilization of algorithms or models that have been trained on annotated language data. Various pre-built POS taggers are available for different Python libraries, such as those provided by NLTK and spaCy.

Code 9.7 shows how to extract the POS of a sentence. The code starts by importing the NLTK library and "word_tokenize" from the NLTK library, then downloads the "averaged_perceptron_tagger" (Line 6), which is essential for tagging words with their parts of speech. The text about Michael Jordan is defined in Lines 9 and 10 and is tokenized into words in Line 13 using the "word_tokenize()" method, with the result stored in the variable "tokens". On Line 16, the "nltk.pos_tag()" method is applied to these tokens to assign a part of speech to each token, storing the output in the "tags" variable. The code concludes by printing these tags on Line 19, which now accurately reflect the parts of speech for each word in the text.

**CODE 9.7 APPLYING POS TAGGING ON A SENTENCE**

```
1. # Import the necessary modules from the nltk library
2. import nltk
3. from nltk.tokenize import word_tokenize
4.
5. # Download the 'averaged_perceptron_tagger' resource from nltk
6. nltk.download('averaged_perceptron_tagger')
7.
8. # Define the text
9. text = "Michael Jordan was born in Brooklyn, New York, and he became a "\
10.        "renowned professional basketball player."
11.
12. # Tokenize the text into individual words
13. tokens = word_tokenize(text)
14.
15. # Perform part-of-speech tagging on the tokenized words
16. tags = nltk.pos_tag(tokens)
17.
18. # Print the list of words with their corresponding part-of-speech tag
19. print("Tags:",tags)
```

The result of executing Code 9.7 is shown in Box 9.1. The "Tags" output displays a list of tuples where each tuple represents a word from the input text and its corresponding POS tag. For example,

### BOX 9.1 RESULTS OF APPLYING POS TAGGING

**Output**

```
Tags: [('Michael', 'NNP'), ('Jordan', 'NNP'), ('was', 'VBD'), ('born', 'VBN'), ('in', 'IN'),
('Brooklyn', 'NNP'), (',', ','), ('New', 'NNP'), ('York', 'NNP'), (',', ','), ('and', 'CC'), ('he',
'PRP'), ('became', 'VBD'), ('a', 'DT'), ('renowned', 'JJ'), ('professional', 'NN'), ('basketball',
'NN'), ('player', 'NN'), ('.', '.')]
```

"Michael" and "Jordan" are tagged as "NNP" (proper noun), "was" is tagged as "VBD" (past tense verb), and "born" as "VBN" (past participle).

## 9.2.2 Chunking

"Chunking" is the process of grouping individual words from unstructured text into larger, meaningful phrases based on patterns or rules, considering their parts of speech. This process is crucial in generating meaningful insights from large volumes of text data and plays a vital role in various NLP applications.

Code 9.8 demonstrates how to chunk a piece of text using Python. In Line 6, a sentence is defined for analysis. The sentence is tokenized (Line 9), and each token is assigned a POS tag (Line 12). Line 15 defines a chunk grammar pattern using a regular expression, where "NP" represents a noun phrase (NP), "<DT>" represents a determiner (DT), "<JJ>" represents adjectives, and "<NN>" represents a noun. This pattern allows for an optional DT, any number of adjectives, and a noun. A "RegexpParser" is initialized with the defined grammar in Line 18, and Line 21 applies this parser to the tagged tokens to identify chunks. Finally, Line 24 visualizes the chunk structure in a tree format, showing how the sentence's tokens are grouped into chunks based on the specified grammar.

### CODE 9.8 CHUNKING A SENTENCE

```
1. # Import the necessary methods and classes from the nltk library
2. from nltk import word_tokenize, pos_tag
3. from nltk.chunk import RegexpParser
4.
5. # Define a sample text
6. sentence = "A swift auburn fox leaps over the dormant hound."
7.
8. # Tokenize the sentence into individual words
9. tokens = word_tokenize(sentence)
10.
11. # Perform part-of-speech tagging on the tokenized words
12. tagged_tokens = pos_tag(tokens)
13.
14. # Define the grammar for noun phrase (NP) chunking
15. grammar = "NP: {<DT>?<JJ>*<NN>}"
16.
17. # Create a RegexpParser object with the defined grammar
18. cp = RegexpParser(grammar)
19.
20. # Parse the tagged tokens to identify noun phrases
21. chunked = cp.parse(tagged_tokens)
22.
23. # Draw the chunked structure
24. chunked.draw()
```

The output of Code 9.8 is presented as a tree diagram, with the root node denoted by the letter S representing the entire sentence (Figure 9.1). The tree has five child nodes, of which three are NPs. The

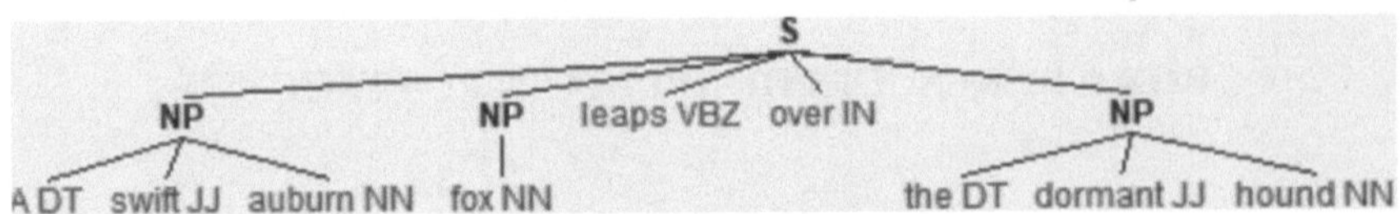


**FIGURE 9.1** Tree representation of chunking based on POS tags.

first NP child node has three sub-nodes, with the letter A being classified as a DT, the word "swift" as an adjective (JJ), and "auburn" as a noun. The second NP child node contains the noun "fox", which is classified as a noun (NN). The third child node contains the verb "leaps" and is classified as a third-person singular present verb (VBZ). The fourth child node contains the word "over" and is classified as a preposition or subordinating conjunction (IN). The fifth NP child node has three sub-nodes, with the word "the" being classified as a DT, "dormant" as an adjective (JJ), and "hound" as a noun (NN).

### 9.2.3 Using a Concordance

A concordance is a tool for examining the contextual usage of a specific word. Analyzing the immediate contexts of the word across sentences allows one to gain insights into its meaning and associated words.

Code 9.9 shows an example of using a concordance. Line 3 imports the "Text" class, necessary for analyzing a sample text. In Lines 6 and 7, the sample text is defined. Then, the sample text is split into individual words and punctuation marks using the "word_tokenize()" method, creating a list of tokens (Line 10). Next, the "Text" class is employed to structure the text for analysis (Line 13). Finally, the "concordance()" method is used to locate and display every occurrence of the word "solidarity" within the text and its context (Line 16). Box 9.2 displays two occurrences of the word "Solidarity" with the context in which it is used.

**CODE 9.9 APPLYING CONCORDANCE ON THE WORD SOLIDARITY**

```
1. # Import the necessary methods and classes from the nltk library
2. from nltk import word_tokenize
3. from nltk.text import Text
4.
5. # Define a sample text
6. sample_text = "Solidarity is enduring. It does not harbor jealousy, it does not brag, "\
7.               "it is not arrogant. Solidarity is compassionate."
8.
9. # Tokenize the sentence into individual words
10. tokens = word_tokenize(sample_text)
11.
12. # Create a Text object from the tokenized words
13. text_obj = Text(tokens)
14.
15. # Display the concordance of the word "Solidarity"
16. text_obj.concordance("solidarity")
```

**BOX 9.2 RESULTS OF APPLYING CONCORDANCE**

**Output**

```
Displaying 2 of 2 matches:
Solidarity is enduring . It does not harbor
s not brag, it is not arrogant . Solidarity is compassionate .
```

### 9.2.4 A Dispersion Plot

A dispersion plot helps find a word's frequency and location, compare synonym usage, and identify where words appear in a text or corpus. It also helps determine the co-occurrence of words in a single text or the usage of words over some time in a chronologic corpus of texts.

Code 9.10 analyzes a text called "text1" from the "book" corpus in the NLTK library. First, we import the "pylab" module to generate plots (Line 2). Then, we download the "book" package using the "nltk.download" method (Line 4). Line 6 imports "text1" from the book (which is a novel by Herman Melville). Line 9 uses the "concordance()" method to find all occurrences of the word "ship" in "text1" and displays them with their immediate context. In Line 12, a list of words related to "ship" is created and assigned to the variable "target_words". Line 15 generates a dispersion plot for the words listed in "target_words" in "text1", visually representing the occurrences of these words throughout the text. Finally, Line 18 uses the "pylab.show()" method to display the plot created.

**CODE 9.10 SHOWING THE FREQUENCY OF WORDS USING A DISPERSION PLOT**

```
1. # Import the necessary libraries
2. import pylab, nltk
3. # Download the 'book' collection from nltk
4. nltk.download("book")
5. # Import ‘text1’ from the nltk book collection
6. from nltk.book import text1
7.
8. # Display the concordance for the word ‘ship’ in text1
9. text1.concordance("ship")
10.
11. # Define a list of target words for the dispersion plot
12. target_words = ['ship', 'boat', 'vessel', 'craft']
13.
14. # Generate a dispersion plot for the target dispersion plot
15. text1.dispersion_plot(target_words)
16.
17. # Show the plot
18. pylab.show()
```

The resulting plot shows the frequency of the word "craft" in comparison with less common words such as "ship", "boat", and "vessel" (Figure 9.2).

### 9.2.5 A Frequency Distribution

A frequency distribution is a statistical representation that provides a concise and comprehensive summary of the frequency of occurrence of unique values or categories of a variable in a dataset. In the context of text analysis, a frequency distribution would typically present the frequency of occurrence of each word in a text, offering insights into the most and least frequently used words.

Code 9.11 shows an example of creating a frequency distribution. The code begins by importing the necessary libraries in Line 2. "FreqDist" is imported in Line 3. The "book" corpus is downloaded at Line 5. In Line 7, "text1" is imported from the book corpus. It is a class used to construct a frequency distribution of words. Line 10 creates a frequency distribution instance, "fdist", by applying

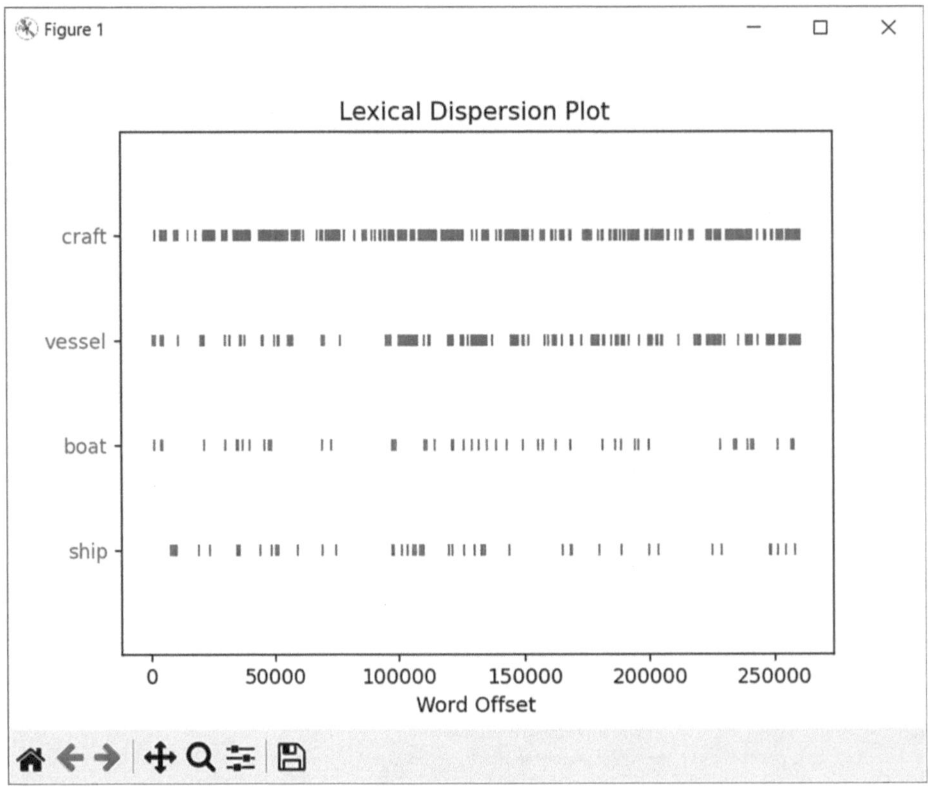


**FIGURE 9.2** Dispersion plot for selected words within the text.

"FreqDist()" to "text1", a collection of words from a book. In Line 13, we print the 20 most common words in "text1" along with their frequencies, using the "most_common(20)" method of the "FreqDist" instance. Finally, Line 16 generates a cumulative plot for the 20 most frequent words in the text, visualizing how the frequencies accumulate over the set of these words, with "cumulative=True" indicating that the plot should display a cumulative count of frequencies. The result is shown in Figure 9.3.

**CODE 9.11 CREATING A FREQUENCY DISTRIBUTION OF 20 MOST COMMON WORDS**

```
1. # Import the necessary libraries and modules
2. import pylab, nltk
3. from nltk.probability import FreqDist
4. # Download the "book" collection from nltk
5. nltk.download("book")
6. # Import text1 from the nltk book collection
7. from nltk.book import text1
8.
9. # Create a frequency distribution of the words in text1
10. fdist = FreqDist(text1)
11.
12. # Print the 20 most common words in a cumulative frequency plot
13. print(fdist.most_common(20))
14.
15. # Show the plot
16. fdist.plot(20, cumulative=True)
```

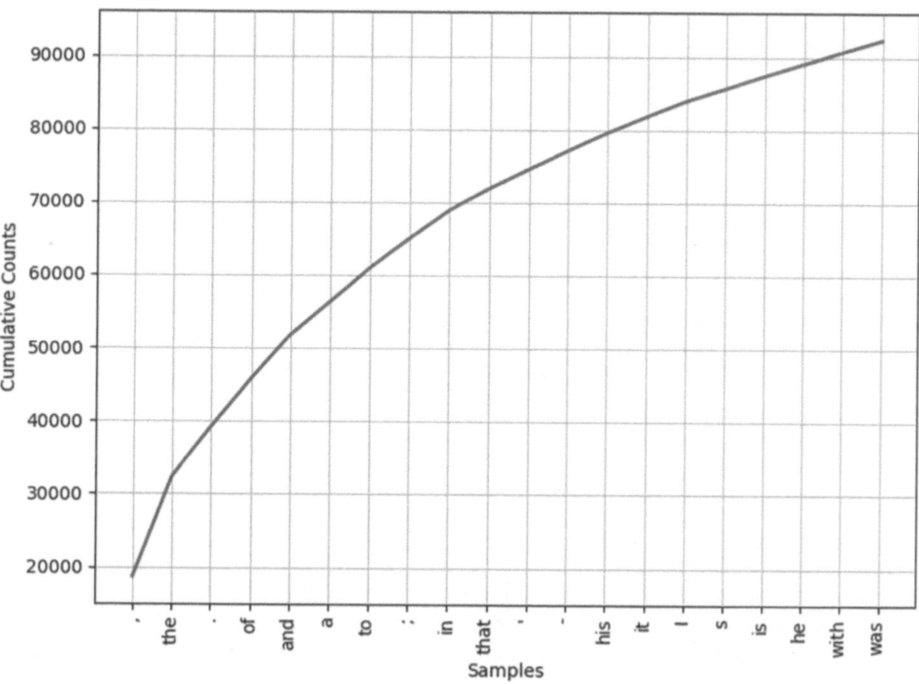


**FIGURE 9.3** Frequency distribution of the most common 20 words in the text.

### 9.2.6 Collocations

Collocations are word combinations that occur frequently together in a language. They convey meanings beyond the individual words, including idiomatic expressions, compound nouns, or other phrases. Collocations are crucial in linguistics and language learning and can improve machine translation and natural language processing (NLP) tasks. Code 9.12 shows an example of extracting collocations from text1. Box 9.3 shows a part of the results of performing collocations on "text1".

**CODE 9.12 PERFORMING COLLOCATION ON TEXT**

```
# Import the necessary NLTK library
import nltk
# Download the "book" collection from nltk
nltk.download("book")
# Import text1 from the nltk book collection
from nltk.book import text1

# Display the collocations found in text1
text1.collocations()
```

**BOX 9.3 OUTPUT OF PERFORMING COLLOCATIONS ON TEXT 1**

**Output**

```
*** Introductory Examples for the NLTK Book ***
Loading text1, …, text9 and sent1, …, sent9
Type the name of the text or sentence to view it.
Type: 'texts()' or 'sents()' to list the materials.
text1: Moby Dick by Herman Melville 1851
text2: Sense and Sensibility by Jane Austen 1811
text3: The Book of Genesis
text4: Inaugural Address Corpus
text5: Chat Corpus
text6: Monty Python and the Holy Grail
text7: Wall Street Journal
text8: Personals Corpus
text9: The Man Who Was Thursday by G . K . Chesterton 1908
```

### 9.2.7 Sentiment Analysis

Sentiment analysis is a method for determining the emotions behind a piece of text. It is useful for understanding people's attitudes, opinions, and emotions expressed online. This technique is often applied to reviews, social media posts, and customer feedback. It helps businesses and researchers understand public sentiment and track brand reputation. This way, companies can develop strategies based on consumer responses.

Code 9.13 utilizes the NLTK library to perform sentiment analysis on two different pieces of text. This analysis is achieved using the "SentimentIntensityAnalyzer" class from the "nltk.sentiment" module, specifically designed to determine the emotional tone of texts (Line 2). Line 4 imports the NLTK library.

The code downloads the VADER lexicon, a finely tuned list of known words and their sentiment intensities, by "calling nltk.download('vader_lexicon')" (Line 7). This lexicon is crucial, as the "SentimentIntensityAnalyzer" uses it to assess the sentiment of given texts. An instance of "SentimentIntensityAnalyzer" is created and assigned to the variable "sia", which is then used to analyze the emotional tone of texts (Line 10). Two strings, "text_1" and "text_2", are defined. They contain distinct sentences that express strong sentiments. "text_1" describes a very unpleasant experience, while "text_2" conveys a highly positive experience (Lines 13–22). The "polarity_scores()" method of the "sia" object is given "text_1" as the argument. This method returns a dictionary containing scores in several categories, including "neg" (negative), "neu" (neutral), "pos" (positive), and "compound" (an aggregated score). These scores, which show the sentiment analysis for "text_1" and "text_2", are then printed (Lines 24 and 26).

**CODE 9.13 PERFORMING SENTIMENT ANALYSIS ON A NEGATIVE AND A POSITIVE SENTENCE**

```
1. # Import the SentimentIntensityAnalyzer class from the nltk.sentiment module
2. from nltk.sentiment import SentimentIntensityAnalyzer
3. # Import the NLTK library
4. import nltk
5.
6. # Download the 'vader_lexicon' from NLTK, which is required for sentiment analysis
7. nltk.download('vader_lexicon')
8.
9. # Create an instance of the SentimentIntensityAnalyzer
```

```
sia = SentimentIntensityAnalyzer()

# Define a negative text for sentiment analysis
text_1= (
    "The service was absolutely horrendous, the staff was rude, and the entire "
    "experience was a complete disaster. I've never felt so unwelcome and "
    "disrespected in my life. Avoid this place at all costs; it's truly a nightmare."
)
text_2= (
    "The service was outstanding, the staff was incredibly friendly, and the whole "
    "experience was delightful. I've never felt so valued and well-treated."
    "Highly recommend this place; it's absolutely wonderful!"
)
# Print the sentiment analysis scores for the negative text
print(sia.polarity_scores(text_1))
# Print the sentiment analysis scores for the positive text
print(sia.polarity_scores(text_2))
```

The result of executing the code in Code 9.13 is in Box 9.4. The first sentiment analysis result indicates a predominantly negative tone with 39.5% negative, 54.9% neutral, and a compound score of -0.9535, reflecting a strong negative sentiment overall. The second result shows a mostly positive sentiment, with no negative expressions, 49% positive content, and a compound score of 0.9723, suggesting an overwhelmingly positive tone. These insights are useful for understanding emotional responses in text data, such as customer feedback or social media comments.

**BOX 9.4 RESULT OF SENTIMENT ANALYSIS**

**Output**

```
{'neg': 0.395, 'neu': 0.549, 'pos': 0.057, 'compound': -0.9535}
{'neg': 0.0, 'neu': 0.51, 'pos': 0.49, 'compound': 0.9723}
```

## 9.3 TEXT MINING SOCIAL MEDIA – TWITTER

An interesting use case of Text Mining is analyzing posts on social media platforms like Twitter. Python provides the Tweepy package to read, write, and analyze Twitter posts. A detailed description is beyond the scope of this book. However, an introduction to Tweepy is provided here to help understand the basic requirements to get started.

### 9.3.1 Twitter to X

Twitter was launched in 2006 and is among the most popular social media platforms. Users on the platform write short messages called tweets, and the brevity of the messages became Twitter's core identity. In 2010, Twitter introduced the logo of a small blue bird to symbolize the act of "tweeting". Some "tweeters" have millions of followers, with dedicated tweeters posting several times daily to keep their followers engaged. The platform took the Internet by storm and became synonymous with breaking news, government discourse, celebrity spats, and the rhythm of online communications.

Then, in 2022, the business magnate Elon Musk acquired Twitter. Twitter was overhauled, with a massive shift in the organization and its online content moderation. The rebranding of Twitter to X in April 2023 sealed the change. The familiar bird logo gave way to a plain "X". Though the platform has a new name, "Twitter" is still the most commonly used name for reference, and therefore, the same is used here to refer to the social media platform.

### 9.3.2 Twitter and Text Mining

Twitter boasts a massive user base, generating millions of tweets daily, with thousands sent every second. While the sheer volume is overwhelming (like "drinking from a fire hose"), it's a goldmine for researchers and businesses. This real-time stream, accessible to programmers, offers valuable insights into current events and public opinion. However, this stream of data is not completely structured.

Twitter provides its users free access to a limited sample of recent tweets. However, through paid partnerships and select businesses, Twitter offers everyone access to a much larger historical dataset. Researchers and businesses process (mine) this data to conduct in-depth analysis, uncover trends and customer preferences, forecast user sentiments, and predict future events. The tweets are accessed through Twitter APIs.

### 9.3.3 Twitter APIs

Twitter application programming interfaces (APIs) provide developers with various ways to interact with Twitter's data and functionality. There are two types of APIs, the REST APIs and the streaming APIs, each with different purposes.

#### 9.3.3.1 REST APIs (Representational State Transfer APIs)

- **Query**: A REST API is primarily used for querying and retrieving specific, predefined sets of data from Twitter, such as user profiles, tweets, lists, and trends.
- **Request-response model**: REST APIs are inherently based on the request-response model, in which the developer application sends a request to Twitter's servers and receives a response.
- **Sampling**: A REST API is executed to retrieve updated information. Twitter responds with the most recent sample of the data based on the query parameters provided.
- **Rate limits**: The REST API has rate limits that restrict the number of requests a client can make within a specific time frame. Developers need to manage these limits to avoid being rate-limited by Twitter.

#### 9.3.3.2 Streaming APIs

- **Real-time data**: The streaming API allows developers to access a continuous flow of data in real time and receive tweets as they occur on Twitter.
- **Pub-sub model**: In the pub-sub (publishers and subscriber) model, Twitter pushes data to the client as soon as it becomes available. Publishers are responsible for sending messages to the system, while subscribers are responsible for receiving those messages.
- **Data filtering**: Developers can set up filters to receive only the data that matches specific criteria.
- **Connection-based**: Client applications establish a persistent connection to the Twitter servers to receive a continuous data stream. The connection remains open until explicitly closed by the client or terminated by Twitter.

#### 9.3.3.3 Important Restrictions

Each Twitter API method has a rate limit, i.e., the maximum number of requests that can be made during a window of time. Twitter may block access to its APIs if the rate limit is crossed. It is advisable to read the following documentation before mining Twitter data.

- https://developer.x.com/en/docs/rate-limits
- https://developer.x.com/en/developer-terms/agreement-and-policy
- https://developer.x.com/en/developer-terms/more-on-restricted-use-cases

### 9.3.4 Creating a Twitter Developer Account

Twitter requires programmers to register to access their data. As part of the registration process, the user is required to provide the reasons for accessing Twitter feeds. After submission, the request is reviewed based on Twitter's terms and conditions. Approval is not assured. Once the request is approved, the Developer Portal with a default project is presented to the user.

The following link can be used to register a developer account: https://developer.x.com/

#### 9.3.4.1 Twitter Projects and Apps

Twitter projects help organize, manage, and monitor the usage of Twitter APIs. Each project can contain one or more Apps. Depending on the type of account and access level that a user registers for, Twitter provides consumption caps. For the free version, only one project with one App and a maximum of 1500 monthly Tweets can be accessed. This may change in the future. The other account types provide more access privileges.

The free version also provides OAuth 2.0 with PKCE authorization. OAuth 2.0 is Open Authorization, a standard that allows a website or application to access resources hosted by other Web applications on behalf of a user. Version 2.0 replaced OAuth 1.0 in 2012 and is now the industry standard for online authorization. Proof Key for Code Exchange (PKCE) provides an additional security layer that helps prevent code interception attacks.

The App listed below is associated with a project. The App details include a name, App ID, App icon, and description. The App ID is not editable, but the other fields are. Check the following link for more information on the access levels and versions.

- https://developer.x.com/en/docs/twitter-api/getting-started/about-twitter-api#Access

Once the Twitter App is created, the "User Authentication Settings" need to be set up.

- Set the App permissions to "Read and Write".
- Enable the OAuth 2.0 Authentication.
- Create callback Uniform Resource Identifiers (URIs). Callback URIs are destinations that OAuth redirects to after the authentication process. This is important for OAuth to recognize the URIs as valid.

### 9.3.5 Client Program to Access Twitter

Python programs can access the Twitter APIs to interact with the social media platform. This requires installing the Tweepy package and writing code that uses the classes defined in the package. The steps here will help set up and write some basic code.

#### 9.3.5.1 Installing Tweepy

The "tweepy" library package provides object-oriented wrappers to access the Twitter APIs. Tweepy supports both Twitter API v1.1 and Twitter API v2.

It can be installed from the console using pip, or in an IDE like PyCharm, the built-in installer can be used to include the package in the Python project.

```
pip install tweepy
```

The following link can be used to refer to the documentation on Tweepy: https://docs.tweepy.org/

#### 9.3.5.2 Authentication Credentials

Twitter requires developers to use specific credentials (keys) to authenticate access to tweets. These keys are initially generated when the App is created and then regenerated whenever the user requires them to be updated. Typically, these keys are regenerated whenever there is a change to the App settings. Creating a separate Python file containing these keys and updating them when needed is advisable for the Python programs. For instance, Box 9.5 provides an example of how these keys can be stored. The "XXXXXXXXXXXXXXXXXXXX" must be replaced with the values of keys copied from the developer platform.

**BOX 9.5 PYTHON FILE "TWITTER_CREDENTIALS.PY" WITH KEYS TO AUTHENTICATE TWITTER API ACCESS**

```
# Define consumer Keys and Tokens to access Twitter APIs

# Consumer Keys for Twitter API
API_KEY = "XXXXXXXXXXXXXXXXXXXX"
API_KEY_SECRET = "XXXXXXXXXXXXXXXXXXXX"

# Authentication Tokens for Twitter API
BEARER_TOKEN = "XXXXXXXXXXXXXXXXXXXX"
ACCESS_TOKEN = "XXXXXXXXXXXXXXXXXXXX"

# Client credentials for OAuth authentication
CLIENT_ID = "XXXXXXXXXXXXXXXXXXXX"
CLIENT_SECRET = "XXXXXXXXXXXXXXXXXXXX"
```

### 9.3.6 Create a Tweet

Code 9.14 provides a class that inherits from "tweepy.Client". The class is a wrapper for methods that Tweepy provides for accessing Twitter.

**CODE 9.14 A CLASS TO USE TWEEPY.CLIENT TO WRITE A TWITTER POST**

```
# Import the tweepy library for Twitter API interaction
import tweepy
# Import twitter_credentials module containing API keys and tokens
import twitter_credentials


class TweetWriter(tweepy.Client):
    """Wrapper Class for posting tweets."""

    def __init__(self):
        super().__init__() # Initialize parent class
        self.client = None # Client to access Twitter
        self.auth = None # Authorization of consumer key

    # Create a client to post tweets
    def createClient(self):
        self.client = tweepy.Client(twitter_credentials.BEARER_TOKEN,
                                    twitter_credentials.API_KEY,
                                    twitter_credentials.API_KEY_SECRET,
                                    twitter_credentials.ACCESS_TOKEN,
                                    twitter_credentials.ACCESS_TOKEN_SECRET)

```

```
23.     # Create authentication for the client access
24.     def createAuth(self):
25.         self.auth = tweepy.OAuthHandler(twitter_credentials.API_KEY,
26.                                         twitter_credentials.API_KEY_SECRET,
27.                                         twitter_credentials.ACCESS_TOKEN,
28.                                         twitter_credentials.ACCESS_TOKEN_SECRET)
29.
30.     # Post a tweet on Twitter
31.     def writePost(self, post):
32.         self.client.create_tweet(text=post)
33.
34.     # Reply to a post on Twitter
35.     def replyPost(self, tweetid, reply):
36.         self.client.create_tweet(in_reply_to_tweet_id=tweetid, text=reply)
37.
38. # Create TweetWriter object and post a tweet
39. mytweeter = TweetWriter()
40. mytweeter.createClient()
41. mytweeter.createAuth()
42. mytweeter.writePost("This is a new post")
```

Lines 2 and 4 import the "tweepy" package and the file with the credentials, respectively. Lines 10–13 initialize the required attributes for creating and authenticating a client. The "createClient()" method in Lines 16–21 creates a client using the bearer token, API key, and access token. The "createAuth()" method from Lines 24 to 28 ensures the client's authentication using the API key and access token as parameters to the "tweepy.OAuthHandler()" method. The "mytweeter" object is created as an instance of the class "TweetWriter" in Line 39. The object creates and authenticates a client in Lines 40 and 41. After this, the "writePost()" method is invoked, defined in Lines 31 and 32. The "writePost()" method is a wrapper for the "create_tweet()" method of the Tweepy. After executing this program, the string "This is a new post" will appear on the associated client's Twitter home page, as shown in Figure 9.4.

When clicking on the post, the post's ID is displayed in the URL of the post, indicated by the red arrow in Figure 9.4. This ID can be used to reply to the post. The wrapper method to reply to a post is written in Lines 35 and 36 of Code 9.14. This "replyPost()" method calls the "create_tweet()" method of the "tweepy.Client" class with the ID provided as the parameter value for the

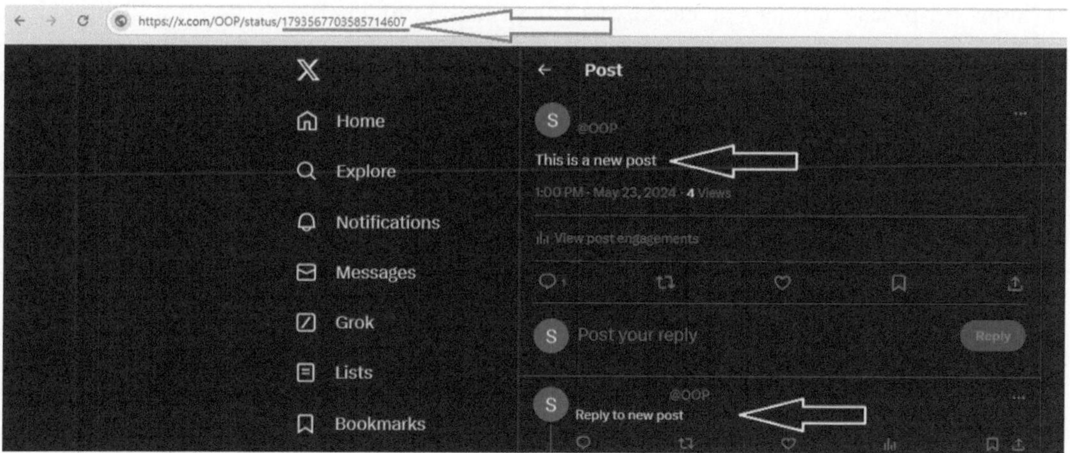


**FIGURE 9.4** Tweet ID in the URL of the tweet.

"in_reply_to_tweet_id". The second parameter, "text", contains the message. The sample parameter values to call the "replyPost()" wrapper method are shown in Code 9.15.

**CODE 9.15 PARAMETER VALUES FOR THE REPLYPOST() METHOD**

```
1. mytweeter.replyPost("1793567703585714607", "Reply to new post")
```

Free developer access to the X API only facilitates writing tweets to an account. Paid access is required to mine text data from Twitter. Therefore, in the following section, historical Twitter data from Kaggle (https://www.kaggle.com/) is used to showcase how to perform natural language processing tasks on social media data.

### 9.3.7 Topic Modeling with Twitter Data

The exponential growth of data production on social media necessitates efficient methods for gleaning quick insights. Machine learning solves this challenge, enabling the extraction of valuable knowledge from vast amounts of data. One such technique is topic modeling, which is applied to textual data to find the most occurring terms in multiple documents and determine themes and trends.

Topic modeling can determine social media trends, analyze customer feedback, evaluate quantitative research, and more. In this section, we apply topic modeling to Twitter data to find the most repeating resolutions made in the new year, illustrated in Code 9.16. The New Year's resolutions dataset is adapted from the Kaggle website and contains 5011 tweets. Lines 1–9 import the necessary libraries. Two packages introduced in this code are: "re" and "genism". The "re" package refers to the regex package, which is used for finding patterns within strings, and the genism package is used for applying two important natural language processing tasks, vectorization and topic modeling. Lines 12–14 download the resources for the NLTK library introduced earlier. Lines 16–80 define a class, "NLPTweets". This class is created to abstract critical NLP processes and their functionality for clarity and reuse. The class constructor from Lines 20 to 23 stores the Pandas DataFrame and column that contains the tweets for later access. It also creates an empty list to store "processed_tweets" for further analysis.

The "pre_process_tweets()" method, Lines 25–57, removes unwanted data and performs tokenization and lemmatization. Line 28 drops rows from the data frame containing no data in the "text" column. Line 31 extracts the "text" column as a list. A "for" loop iterates through each tweet and performs per-processing steps on each tweet from Lines 33 to 55. Tweets are most likely to have hashtags starting with a "#" symbol, mentions or references to other users beginning with an "@" symbol, website links, and more. As we are interested in users' goals for the new year, we consider hashtags, mentions, website links, dates, and punctuation to be noise in our data. Therefore, Lines 41–44 apply regex patterns to substitute noisy data with an empty string using the "sub()" method. For example, in Line 41, the "#\w+" pattern finds all words starting with a # and removes them from the tweet. Line 45 iterates through each character in the tweet and extracts only those characters without punctuation, resulting in tweets without punctuation. Line 48 performs tokenization, and Lines 49–52 perform lemmatization. Line 55 adds the tweet to the "processed_tweets" list. Line 57 outside the for-loop returns this list.

The "vectorize_tweets()" method from Lines 59 to 67 converts textual data to numeric representation to apply statistical analysis. Line 62 uses the Gensim's corpora to identify unique words in the tweets and assign a unique integer value to represent each word. In Line 65, this "self.dictionary" dictionary is used to convert all words to their integer value representation, along with their frequency of occurrence in the text. This strategy is called the "bag of words" strategy and helps create a corpus of text in numeric format. Numeric representations of data are referred to as vectors and are necessary for applying statistical analysis. Following is a sample sentence, converted to integer values and frequency of occurrences as tuples.

*Pre-processed tweet= ["Eat", "less", "quit", "lying"]*
*Numeric word representations with frequency = [(33, 1), (34, 1), (35, 1), (36, 1)]*

The "apply_topic_modelling()" method from Lines 70 to 76 applies an unsupervised learning algorithm called the Latent Dirichlet Allocation (LDA) algorithm. This algorithm finds the most frequently occurring words within different texts to identify trending topics. The LDA instance is initialized at Lines 73–76 with four parameters. The first is the vector representation of the text, "self.corpus". The second parameter is the number of trending topics to retrieve from the text. Followed by the parameter "id2word=self.dictionary", which converts the integer representation of the words to their original text format after performing the statistical inference. The last parameter, "passes=10", is the number of times the model is trained to improve performance. "lda_model.print_topics()" method at Line 79 finds the top frequently co-occurring words within sentences based on the number of words specified, which in this code are "6" words. Line 84 creates a Pandas dataframe after retrieving the data from a CSV file. Line 86 creates the "NLPTweets" instance to apply natural language processing to the data. Line 87 applies the "pre_process_tweets()" method to clean the data. Line 89 applies the "vectorize_tweets()" method to convert text to a vector format. Line 91 applies the LDA algorithm to retrieve top-5 topics. Lines 92–93 iterate through the topics and print them, as shown in Box 9.6.

### CODE 9.16 TOPIC MODELING ON TWITTER DATA

```
1. import pandas as pd
2. import nltk
3. import string
4. from nltk.corpus import stopwords
5. from nltk.stem import WordNetLemmatizer
6. from nltk.tokenize import word_tokenize
7. import re
8. import gensim
9. from gensim import corpora
10.
11. # Download necessary NLTK resources
12. nltk.download('stopwords')
13. nltk.download('punkt')
14. nltk.download('wordnet')
15.
16. class NLPTweets:
17.     """A class to apply NLP techniques on Tweets"""
18.
19.     # Storing the dataframe and the column for processing the tweets
20.     def __init__(self, df,col_name):
21.         self.df= df # DataFrame containing the tweets
22.         self.col_name=col_name # Column name with the tweets
23.         self.processed_tweets=[] # List of pre-processed tweets
24.
25.     def pre_process_tweets(self):
26.
27.         # Dropping rows with null values in the column with tweets
28.         self.df.dropna(subset=[self.col_name], inplace=True)
29.
30.         # The column with textual data
31.         tweets = self.df[self.col_name].values
32.
```

```
        for tweet in tweets:
            # Using stopwords from the english dictionary to remove
            stop_words = set(stopwords.words('english'))

            #Using a Lemmatizer to find and replace with root words
            lemmatizer = WordNetLemmatizer()

            # Using pattern matching to remove noise from the data
            tweet = re.sub(r"#\w+", "", tweet) # Removing any  hastags
            tweet = re.sub(r"@\w+", "", tweet)  # Removing mentions
            tweet = re.sub(r"https?://\S+|www\.\S+","",tweet)  # Removing URLs
            tweet = re.sub(r"\d+", "", tweet) # Removing numbers such as a date
            tweet=''.join([char for char in tweet if char not in string.punctuation])

            # Tokenizing and Lemmatizing each tweet
            words = word_tokenize(tweet)
            words = [
                lemmatizer.lemmatize(word)
                for word in words
                if word not in stop_words ]

            #Adding the processed tweets to a list
            self.processed_tweets.append(words)

        return self.processed_tweets

    def vectorize_tweets(self):

        # Creating a dictionary of unique words from tweets, each given an integer value
        self.dictionary = corpora.Dictionary(self.processed_tweets)

        # Bag of words strategy, each word as integer its frequency in the tweet
        self.corpus = [self.dictionary.doc2bow(text) for text in self.processed_tweets]

        return self.corpus


    def apply_topic_modelling(self,num_topics):

        # applying LDA
        lda_model = gensim.models.LdaModel(
            self.corpus,
            num_topics=num_topics,
            id2word=self.dictionary, passes=10)

        # Retrieving the topics
        topics = lda_model.print_topics(num_words=6)
        return topics


# Retrieving the dataset
df = pd.read_csv(
    "https://raw.githubusercontent.com/Object-Oriented-Programming-2024/" \
    "Object-Oriented-Programming/main/Chapter9/newyear_resolutions.csv"
)

nlp_tweets=NLPTweets(df,"text") #Creating the NLPTweets Instance
processed_tweets=nlp_tweets.pre_process_tweets()

corpus=nlp_tweets.vectorize_tweets()

topics = nlp_tweets.apply_topic_modelling(5)
for topic in topics:
    print(topic)
```

**BOX 9.6 OUTPUT OF TOPIC MODELING APPLIED IN CODE 9.16**

**Output**

```
(0, '0.075*"I" + 0.041*"year" + 0.019*"new" + 0.011*"get" + 0.010*"Im" + 0.010*"want"')
(1, '0.095*"New" + 0.080*"Years" + 0.080*"resolution" + 0.051*"My" + 0.020*"year" + 0.020*"Resolution"')
(2, '0.032*"I" + 0.029*"My" + 0.026*"make" + 0.023*"money" + 0.019*"possible" + 0.017*"Im"')
(3, '0.022*"get" + 0.018*"I" + 0.008*"like" + 0.007*"back" + 0.007*"day" + 0.007*"dont"')
(4, '0.017*"I" + 0.017*"time" + 0.016*"My" + 0.016*"stop" + 0.007*"smoking" + 0.007*"right"')
```

Reviewing the output in Box 9.6, we observe that the most frequently co-occurring words in the tweets are given with their weight within the sentence. The words new year resolution are in the top-most results with high weights in topic 2, as most people added these words to their tweets. Words like “make”, “possible”, “want”, and “stop” suggest that people want to change their lives. Moreover, making money is the third topic, with stopping smoking being the fifth.

## 9.4 CASE STUDY

In this case study, we will use Python libraries to analyze customer reviews of tablets. The dataset can be found here on GitHub.[2] Table 9.1 is an example of a few data rows.

We will mainly focus on the review text column and perform different types of text preprocessing and analysis: tokenization, text normalization, POS tagging frequency distribution, and sentiment analysis.

Code 9.17 aims to preprocess and analyze the reviews stored in the dataset on customer reviews. The code begins by importing the necessary libraries (Lines 1–6). Then, a dataset is imported from a CSV file named customer_reviews.csv’ (Line 8). The Pandas library is a Python library that is useful for processing datasets. A CSV file is a format that structures data tables into delimiter (e.g., colon and semicolon) separated fields. The text data is extracted from the “ReviewText” column and stored in a list called review_texts (Line 11). The code defines several methods: “tokenize_text()” to break text into tokens (Lines 13–15), “normalize_text()” to filter and lowercase tokens (Lines 17–21), “tag_tokens()” to assign POS tags (Lines 23–25), and “get_sentiment()” to analyze sentiment (Lines 27–30). Following these definitions, the script iterates through each text in the list, performing tokenization, normalization, tagging, frequency distribution of tokens, and sentiment analysis, outputting the results for each step to the console (Lines 32–51). This approach systematically processes the text for analysis, cleaning, and tagging the data before assessing sentiment, making it effective for evaluating and understanding complex textual data in a structured format.

**TABLE 9.1**
**A sample of reviews for computer tablets**

| Review ID | Date | Reviewer Name | Tablet Model | Rating | Review Text |
|---|---|---|---|---|---|
| 1 | 2023-02-08 | Tara Riley | Pixel Slate | 2 | Lacks durability and has a confusing interface, making it a poor choice. |
| 2 | 2024-04-11 | Mrs. Brittney Phillips | Surface Pro | 1 | Slow performance and frequent crashes make it frustrating to use. |
| 3 | 2023-02-12 | Nancy Bailey | Tab S6 | 4 | Easy to use |

**CODE 9.17 SENTIMENT ANALYSIS OF CUSTOMER REVIEWS**

```
import pandas as pd
import nltk
from nltk.tokenize import word_tokenize
from nltk.corpus import stopwords
from nltk.sentiment import SentimentIntensityAnalyzer
from nltk.probability import FreqDist
 # Load the customer review dataset
df = pd.read_csv('https://raw.githubusercontent.com/Object-Oriented-Programming-2024/'\
     'Object-Oriented-Programming/main/Chapter9/customer_reviews.csv')
# View the column information
print(df.info())
review_texts = df['Review Text'].tolist() # Extract text and store it into a list

def tokenize_text(text):
    """Tokenize the input text."""
    return word_tokenize(text)

def normalize_text(tokens):
    """Normalize text by converting to lowercase and remove stopwords."""
    tokens = [token.lower() for token in tokens if token.isalpha()]
    stop_words = set(stopwords.words('english'))
    return [token for token in tokens if token not in stop_words]

def tag_tokens(tokens):
     """Tag tokens with their parts of speech."""
     return nltk.pos_tag(tokens)

def get_sentiment(text):
    """Analyze the sentiment of the input text."""
    sia = SentimentIntensityAnalyzer()
    return sia.polarity_scores(text)

for text in review_texts:
    # Tokenize the input text
    tokens = tokenize_text(text)
    print(f"Tokens: {tokens}")

    # Normalize the tokens
    normalized_tokens = normalize_text(tokens)
    print(f"Normalized Tokens: {normalized_tokens}")

    # Apply Parts-of-Speech Tagging
    pos_tags = tag_tokens(normalized_tokens)
    print(f"POS Tags: {pos_tags}")

    # Apply Frequency Distribution
    freq_dist = FreqDist(normalized_tokens)
    print("Most Common Words:", freq_dist.most_common(10))

    # Apply Sentiment Analysis
    sentiment = get_sentiment(text)
    print(f"Sentiment: {sentiment}\n")
```

## 9.5 CHAPTER SUMMARY

Natural language processing is revolutionizing human interaction with technology, resulting in the seamless integration of systems that support various aspects of our lives. This chapter briefly introduces basic preprocessing tasks involved in NLP, including tokenization, normalization, stop-word removal, and POS tagging. These tasks play an important role in preparing textual data for further analysis. In addition, sentiment analysis and topic modeling were introduced as applications of

NLP. Sentiment analysis helps determine the emotional tone in textual content, a valuable technique for analyzing customer reviews, social media content, and feedback. Topic modeling identifies themes within the text and assigns probabilities to co-occurring words to perform document classification or trend analysis. Therefore, this chapter provides the first step toward understanding how to perform basic analysis to gain insights from textual data.

## 9.6 EXERCISES

### 9.6.1 Test Your Knowledge

1. Discuss the importance of text preprocessing in NLP. Why is it necessary to transform raw text data before analysis?
2. Explain the process and significance of tokenization in NLP.
3. Describe text normalization and its components.
4. Compare and contrast stemming and lemmatization.
5. Describe the process of Chunking.
6. Explain the purpose of a dispersion plot in text analysis.
7. How does a frequency distribution help in understanding text data?
8. What is the main objective of POS tagging in NLP?
9. Describe the pub-sub model.
10. Elaborate on the steps in pre-processing tweets for natural language processing tasks. Why is each step necessary, and what challenges might arise?

### 9.6.2 Multiple Choice Questions

1. Which of the following is not a standard text preprocessing step?
   a. Tokenizing
   b. Filtering Stop Words.
   c. Stemming.
   d. Compiling.
2. What does the "punkt" tokenizer model do in NLTK?
   a. Translates text into multiple languages.
   b. Splits text into sentences and words.
   c. Convert text to uppercase.
   d. Removes punctuation from the text.
3. Which Python library is commonly used for NLP tasks?
   a. Matplotlib.
   b. Scikit-learn.
   c. NLTK.
   d. TensorFlow.
4. Which of the following is a technique for reducing words to their base form by removing suffixes?
   a. Tokenization.
   b. Stemming.
   c. Chunking.
   d. Stop Words Removal.
5. What does the "NNP" tag represent in POS tagging?
   a. Noun phrase.
   b. Proper noun.
   c. Past tense verb.
   d. Adjective.

6. Which of the following is the primary purpose of POS tagging?
   a. Summarizing the text.
   b. Assigning grammatical tags to words.
   c. Translating text.
   d. Generating word clouds.
7. Which of the following is the purpose of Twitter's REST APIs?
   a. Accessing a continuous flow of real-time data
   b. Querying and retrieving specific sets of data
   c. Publishing tweets directly to users
   d. Setting up OAuth 2.0 authentication
8. Which package is used to read, write, and analyze Twitter posts in Python?
   a. Tweepy
   b. NumPy
   c. Pandas
   d. Scikit-learn
9. In topic modeling, which method converts text data into numeric presentations?
   a. Tokenization
   b. Lemmatization.
   c. Vectorization
   d. Filtering
10. What does the "OAuth 2.0 with PCKE" authorization provide?
    a. An alternative to REST APIs.
    b. An additional security layer.
    c. Enhanced data filtering options.
    d. Real-time data streaming.

### 9.6.3 Short Answer Questions

1. Name the process of transforming text by converting all characters to lowercase, stripping punctuation, and Tokenizing the text into discrete pieces for enhanced analysis.
2. Name the technology that has emerged besides NLP to extract meaningful information from unstructured text data.
3. Name the Python library used for performing various NLP tasks.
4. What is the use of "mltk.pos_tag()" method.
5. What is the primary purpose of using Twitter APIs in Text Mining?
6. What does the "pre-process_tweets" method do?
7. Why are callback URIs important in OAUth 2.0 authentication?
8. What does the OAuth stand for?
9. What is the "re" package in Python used for?
10. What is required from developers to authenticate access to tweets?

### 9.6.4 True or False Questions

1. Stop words are common words that do not add significant meaning to a sentence and are often removed during text preprocessing.
2. Lemmatization always produces words that are found in the dictionary.
3. A concordance can show every occurrence of a word and its context within a text.
4. A dispersion plot visually represents the frequency and location of words in a text.
5. The frequency distribution in text analysis provides a statistical representation of word occurrences.
6. Collocations are essential in linguistics and language learning because they help improve machine translation and natural language processing tasks.

7. The Streaming API allows developers to access real-time data as tweets occur in Twitter.
8. The Streaming API uses a request-response model to fetch data.
9. A rate limit is the maximum number of requests that can be made to a Twitter API within a specific time frame.
10. The Tweepy package supports Twitter API v1.1 and Twitter API v2.0.

### 9.6.5 Fill in the Blanks

1. Complete the code below with the correct methods to tokenize and perform POS given sentence using NLTK.

```
# Import the necessary modules
from nltk import word_tokenize, pos_tag
from nltk.chunk import RegexpParser

# Define the sentence to be chunked
sentence = "The Burj Khalifa stands as a testament to the nation's ambition and innovation."

# Tokenize the sentence into words
tokens = ___________(sentence)

# Perform part-of-speech tagging on the tokenized words
tagged_tokens = _________(tokens)
```

2. Fill in the blanks in the code snippet below that is a continuation of the above code to remove stopwords from the above text:

```
from nltk.corpus import stopwords

# Download the 'stopwords' data package from nltk, if not already downloaded
nltk.download('__________')

# Create a set of English stopwords
stop_words = set(___________.words('english'))

# Filter out the stopwords from the tokenized words
filtered_text = [word for word in tokens if not word.lower() in stop_words]

# Print the filtered text, which excludes the stopwords
print(filtered_text)
```

3. Fill in the blanks in the code snippet below that is a continuation of the first code in this exercise to perform lemmatization on the given text:

```
from nltk.stem import WordNetLemmatizer

# Download the 'wordnet' resource from nltk
nltk.download('wordnet')

# Create an instance of the WordNetLemmatizer
lemmatizer = _____________()

# Lemmatize each tokenized word using the WordNetLemmatizer with part of speech as 'verb'
lemmatized_words = [__________.lemmatize(word, pos='v') for word in tokens]

# Print the list of lemmatized words
print(lemmatized_words)
```

### 9.6.6 Coding Problem

You are tasked with implementing various text preprocessing operations using Python's NLTK library. The goal is to clean and prepare raw text data for further analysis by performing tokenization, stop word removal, stemming, lemmatization, and POS tagging.

#### 9.6.6.1 Requirements

1. **Tokenization:** Write a method *tokenize_text* that takes a text string as input and returns a list of word tokens. Use the following text for this requirement: "Tokenization breaks text into words or sentences, making it easier to work with manageable segments".
2. **Stop Words Removal:** Write a method *remove_stop_words* that takes a list of word tokens and returns a list with the stop words removed. Use the following text for this requirement: "Text data can be messy, unstructured, and filled with noise, so preprocessing is essential".
3. **Stemming:** Write a method stem_words that takes a list of word tokens and returns a list of stemmed words. Use the following text for this requirement: "Stemming is a process that reduces words to their root form, such as 'running' to 'run'".
4. **Lemmatization:** Write a method lemmatize_words that takes a list of word tokens and returns a list of lemmatized words. Use the following text for this requirement: "Lemmatization reduces words to their base or root form, ensuring that the resulting word is valid".
5. **POS Tagging:** Write a method pos_tagging that takes a list of word tokens and returns a list of tuples where each tuple contains a word and its corresponding POS tag. Use the following text for this requirement: "Natural Language Processing is a fascinating field that involves analyzing and understanding human language".

## NOTES

1. https://www.nltk.org/install.html.
2. https://raw.githubusercontent.com/Object-Oriented-Programming-2024/Object-Oriented-Programming/main/Chapter9/customer_reviews.csv.

# Appendix A
# Installing Python and Environment Setup

## A.1 INSTALLATION FOR WINDOWS OS

Visit the following link to download the latest version of Python for the Windows operating system at https://www.Python.org/downloads/, as shown in Figure A.1.

Once the file downloads on your machine, run the executable file and the installation wizard will start, as shown in Figure A.2.

Select the *Add Python.exe to PATH* checkbox (this option ensures that your system knows where to find Python). Select the *Install Now* option to begin the installation. Once the installation is complete, you can program in the Python terminal.

### A.1.1 STARTING THE PYTHON TERMINAL

To run your Python code, you can access the Python interpreter using the command prompt or download a development environment covered in more detail in Appendix B. In the Windows search bar, type *cmd* and the command prompt option will show. Open the command prompt and type Python, and your Python terminal prompt will open, as shown in Figure A.3.

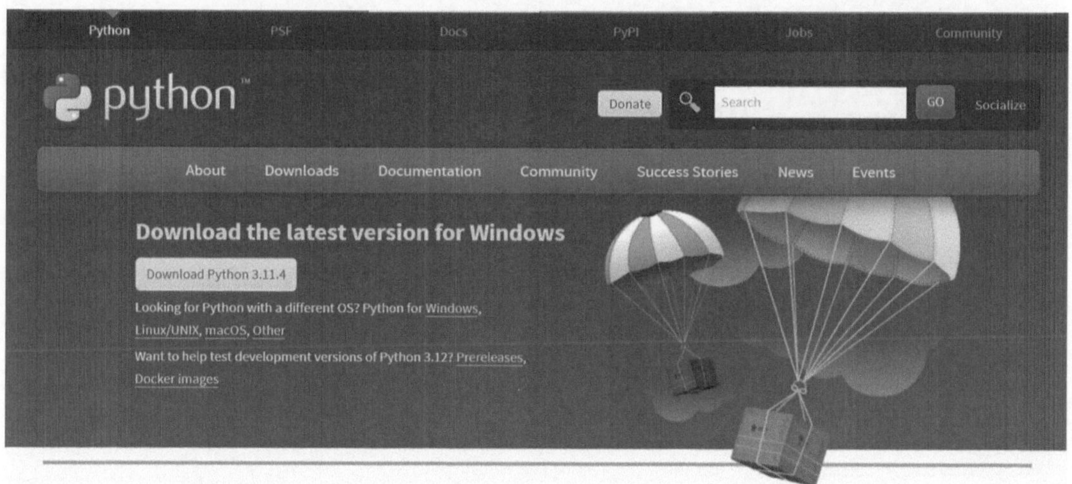


**FIGURE A.1** Python download for Windows users.

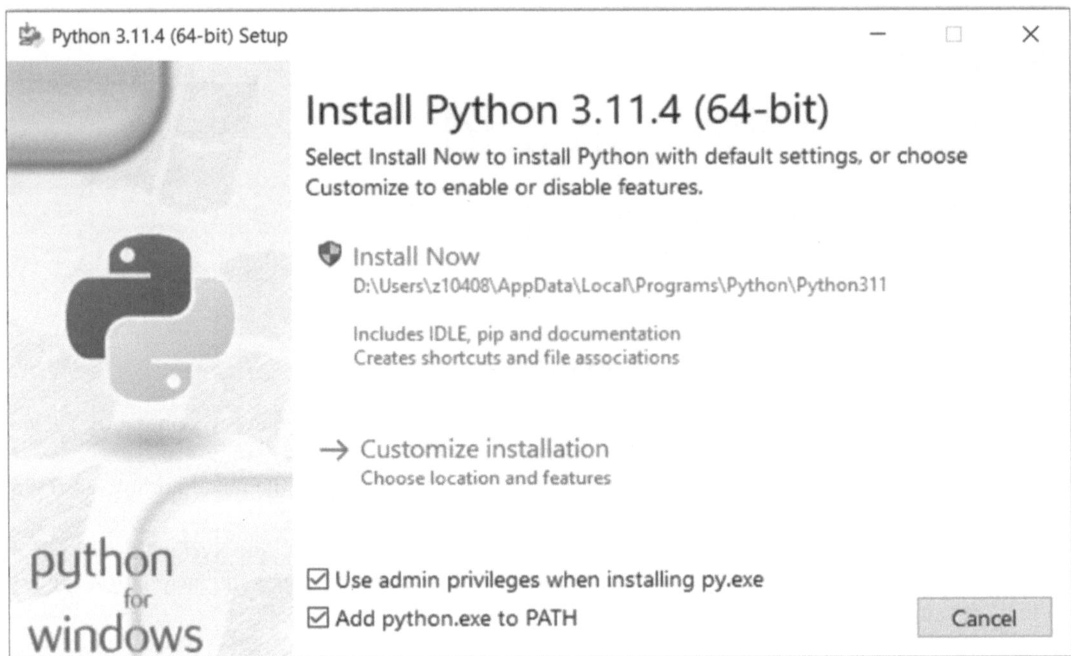


**FIGURE A.2** Python installation wizard.

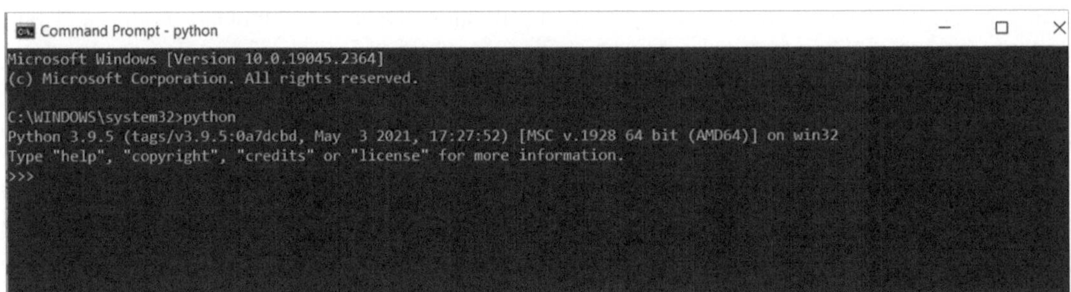


**FIGURE A.3** Python Terminal.

To test your Python terminal, type a simple program such as the one given below:

**CODE 1 TEST PYTHON COMPILER USING PRINT() FUNCTION**

```
>>>a= 2
>>>print(a)
```

To quit the Python terminal, type *quit()* or select the keys *ctrl+z*.

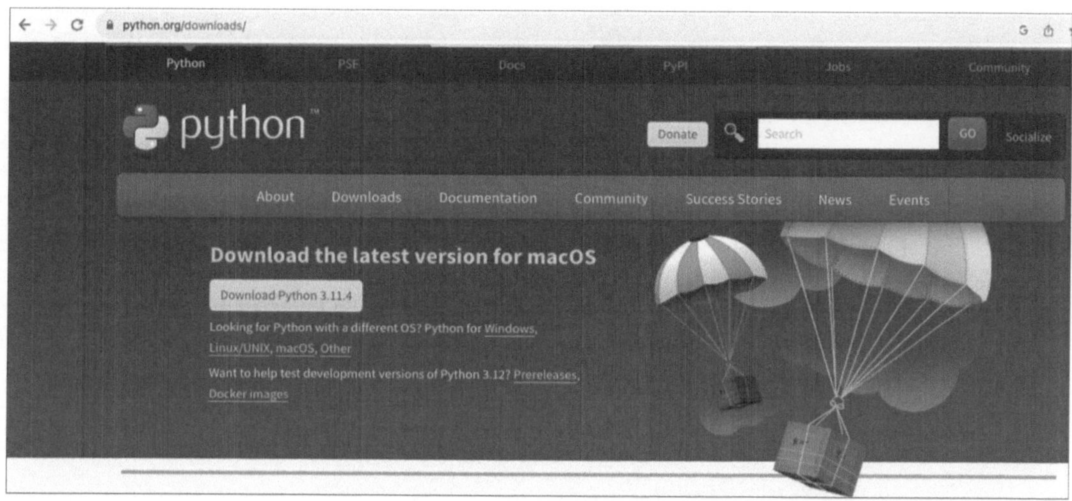


**FIGURE A.4** Python download for Mac users.

## A.2 INSTALLATION FOR MAC OS

Mac, by default, has the Python 2 version installed. However, we do need the latest version of the Python compiler, which you can download from the following link: *https://www.Python.org/downloads/*, as shown in Figure A.4. Once the file is installed on your computer, run the installation wizard with the default settings, and you are ready to begin programming in Python.

### A.2.1 Starting the Python Terminal

Go to *Finder*, select *Applications* and then *Utilities*, as shown in Figure A.5. Within the utilities folder, look for the terminal application and open it.

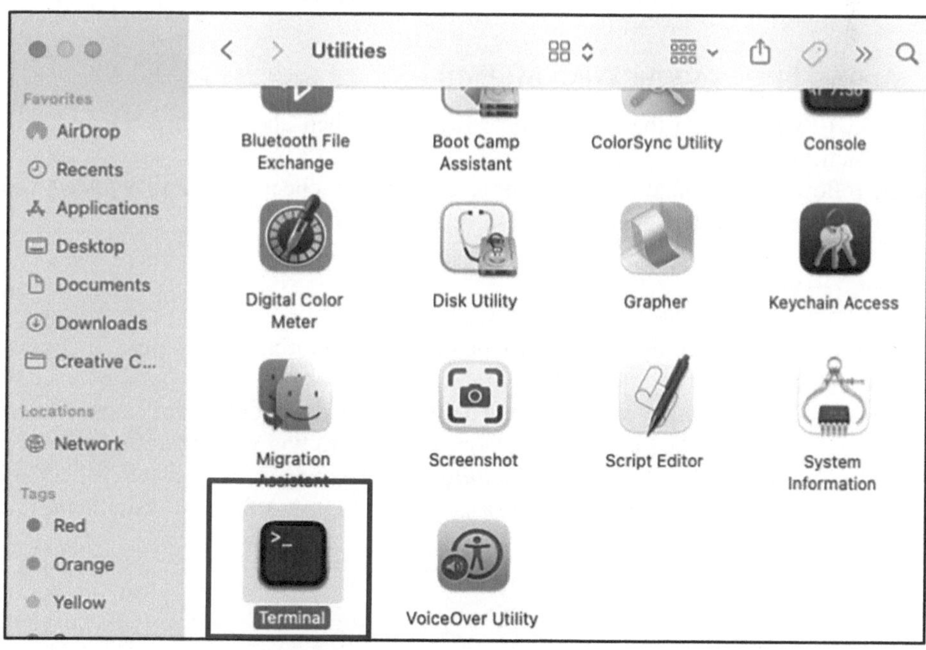


**FIGURE A.5** Selection of "Terminal" within the Utilities folder.

```
avuser — -bash — 80×24
ML086065-avuser:~ avuser$
```

FIGURE A.6 Terminal command window in Mac.

The terminal window will open with an input prompt, allowing you to type commands, as shown in Figure A.6. In this figure, the last line, *ML086065*, is the computer's name and *avuser* is the directory created based on a login username. These details will depend on the system used to run Python.

Type *Python3* in the terminal window. The Python terminal prompt will show as "**>>>**". To test the Python terminal, type a simple program such as the one given below:

```
>>>a= 13+26
>>>print(a)
```

To quit the Python terminal, type *quit()* or select the *keys ctrl+d.*

## A.3 ADDING PYTHON TO THE PATH VARIABLE

First, let's understand the path of the operating system. To give commands to the operating system, we create executable files stored in directories in our systems. The "Path" is a variable the operating system uses to find executable files, such as Python. When you run a command in the command prompt, the operating system searches for an executable file of the same name in the "Path" variable. Therefore, the "Path" variable must contain the complete location of the executable file. After Python is added to the "Path" variable, you can access Python even if your command prompt or terminal is currently in a different directory from Python. Moreover, you can install Python packages using the "PIP" command discussed in more detail in Section A.4.

### A.3.1 Adding Python to the Windows Path

If you installed Python using the installation wizard described above, it has been added to your operating system path.

Type the following command in your command prompt to check if Python is added to the Path variable. The command prompt will display the version of Python.

```
python --version
```

If an error message shows, Python is not added to your operating system path. In that case, you can follow the steps below.

First, find the complete path to where Python is installed in your folders, for example, c:\desktop\Python3.9. Once you know the location of Python, you need to find your operating system environment variables. Search for *Advanced system settings* and select open, as shown in Figure A.7.

Once the system properties window opens, as shown in Figure A.8, select the *Advanced* tab, and then *Environment Variables*.

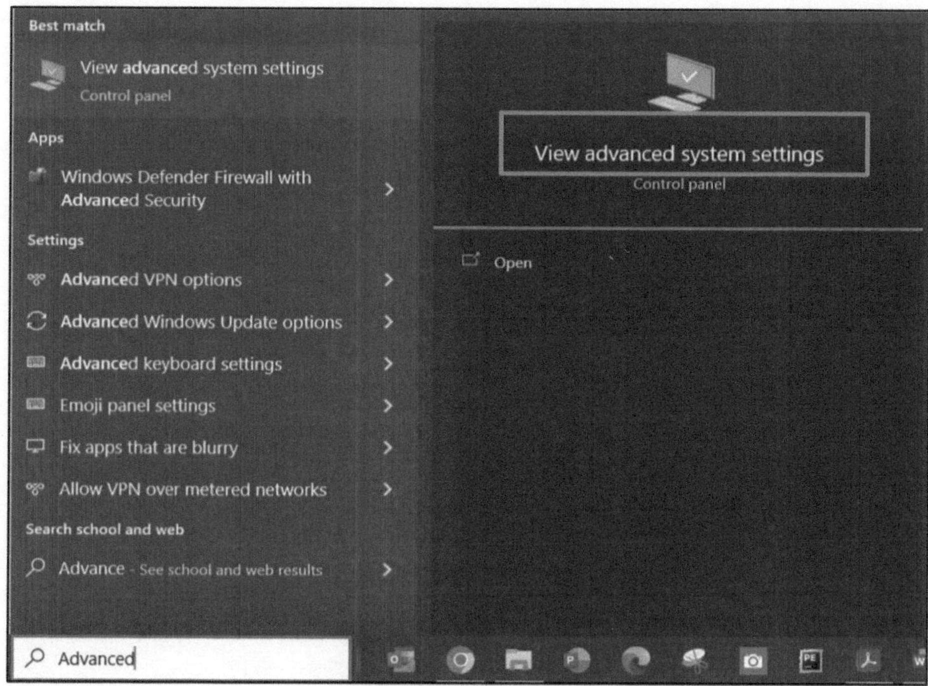


**FIGURE A.7** Opening advanced system settings.

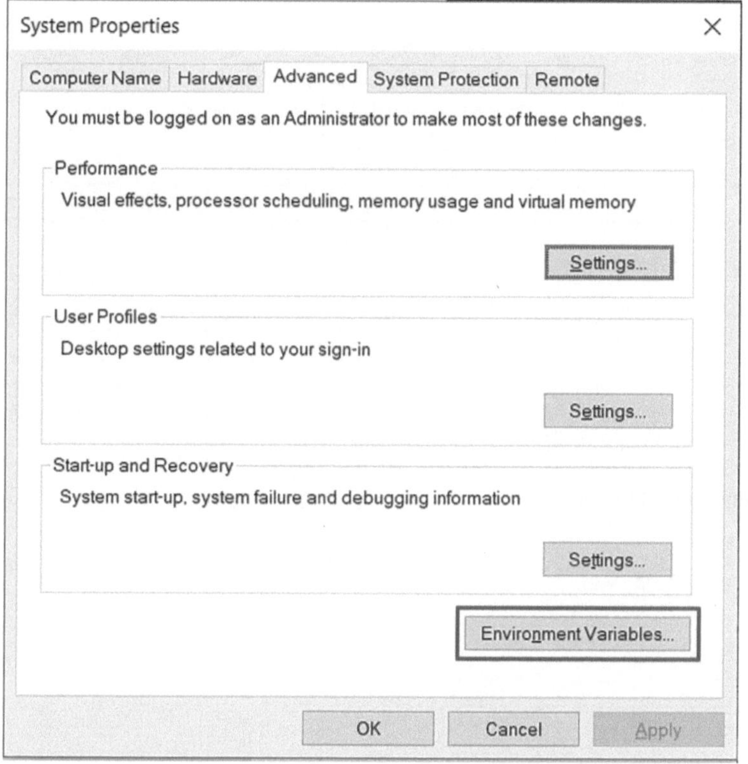


**FIGURE A.8** Selecting environment variables under the Advanced tab.

On this new window, system variables are listed in the lower list box. Select the *path* and click *Edit*. You will find the list of executables added to the Path variable, as shown in Figure A.9.

After clicking *Edit*, a new window will show the list of executables that you can directly run from the command prompt without knowing their complete path, as shown in Figure A.10.

Now select *New* and add the location where Python is installed on your computer. As you can see in Figure A.10, my computer has Python installed in *C:\Program Files\Python38*.

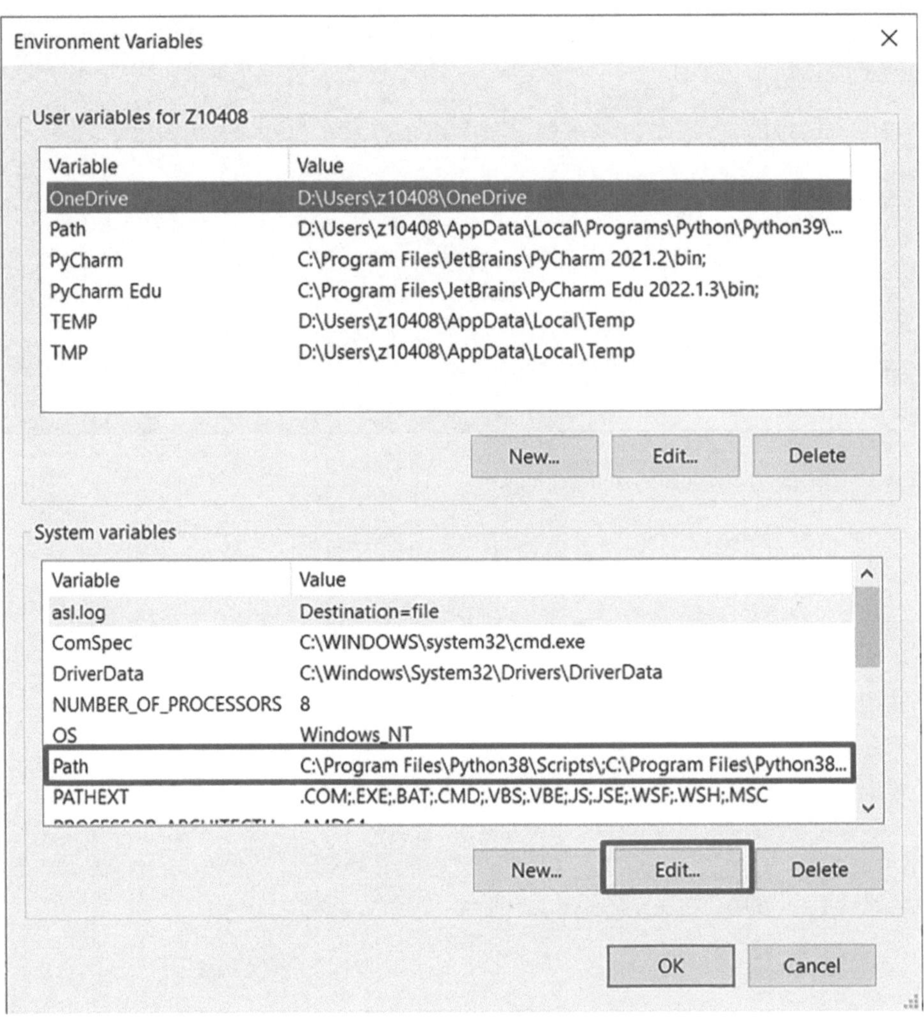


**FIGURE A.9** Environment variables.

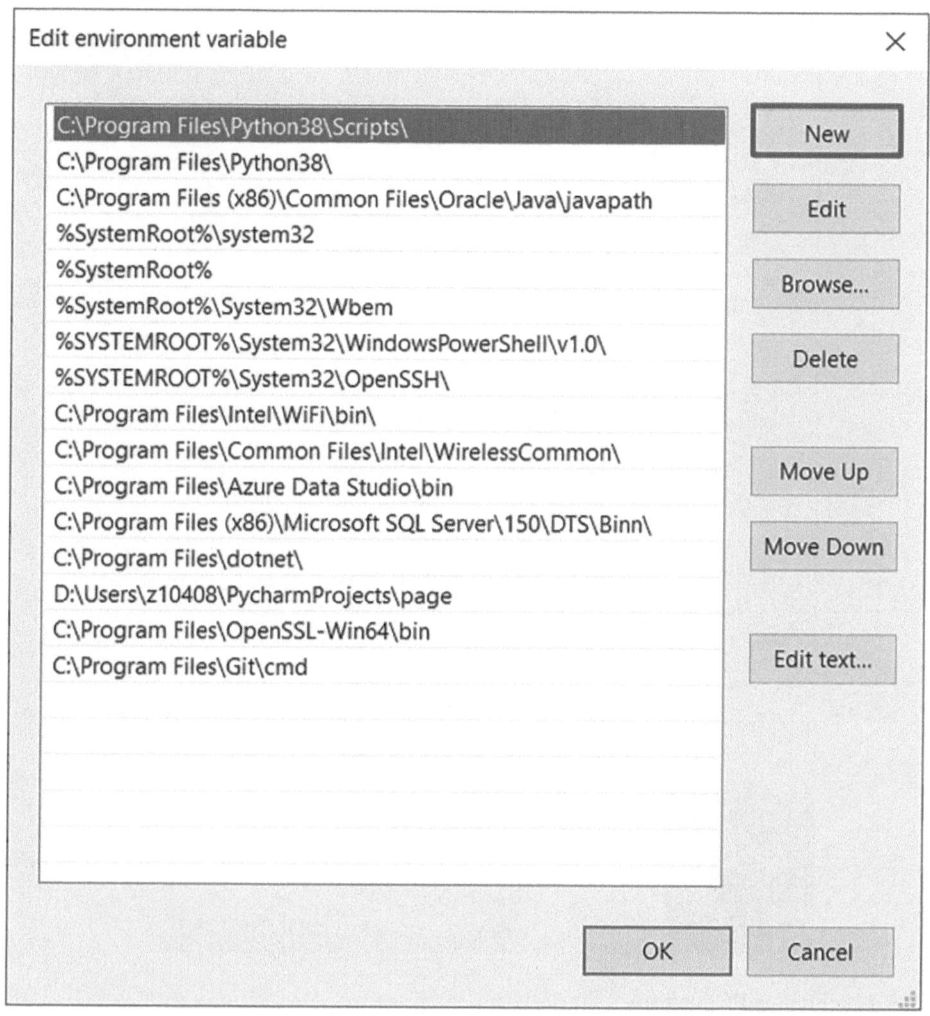


**FIGURE A.10** Executable files added to the Path variable.

### A.3.2 Adding Python to the Mac Path

To check if Python is added to the Path variable on your MAC computer, type *Python3* in your terminal. This will open the Python prompt in the terminal window as >>>.

If the above command gives an error message, you must add Python to the Path variable. First, find where Python is installed on your Mac. You can do that by searching for Python in *Finder*. When you select Python Launcher, the complete directory path to where Python is installed will show, as shown in Figure A.11.

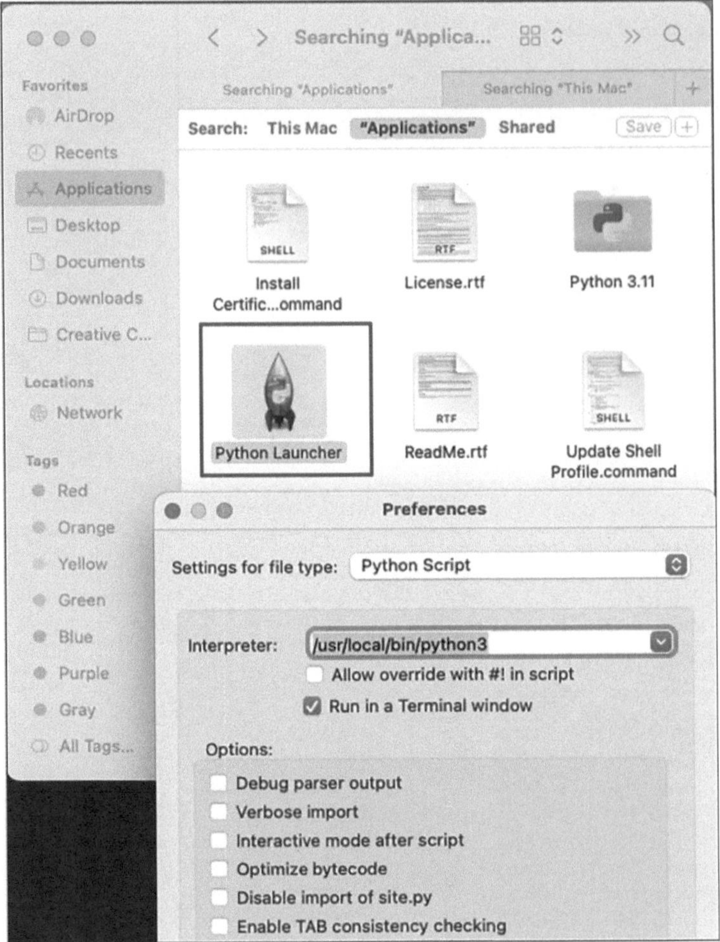


**FIGURE A.11** Finding Python path.

Now open the terminal from Launchpad. Type the following command in your terminal:

*sudo nano /etc/paths.*

The terminal will ask for your password to access the system, as shown in Figure A.12.

After you provide the password, the list of directories stored in the Path variable is displayed, as shown in Figure A.13. Now, enter the complete path to where Python is installed on your computer.

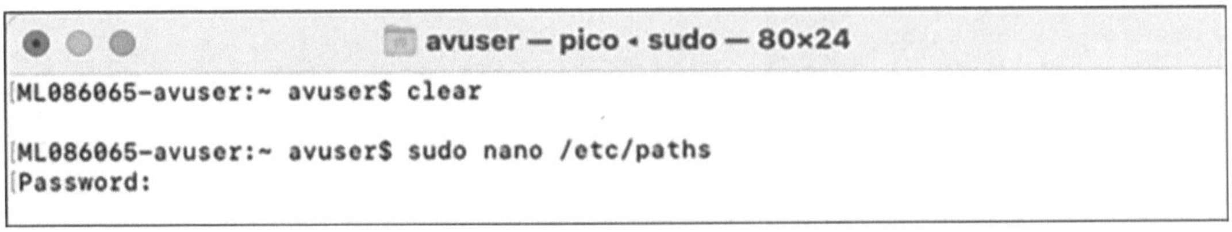

```
avuser — pico ◂ sudo — 80×24
ML086065-avuser:~ avuser$ clear

ML086065-avuser:~ avuser$ sudo nano /etc/paths
Password:
```

**FIGURE A.12** Finding directories added to the Path variable.

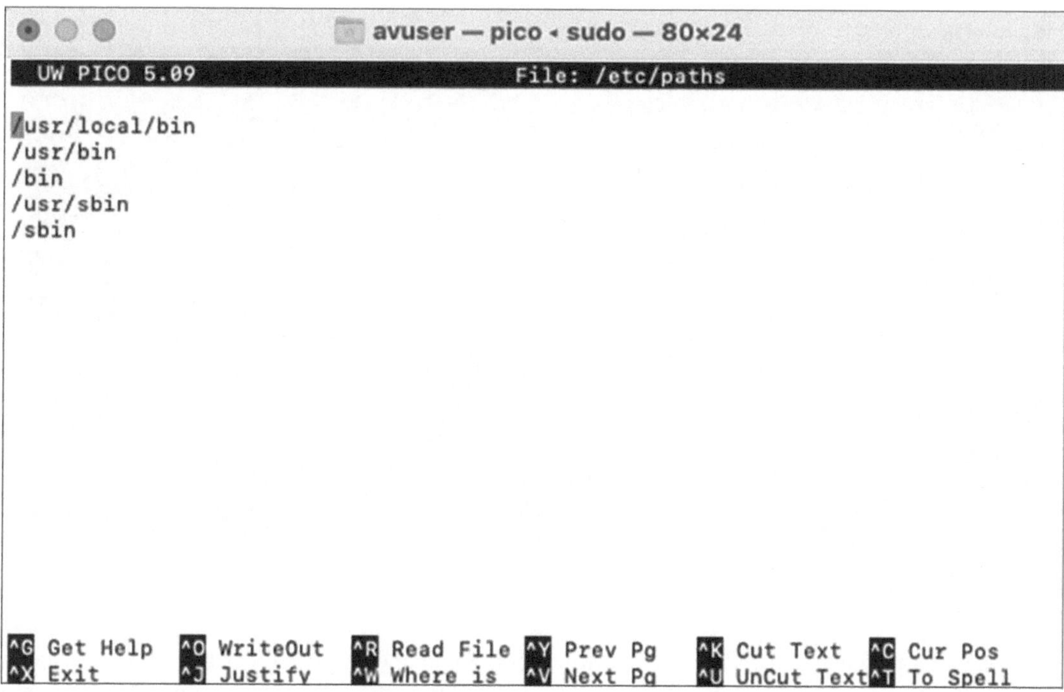


**FIGURE A.13** Adding Python to the Path variable.

For example, based on Figure A.13, Python is installed in *user/local/bin* which is already added to the Path variable. After adding the path, press *ctrl+X* to quit and select ***Y*** to save changes.

## A.4 PYTHON PACKAGE MANAGER – PIP

PIP is a package manager that allows easy installation and inclusion of Python packages to your project. Packages contain existing code libraries for reuse. Python 3.4 or later versions have PIP installed by default. If you installed the latest version of Python, then PIP is already on your system. To test if PIP is installed on your system, type the following command in your command prompt/terminal: *pip3 --version*, as shown in Figure A.14. The version of PIP is displayed with the complete path to the directory in which it is installed. The command is the same for both MAC and Windows users.

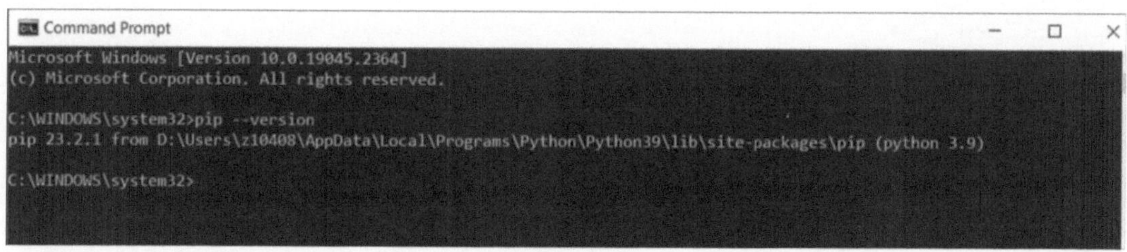


**FIGURE A.14** Confirming PIP version and installation.

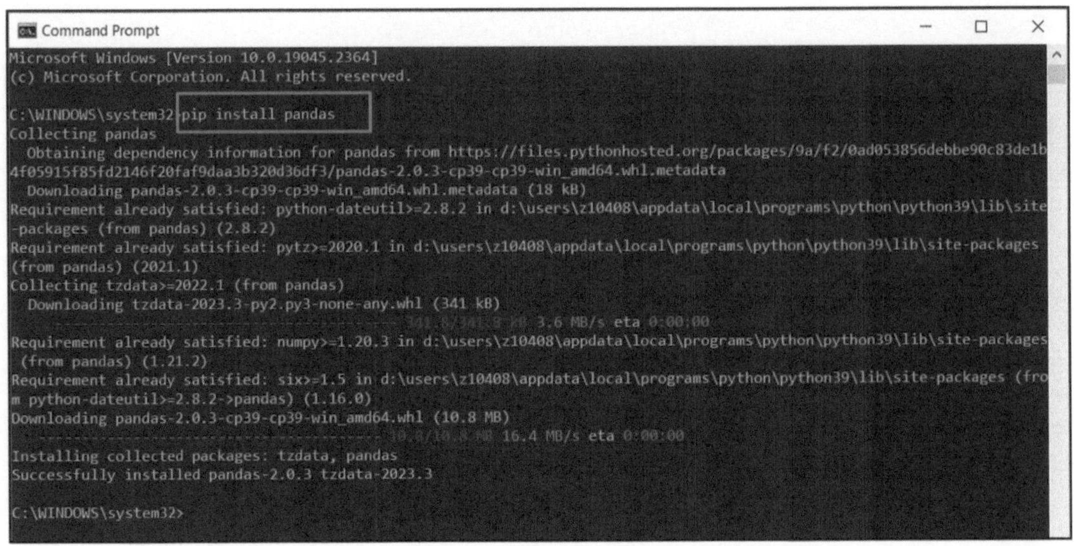


**FIGURE A.15** Installing the Pandas package using PIP.

To install a package using PIP, type the following command, where ***XX*** is the package's name: *pip install XX*. For example, to install the *pandas* library, type the following command, as shown in Figure A.15:

*pip install pandas.*

You should be able to see a successful message once your package has been installed.

# Appendix B
# Choosing an IDE

## B.1 PYCHARM

PyCharm is a popular integrated development environment (IDE) for Python programming. It was developed by JetBrains, a software development company. It is very popular due to its many features, continuous updates, and user-friendly interface.

There are two editions of PyCharm: a community edition and a professional edition. The community edition is free and open-source and provides many key functionalities for Python development. The professional edition is not free and includes more sophisticated features and tools. It is appropriate for developers of all levels of expertise, from novices to seasoned professionals. Key features of PyCharm are listed here.

### B.1.1 Code Editor

The editor is the main part of the IDE, which is used to create, read, and modify code. Its features include code completion and inspections, syntax error highlighting, and intelligent code analysis. It helps fix bugs and write better code with its assistance.

The editor is divided into four sections: the scrollbar displays Errors and warnings in the current file, the gutter displays line numbers and annotations, breadcrumbs help the user navigate the code in the current file, and tabs display the names of the open files.

### B.1.2 Code Navigation

PyCharm facilitates code navigation, which helps move directly to code definitions, find usages, and navigate the code files easily.

### B.1.3 Debugger

The debugger in PyCharm helps locate and correct bugs in the code. To find and fix issues, the programmer can step through the code, check variables, and set breakpoints.

### B.1.4 Web Development Support

PyCharm provides support for web development frameworks like Django and Flask. It provides project setup, code completion, and templates specific to these frameworks.

### B.1.5 Database Tools, Data Science, and Machine Learning support

PyCharm includes tools for working with databases, making it easier to manage database connections, run queries, and view data. It also contains features for data science and machine learning tools, such as a scientific calculator, a plotting tool, and a data viewer.

### B.1.6 Multi-technology Development and Google App Engine

The PyCharm Professional Edition can create projects combining different programming languages, including Python, Java, and JavaScript. PyCharm, in its professional edition, also supports the Google App Engine, a platform for deploying web apps.

### B.1.7 Code Templates and Cross-Platform Features

Frequent code snippets can be easily written using PyCharm's collection of code templates. It is a cross-platform IDE as it is accessible for Linux, macOS, and Windows.

## B.2 JUPYTER NOTEBOOK

Jupyter Notebook is not a traditional desktop-based IDE but a web-based application that provides an interactive computational environment for creating and sharing documents containing code, equations, visualizations, and narrative text.

It can compile code and display interim results. While Jupyter Notebook is used for Python programming, it is widely used for data science, data analysis, and machine learning projects. Its ability to integrate code, text, and graphics into one document makes it a popular choice for education and documentation.

Jupyter Notebook also has several features, some of which are discussed here.

### B.2.1 Data Visualization

Jupyter Notebook enables the integration of several visualization tools, including Matplotlib, Seaborn, and Plotly. This allows users to generate charts and graphs directly within the notebook.

### B.2.2 Interactive Environment, Data Analysis, and Machine Learning

Users can write and run code in a cell-based environment with Jupyter Notebook. Each cell may include text formatted using Markdown, code written in Python or another language (such as R or Julia), or mathematical equations written in LaTeX.

Jupyter Notebook's interactive features and support for libraries like Pandas, NumPy, and scikit-learn make it a popular choice for data cleansing, transformation, model training, and evaluation tasks.

### B.2.3 Live Code Execution

Code cells can be run individually. There is no need to run the complete code. Results and output are immediately seen after running the code. This feature is useful for data exploration and analysis, making it simple to test various codes and view the results in real time.

### B.2.4 Collaboration

Multiple users can collaborate on Jupyter Notebooks, facilitating group projects. Platforms like Jupyter Notebook Viewer and GitHub allow users to share and view notebooks online without requiring the Jupyter environment.

### B.2.5 Extensibility and Portability

New features and functionality can be added to Jupiter Notebook, which makes it extensible; it is also portable as a Notebook file can run on any web browser.

### B.2.6 JupyterLab

JupyterLab is an upgraded version of Jupyter Notebook that offers a more flexible and integrated user interface. It supports various document formats within a single workspace, including text files, Jupyter notebooks, and more.

## B.3 VISUAL STUDIO CODE

Microsoft developed Visual Studio code that works on Windows, Linux, and Mac operating systems. VS Code is a free, open-source code editor with Pylance as Python's default language server. However, VS Code is for Python and other programming languages such as C++ and Java. VS Code also has a large ecosystem of extensions and plugins, which makes it a fully featured IDE, popular for coding and development work due to its reputation for being lightweight and adaptable, as well as the several features discussed below.

### B.3.1 Version Control and Integrated Terminal

VS Code interfaces with well-known version control tools like Git, so developers can control their code repositories directly from the editor. VS Code also has a built-in terminal that enables programmers to execute command-line utilities and scripts without leaving the editor.

### B.3.2 Integrated Debugger and Syntax Highlighting Feature

VS Code has a built-in debugger that can be used to walk through the code and troubleshoot bugs. It supports several languages and computing platforms. It also offers syntax highlighting for the different programming languages supported, improving the readability of the code.

### B.3.3 Live Share and Live Preview

Live Share is a feature of Visual Studio Code that enables developers to work together in real time while sharing their code and debugging sessions with others. VS Code also supports live preview, where users can preview their code in a web browser without compiling and running it first.

### B.3.4 Customization and Task Automation

Users can alter the editor's look and behavior through themes and settings, in addition to keyboard shortcuts that are extensive and editable. VS Code can also be customized to support multiple cursors for fast simultaneous edits. VS Code has a built-in task runner, a tool that automates repetitive tasks in a development workflow. It performs various tasks, such as compiling, linting, and testing code.

### B.3.5 Code Navigation

VS Code has several code navigation features that can help you quickly and easily find the required code when the cursor is placed over a symbol. There are navigation features like "Go to Definition", which navigates to the definition of the symbol under the cursor; "Go to Symbol", which navigates to the symbol that the user specifies; "Find All References", which locates all references to the symbol under the cursor, and other navigation features like "Peek definition and peek references", which enables the preview of the definitions and references. There are also navigation features like "breadcrumbs", which display the current location in the document.

### B.3.6 Code Snippets

VS Code has a built-in code snippet feature for Python, allowing programmers to add their code snippets. Those are templates that can be used to insert common code blocks quickly. By automating time-consuming operations and providing a starting point, code snippets help developers write more consistent code, save time, and provide the ability to learn new code with provided examples.

# Appendix C
# Debugging Your Python Program

## C.1 DEBUGGING USING TRACEBACK

One common method of debugging code is to trace it line by line and check the values of your variables. Eventually, you will find the step that caused the error.

When an exception or error occurs, the PyCharm console displays the flow or, in other words, traces *back* to events that led to the exception. This information is very useful in identifying the error in your program. For example, let us consider Code C.1. At Line 7 in the code, the Student object is created, and the __init__ function in the Student class is called. This function assigns the id, name, and age to class variables. However, before assigning the age, we convert it to an integer to store the value in the correct format if we need to apply mathematical operations later.

**CODE C.1 CODE WITH ERROR**

```
1. class Student:
2.     def __init__(self, id, name,age):
3.         self.value = id
4.         self.name = name
5.         self.age = int(age)
6.
7. student1 = Student("M20220405","Brian","Twenty")
```

When you run this code, a ValueError is displayed in the console, as shown in Figure C.1. In this figure, look closely at the Traceback in the console. It highlights that on executing Line 7 with the code *obj1 = Student("M20220405", "Brian", "Twenty")*, a call was made to a function, the *__init__* function. Within this function, the error occurred in Line 5 with code *self.age = int(age)*. Comparing these two lines of code helps us identify that we are trying to convert the string "Twenty" to an integer age at Line 5, causing the error. Therefore, tracing your code step by step with how values are processed and passed within your program helps identify errors in the code.

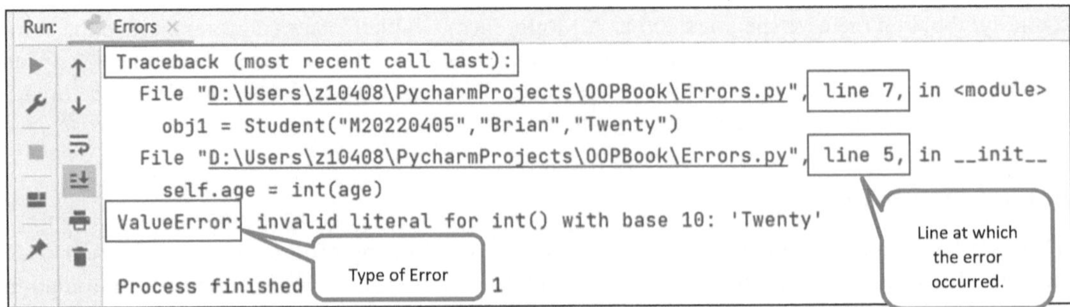


**FIGURE C.1** Exception with a traceback of the steps that caused the exception.

During your programming journey, sometimes your program throws an error, and in other cases, the program runs successfully but does not display the desired output. Therefore, it is important to understand the types of errors that programmers encounter first and then look into how to identify and fix these bugs.

## C.2 TYPES OF ERRORS

There are three major categories of errors that you will come across in your programming journey: syntax errors, runtime errors, and logical errors. An error can fall under more than one of these categories, as explained in the following sections.

### C.2.1 Syntax Errors

Syntax errors occur when you do not follow the rules or grammar of the programming language. Even without running your code, PyCharm underlines the code in red, which does not follow the proper syntax of Python programming. As shown in Figure C.2, the class keyword with the capital letter C is underlined because all keywords are small letters.

If you try to run your code with syntax errors, PyCharm displays an error in the console. The error messages shown in the console are very helpful in identifying the problem with your code. For example, in Figure C.2, the error occurred in Line 1, and the type of error was a syntax error.

### C.2.2 Runtime Errors

Runtime errors occur after your program compiles and while the program is running. They usually occur due to incorrect input or unexpected conditions encountered while running the program. For example, if we run the program given in Figure C.3, an Attribute Error is displayed in the console. The code is trying to access the "address" attribute of the class Student at Line 10. If you observe

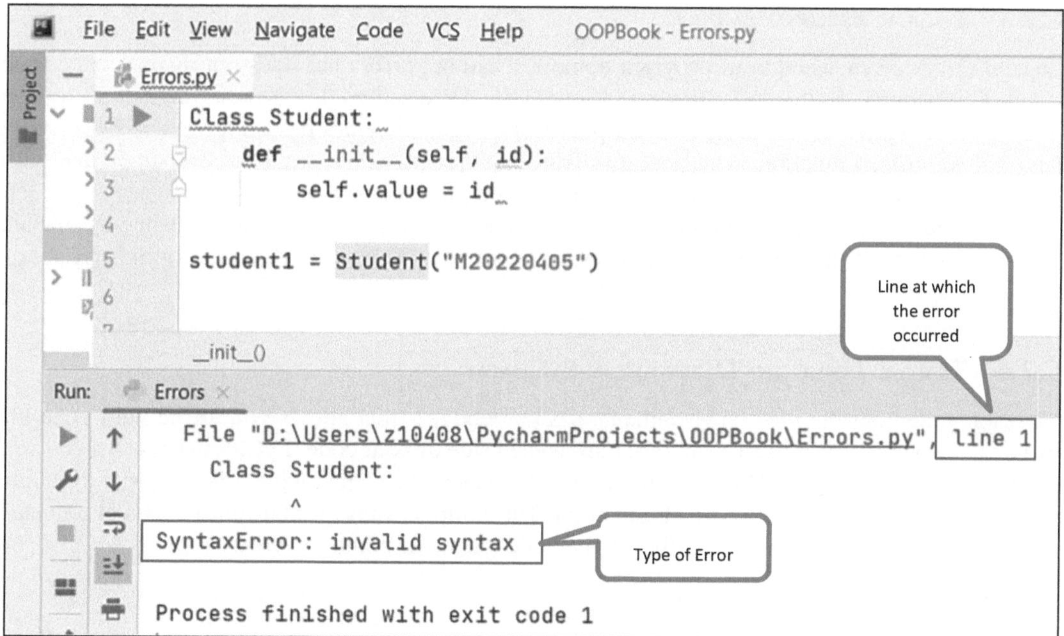


**FIGURE C.2** A syntax error is displayed in the console window.

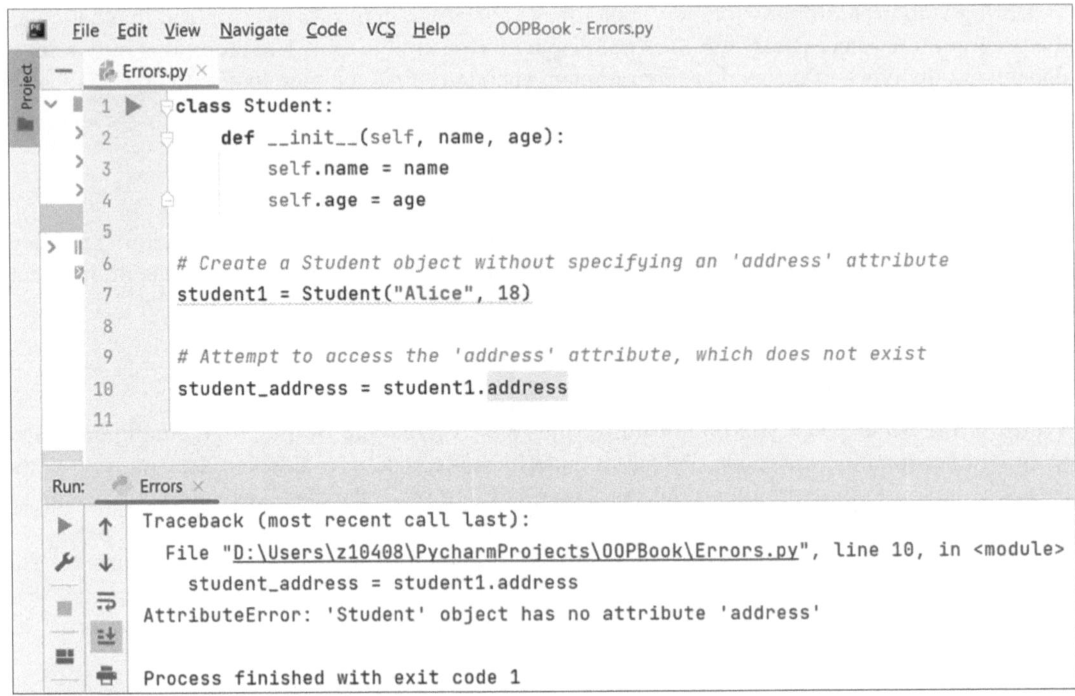


**FIGURE C.3** A runtime error due to access of an attribute that does not belong to the Student Class.

the Student class's __init__() function, you will only find the "name" and "age" attributes, and there is no attribute called "address". Therefore, the compiler cannot find the attribute *"address"*, resulting in an error during runtime.

### C.2.3 Logical Errors

Logical errors occur when your program compiles without errors but does not display the correct result. For example, in the code shown in Figure C.4, Alice scored 45 on the exam and 50 was the passing score. However, the program's output displays that Alice has passed the exam. Observe the *has_passed_exam()* function, which returns True for students with a grade lower than 50 and False for those who scored more than 50. Hence, there is a logical error in the code.

If you cannot get the desired output from your code, it is best to trace your steps line by line and check the values of the variables at each line of your code. This will help identify the step that is causing the error.

### C.2.4 Tracing Using the Debugger in PyCharm

Most integrated development environments have a debugger that allows you to run your code line by line and check the values of your variables at each step of your code. PyCharm provides you with a powerful debugging tool. However, first, you must select the step at which the program execution must pause. This is the selection of a breakpoint. The compiler runs the code until the breakpoint and then pauses, allowing you to manually run the code from that point onwards.

#### C.2.4.1 Starting the Debugger

To illustrate how to pause your compiler until a certain line of code in PyCharm, click in the margin next to a line of code, and a dot will appear in the margin. For example, by clicking in the margin next to Line 8, as shown in Figure C.5, a dot appears. Now you can right-click anywhere in your

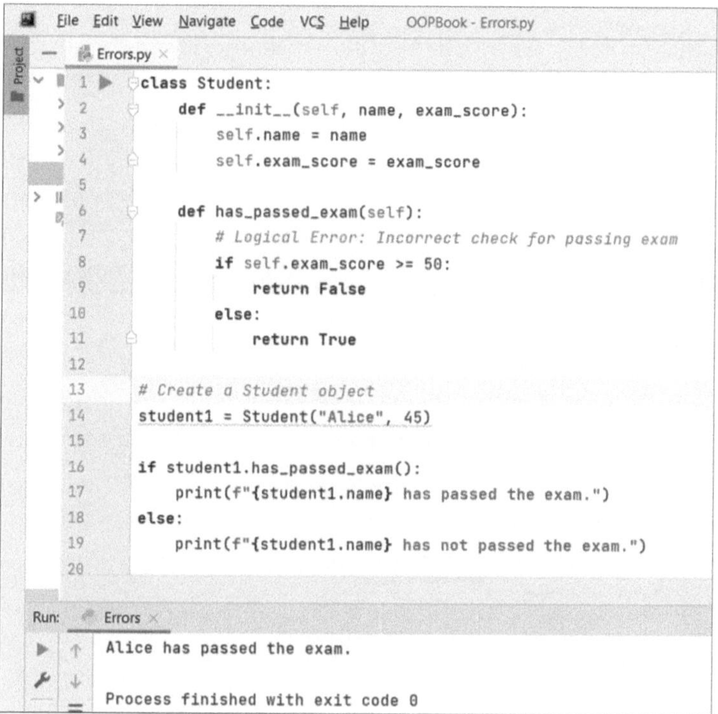


**FIGURE C.4** A logical error: the program shows that Alice passed with a grade of less than 50.

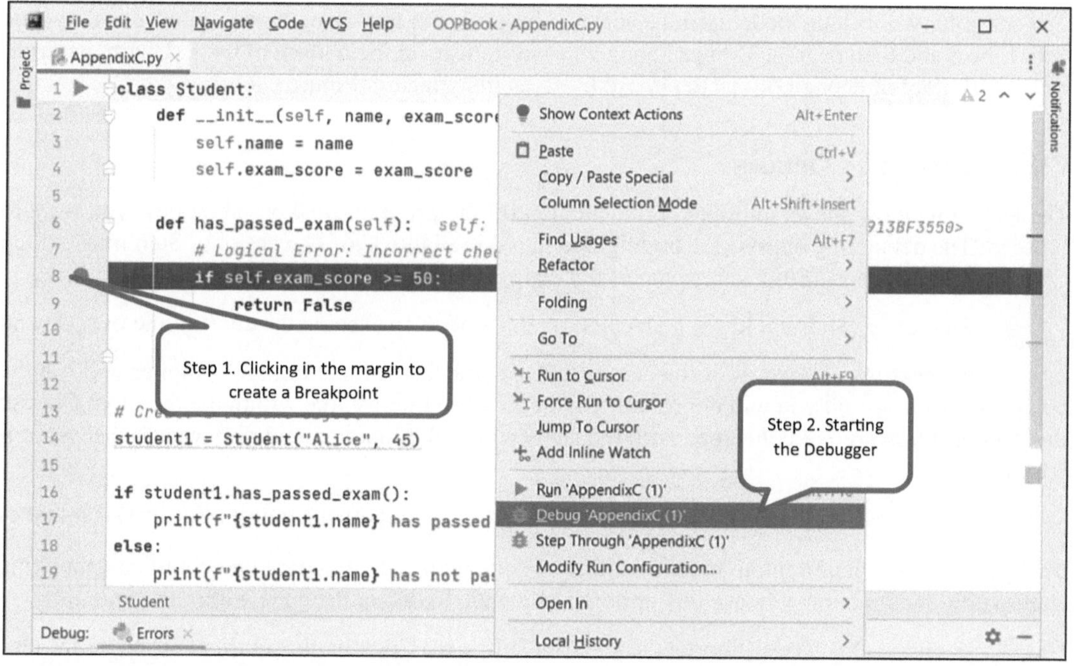


**FIGURE C.5** How to start the Debugger.

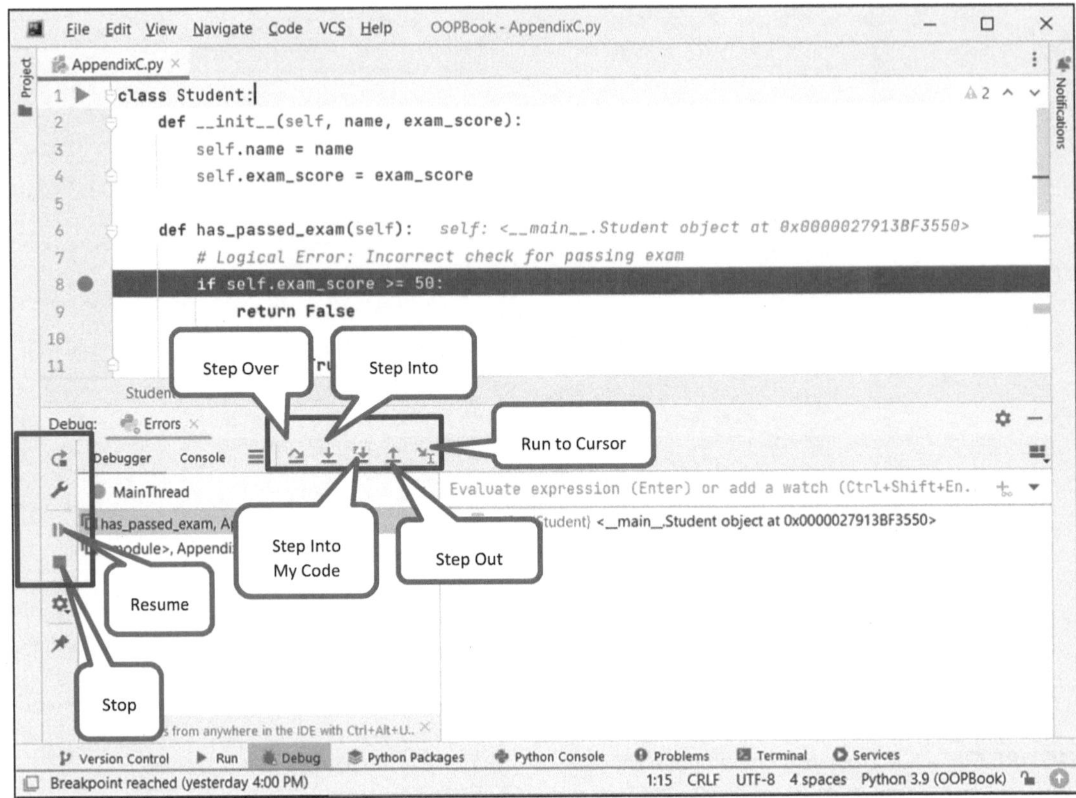


**FIGURE C.6** Debugging options.

code and select *Debug*, instead of *Run*, and the Debug window will appear as shown in Figure C.5. The compiler will run the code until the selected breakout; in this case, the compiler runs that code until Line 8 and then pauses. Then, a debug window appears at the bottom of the PyCharm environment, with a set of debug options to choose from, as illustrated in Figure C.6.

### C.2.5 Debugging Options

The Debug window shows multiple options to control the program flow based on how you want to proceed. The debugging options are highlighted in red in Figure C.6. The options "Step Into", "Step Over", "Resume", and "Stop", are some of the controls that allow you to run your code per line.

Step Over (Shortcut key – F8): Click on this option to execute the current line of code and move to the next line. However, if the current line of code has a function, the compiler will run the complete function, and you will not be able to run the code line-by-line within the function. Choose this option If you are not concerned with the steps within the function and only want to see what it returns.

Step Into (Shortcut key – F7): Select this option to execute the current line, and if it makes a call to another function, jump to the first line of code within the function and continue debugging from within the function. Choose this option if you want to debug the code within the function.

Step Into My Code (Shortcut key – Alt+Shift-F6): Click on this option to execute the current line, and if it makes a call to a function created within your code, then the compiler jumps to the first line of that function. However, suppose it is a built-in function (provided by Python libraries).

In that case, the compiler executes the complete function, returns the result, and does not show you the function's workings. Choose this option if you want to focus on your code base and not be distracted by details of built-in library functions or external modules.

Step Out (Shortcut key – Shift+F8): Click on this option to make the compiler run the rest of the function you are in and return to the line of code where the current function was called.

Run to Cursor (Shortcut key – Alt+F9): Select a line of code and then select this option. The compiler will execute the program until the cursor is in your code editor.

Stop (Shortcut key – Ctrl+F2): Select this option to stop debugging and return to working on your code.

Resume ( Shortcut key – F9): Select this option if you want the compiler to continue running the rest of the program without using the debugging options.

## C.2.6 Tracing Variables

While debugging, as you step into each line of your code, you will mostly need to see the values of the variables and how they are processed. In the "Variables" panel, highlighted in red in Figure C.7, you'll see a list of variables within the scope of the code you are running and their current values.

In Figure C.7, you can observe that the debugger is currently at Line 8. The variables within this scope are name and exam_score, with values Alice and 45, respectively. You can right-click on a variable and select "Add to Watches" to monitor its value as you enter the code.

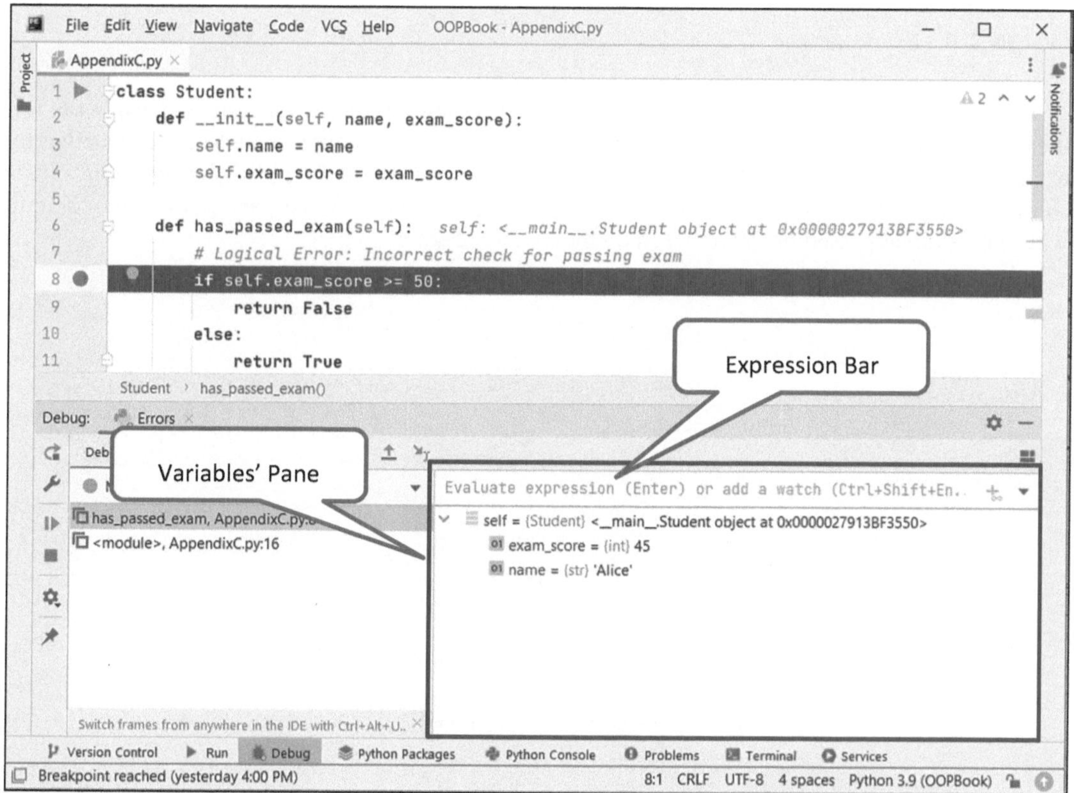


**FIGURE C.7** Variables Pane.

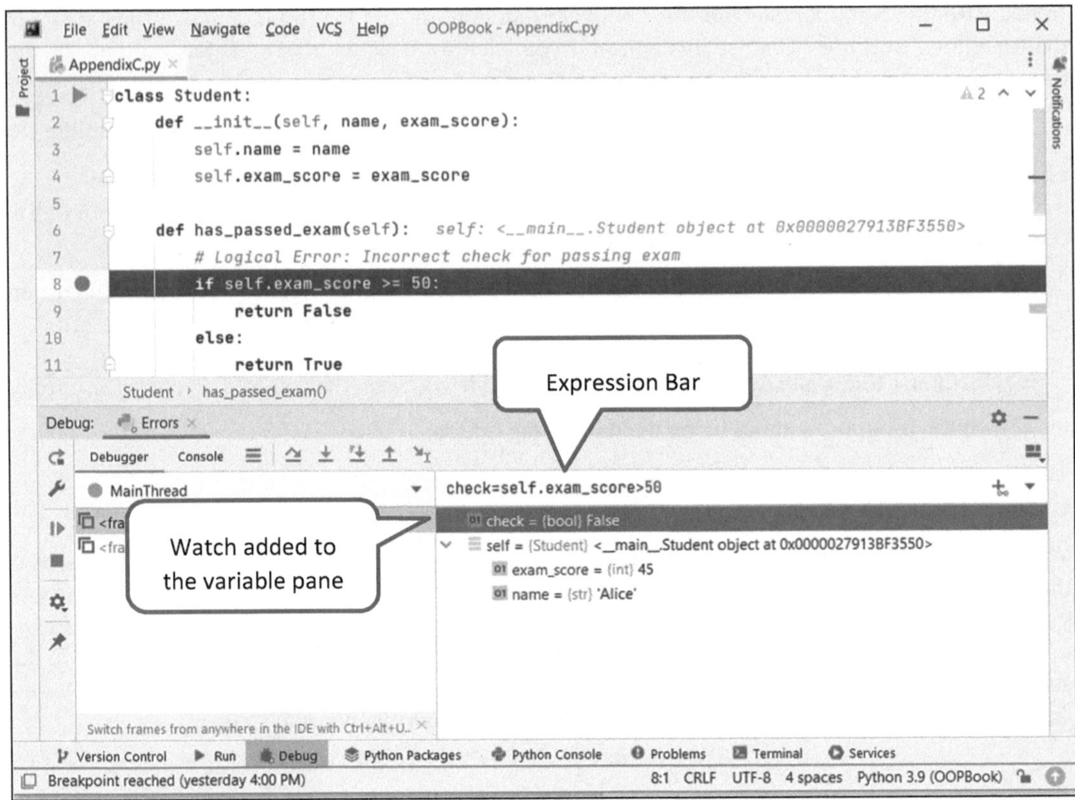


**FIGURE C.8** Expression bar.

If you need to evaluate an expression based on the variables within your code, you can write it in the expression bar and press enter. For example, you can create an expression to check whether the exam_score of a student is above 50 by typing in the expression bar: *check=self.exam_score>50*, as shown in Figure C.8. The watch will permanently show in the variable pane below and compute whether a student passed or failed based on their exam score.

# Appendix D

# PEP Style Guide-Coding Standard and Conventions

## D.1 NAMING CONVENTIONS

In programming, a token is your code's smallest meaningful building block. Like words in a sentence, tokens are formed by grouping characters based on their function within the code. These tokens fall into five main categories, each with a specific role:

- **Operators**: Symbols that perform calculations or comparisons (e.g., +, ==, <). The Python language predefines these.
- **Reserved words**: Keywords with special meanings within Python (e.g., if, for, and while). These are also predefined and cannot be redefined.
- **Constants**: Fixed values like numbers or text. Some constants might be predefined (e.g., string literals), while others can be defined by developers (e.g., mathematical constants).
- **Identifiers**: User-defined names given to variables, functions, or classes. Developers choose these names following Python's naming conventions for better readability.
- **Separators**: Characters that structure the code and separate tokens (e.g., commas, semicolons, white space, and parentheses).

Establishing a standard for naming identifiers (developer-defined names) is beneficial for consistency across this book with multiple programs. Additionally, some external libraries might have their own naming conventions for their components. It's important to follow such libraries' specific naming styles to ensure code correctness and maintainability.

### D.1.1 Name Selection

Descriptive names are recommended for identifiers to show what every name represents and to enhance readability. It's important to avoid confusing naming styles, such as single lower-case letters like "a", except for mathematical equations, to prevent confusion and reduce the difficulties of understanding the role of each name in a long code. Letters like "l" (lower case L), "O" (upper case O), and "I" (upper case i) are not used even in mathematical representations to avoid confusing them with numerals one and zero.

#### D.1.1.1 Naming Styles

Table D.1 summarizes the naming styles used for the different requirements with examples.

### D.1.2 Code Layout

Code organization plays an important role in code readability.

**TABLE D.1**
**Naming conventions**

| Type | Naming Convention | Examples |
|---|---|---|
| Variable | The recommendation is to use a lowercase letter (for mathematical purposes) or a lowercase word. Multiple words may be separated with underscores. | student, cis_student, student_first_name, x, y |
| Constant | The recommendation is to use an uppercase letter (for mathematical purposes) or a lowercase word. Multiple words may be separated with underscores. | TOTAL, MAX_NUMBER, Z |
| Functions, class attributes, instance variables, methods, and global variables. | A lowercase word or words are used. Underscores separate words. Hidden internal methods (or properties) that the user shouldn't use should start with the underscore "_". | max(), calculate_average() class_method () |
| Method arguments | Use *self* for the first argument to class methods. | __init__(self), setaddress(self) |
| Classes and exceptions | A capitalized word or words are used. When multiple words are used, each word is capitalized without an underscore, i.e., camel case. For Exceptions, use the suffix "Error". | Class example: Vehicle, MyClass Exception example: VehicleError, AgeError |
| Modules | A short lowercase word or words are used. Separate words with underscores in module names only if it improves readability. | module.py, my_module.py |
| Package | A short lowercase word or words are used. Multiple words are used without underscores. | package, mypackage |

### D.1.3 Maximum Line Length

Following the Python standard library requirement, code lines in this book are limited to 79 characters, and docstrings and comment lines are limited to 72 characters, wherever possible. Longer code lines are handled using line continuation or separated into different lines inside parentheses.

### D.1.4 Line Break

Longer code lines with arithmetic operators such as '+' and '*' are split into different lines to enhance readability. Python line breaking uses the mathematician concept of math expression line breaking by starting the continuation line with the operator within the parentheses as in the below example:

```
total_salary = (basic_salary
                + housing_allowance
                + transportation_allowance
                - late_attendance_penalty)
```

Long textual content, like URLs or sentences, is split into separate lines using the line continuation backlash ('\') character, as shown in the example below.

```
text = "Michael Jordan was born in Brooklyn, New York, and he became a "\
       "renowned professional basketball player."
```

### D.1.5 Blank Lines

Blank lines are a powerful method for improving readability, and they are also used in Python to separate important code parts. In this book, as recommended, two blank lines were used to surround top-level functions and class definitions, a single blank line was used to surround method definitions and groups of import statements, and blank lines were used in lengthy functions to separate different sections.

### D.1.6 Imports

Import statements mentioned in the previous paragraph can always be found at the top of the code file just after any module comments and docstrings. The layout considered here was to place each import on a separate line.

### D.1.7 Indentation

Indentations are used as multiples of four spaces. When a line is split into multiple lines, the four spaces rule for indentation was optional, and the indentation was either:

- Aligned with opening delimiter:

```
variable1 = (val1, val2
             val3, val4)
```

- Two levels of indentation (four spaces) from the start to distinguish arguments:

```
def var_function (
        arg1, arg2,
        arg3, arg4):
    argument1 = arg1
```

## D.2 WHITESPACE IN EXPRESSIONS AND STATEMENTS

Table D.2 summarizes how the white spaces were selected to be used and where it was avoided:

**TABLE D.2**
**Whitespace situations**

| White space situation | Wrong use of white space | Correct use of white space |
|---|---|---|
| Inside parenthesis, brackets, and braces | `employ( id[ 1 ], { Salary: 2 } )` | `employ(id[1], {salary: 2})` |
| Immediately before a comma, semicolon, or colon | • `print(x, y)`<br>• for i in string : | • `print(x, y);`<br>• `for i in string:` |
| Between a trailing comma and a following close parenthesis | `number = (1, )` | `number = (1,)` |
| Before the parenthesis of the function call argument list | `calculate_average (2,3,4)` | `calculate_ average(2,3,4)` |
| Around assignment operator when trying to align two variables of different character lengths | `Student_Id=1`<br>`Student_name='Lita'` | `student_Id = 1`<br>`student_name = 'Lita'` |

## D.3 COMMENTS

A well-commented code adds to the program's "readability". Comments are followed using the Python Enhancement Proposal (PEP) 8 style, which requires complete sentences starting with a capitalized word, except for identifiers.

Block comments were made in one or more paragraphs, and each paragraph was separated by a line that starts with #. Block comments were indented to the same level as the corresponding code, where each sentence of the block started with a #. Inline comments were separated by two spaces from the statement and started with a # and a single space.

Documentation strings, or docstrings for short, were written for classes, functions, and methods. They were enclosed with triple-double quotes (") or triple single quotes ('). The triple-double quote was placed on a line for a multiline docstring.

## D.4 STRING QUOTES

When strings are used in the code, their quotes can be single or double. When a string contains a single quote character, double quotes are used for the string, and vice versa.

# Index

Pages in *italics* refer to figures and pages in **bold** refer to tables.